Organic
Photochromes

Organic Photochromes

A. V. El'tsov
Lensovet Technological Institute
Leningrad, USSR

Translated from Russian by
Yu. E. Sviridov
Moscow, USSR

Translation edited by
J. Whittall
UWIST
Cardiff, Wales, United Kingdom

CONSULTANTS BUREAU • NEW YORK AND LONDON

Library of Congress Cataloging in Publication Data

Organic photochromes.

Translation of: Organicheskie fotokhromy.
Includes bibliographies and index.
1. Photochromic materials. I. El'tsov, A. V. (Andrei Vasil'evich) II. Title.
QD715.07413 1987   574.1'35   87-24558

ISBN-13: 978-1-4615-8587-9   e-ISBN-13: 978-1-4615-8585-5
DOI: 10.1007/978-1-4615-8585-5

This translation is published under an agreement with the Copyright
Agency of the USSR (VAAP)

© 1990 Consultants Bureau, New York
Softcover reprint of the hardcover 1st edition 1990

A Division of Plenum Publishing Corporation
233 Spring Street, New York, N.Y. 10013

PREFACE

Only a small part of the numerous photochemical reactions in organic compounds result in reversible structural changes. The latter are accompanied by a change in physical properties, in particular, of electronic spectra. It is tempting to try to use this photochromic effect in various systems for controlling and regulating light fluxes and for data recording. Eventually an independent trend emerged in photochemistry – the study of the photochromism of organic compounds to establish relationships between structure and photochemical behavior, the theoretical analysis of these relationships to predict structures with pre-set photochromic parameters, and, finally, the identification of suitable areas for the application of photochromism. This monograph summarizes the results of recent studies carried out by a number of research institutions in the USSR and the GDR. Devoted to an important aspect of applied photochemistry, this monograph contains a detailed exposition of the chemical photochromes referred to briefly in the earlier editions of "Introduction to the Photochemistry of Organic Compounds" (Khimiya Press, Moscow, 1976) and "Photochemical Processes in Layers" (Khimiya Press, Moscow, 1978).

In Chapter 1 we point out that the theoretical elaboration of the problems of photochromism relies on a wide variety of concepts, and that analysis of changes in the potential energy surfaces of photochromic systems during reaction appears to be the most promising method of research. Whereas Chapters 2 and 3 deal with traditional photochromic systems, Chapters 4 and 5 describe the excellent prospects for the synthesis of photochromes from a series of potential tautomers. Chapter 4 also contains a section on aryloxyquinones written by Yu. E. Gerasimenko of the Scientific-Research Institute of Organic Intermediates and Dyes, Moscow, and a section on acylotropic regroupings written by V. I. Minkin, G. D. Palui, V. A. Bren, and A. E. Lyubarskaya of the Scientific-Research Institute of Physical Organic Chemistry, Rostov-on-don, RSFSR. The possibility of using photochromes for accumulating and storing solar energy through the energy of chemical bonds is of special interest in this context. Chapter 6 contains a brief survey of the entire body of available data on the so-called luminescent photochromes, which can be used to increase sharply the sensitivity of photochromic systems.

A. V. El'tsov

CONTENTS

Chapter 1
THEORETICAL STUDIES OF THE PHOTOCHROMISM OF ORGANIC COMPOUNDS
F. Dietz and A. V. El'tsov

The Use of Qualitative Models and the Woodward–Hoffmann Rules
for Describing Photochromic Systems   4

Quantum-Chemical Calculations of the Spectra of Photochromic
System Components and Statistical and Dynamic Indices
for Estimating Their Reactivity   7

Photochromic Spiropyrans.   9
    Chromenes and related compounds.   19
    Reversible photocyclizations and cycloadditions.   21
    Photochromism of indigoids   24
    Photoisomerization of azines   26
    Hydrogen migration in photochromic systems   27

Discussion of the Mechanisms of Photochromism using Current Ideas
on Potential Energy Surfaces.   29
    E/Z-Photoisomerizations.   30
    Reversible photocyclizations   35
    Hydrogen phototransfer   37

The Influence of Solvent on Photochromism   38

References   40

Chapter 2
PHOTOCHROMIC THIOINDIGOID DYES
M. A. Mostoslavskii

The Structure and Spectra of Thioindigoids.   47

Production, Thermal Stability, and Spectra of Z-Isomers
of the Thioindigoids.   58

On the Mechanism of the Photoisomerization of Indigoid Dyes   75

The Effect of the Heteroatom of an Indigoid Dye on $\phi_{E \to R}$. The
Reasons for the Absence of Photochromism in Indigo.   89

The Influence of Substituents and Environmental Parameters
on $\phi_{E \to R}$ 95

Addendum   98

References   99

Chapter 3
PHOTOCHROMIC COMPOUNDS WITH N=N AND C=N CHROMOPHORES
D. Fanghänel, G. Timpe, and V. Orthman

The Principal Paths of Photochemical and Thermal
E/Z-Isomerization   105
    Direct and triplet-sensitized photoisomerization   108
    Photocatalytic isomerizations.   111

Photochromic Azo Compounds.   112
    Electronic absorption spectra.   112
    Luminescence   123
    The mechanisms of E/Z-photo- and thermal isomerization
    and photo- and thermal dissociation.   128
    The application of the photochromism of azo compounds.   134

Photochromic Aryldiazo Compounds.   136
    Electronic spectra   137
    The mechanisms of Z/E-photo- and thermal isomerization
    and photo- and thermal decomposition   146
    The uses of the photochromic properties of aryldiazo
    compounds   151

The Photochromism of Compounds Containing the C=N Chromophore . . .   153
    Electronic spectra   153
    The mechanisms of the E/Z-photo- and thermal isomerization
    of compounds with the C=N chromophore.   158

References.   164

Chapter 4
PHOTOCHROMIC TAUTOMERIC SYSTEMS
A. V. El'tsov, A. I. Ponyaev, É. R. Zakhs, D. Klemm, and E. Klemm

Photochromism of (Nitroaryl)(hetaryl)alkanes.   178
    Structural features of photochromic (nitroaryl)(hetaryl)-
    alkanes.   178
    The nature of photoinduced forms of (nitroaryl)(hetaryl)-
    alkanes and the mechanism of their formation   185
    The effect of structure on the color of observed
    photochromes.   193
    The decoloration of colored forms.   194
    The quantum yields of photocoloration.   201
    Irreversible photodissociation   205
    The photochromism of (nitroaryl)(hetaryl)alkanes in crystals,
    melts, and polymer films   208

Photochromic peri-Aryloxy-p-quinones.   210
    The structure of the photoinduced form   210

The relation between the structure and photochromism
    of aryloxyquinones 212
Absorption and luminescence. 213
The mechanism of photoisomerization. 216
Potential practical applications 217
Reversible Photoacylotropic Regroupings 218

References. 224

Chapter 5
ORGANIC PHOTOCHROMES FOR SOLAR ENERGY STORAGE
V. I. Minkin, V. A. Bren', and A. É. Lyubarskaya

The Principle of Accumulating Light Energy as the Energy of Strained
    Metastable Structures 229

Efficiency Criteria 232

Various Photochromic Systems for Solar Energy Accumulation
    and Their Mechanisms. 233
Geometric isomerizations 234
Valence isomerizations 236
Dissociation reactions and dimerization. 239

A Basic Scheme of the Engineering Solution to the Problem of Solar
    Energy Storage and Release in the Form of Thermal Energy. 240

Acylotropic Azomethine Systems – A New Type of Solar Energy
    Storage Battery 240

References 243

Chapter 6
LUMINESCENCE OF PHOTOCHROMIC COMPOUNDS
M. G. Kuz'min and M. V. Koz'menko

Polyacenes and Their Analogs. 246

Diarylethylenes 248

Thioindigoids 252

Arylmethane Derivatives 252

Spiropyrans 256

Azomethineimines. 257

Compounds with an Intramolecular Hydrogen Bond. 257

Conclusion. 261

References. 261

Index 267

Chapter 1

THEORETICAL STUDIES OF THE PHOTOCHROMISM OF ORGANIC COMPOUNDS

F. Dietz* and A. V. El'tsov**

*Karl Marx University, Leipzig, GDR
**Lensovet Leningrad Institute of Technology, USSR

Photochromism, one of the more striking phenomena of photochemistry [1-4], is the transition between two chemical compounds*** A and B with mar- kedly different absorption or emission (see Chapter 5) spectra induced in at least one direction by electromagnetic radiation:

Compound A $\xrightarrow{\text{hv } \lambda_1}$ Compound B

(absorbs light $\xleftarrow{\hspace{2cm}}$ (absorbs light

in λ_1 region) hv λ_2 or Δ in λ_2 region)

Changes in light absorption and luminescence always entail an abrupt alteration of the chromophore system: in the case of chemical photochromes the A ⇌ B transitions usually result in substantial molecular restructuring.

Photochromic systems are distinguished by the following:

a) Structural changes are reversible. Transition to metastable com- pounds are photochemically induced from thermodynamically stable compounds.

b) Associated with structural changes A ⇌ B is a change in physical and chemical properties, of which color change is, at present, of the greatest scientific and practical interest.

c) Depending on the properties of photochromic systems, forward and reverse reactions may proceed through electronically excited singlet or triplet states. Metastable form B may have a lifetime of from a few micro- seconds to hours, days, or weeks. This factor is of decisive importance for potential practical uses of photochromic systems [5,6].

d) Photochromic systems are often also thermochromic; i.e., reversible changes in them proceed thermally. In many cases the back reaction in photochromic systems is also induced thermally.

***We are not concerned here with physical photochromes whose photoinduced coloration is due to the light absorption of the resulting excited atoms, molecules, and ions.

1

e) In theory the forward and reverse reactions can be repeated indef-
initely. Real photochromic systems show signs of "aging"; the number of
conversion cycles from A to B and back into A decreases on account of side
reactions. The practically achieved number of cycles ranges from 1 to about
10^6.

Most photochromic systems are based on bond cleavage, isomerization, or
oxidation—reduction reactions.

Bond cleavage proceeds either homo- or heterolytically. Thermal recombination of the radicals in the reverse reaction corresponds to homolytic
bond cleavage. The light energy necessary for homolysis corresponds to the
energy of dissociation of the bond being split. The relative stability of
the resultant radicals is heavily dependent on the electronic structure of
the given compound (for instance, "mesomer stability" through delocalization
of the unpaired electron is possible). C—C bonds (for instance, in
bistetraarylimidazoles), as well as C—N bonds (in hexaarylbisimidazoles),
N—N bonds (in nitroso dimers), and S—S bonds (in diaryl disulfides), are
homolytically broken.

Heterolytic bond cleavage and, accordingly, thermal recombination of
resultant ions in the back reaction are to be found particularly frequently
in photochromes whose structural and electronic factors favor the formation
of relatively stable ionic or intraionoid reaction products; the latter have
structural elements of polymethines noted for their deep color (for instance, triarylmethyl cations, merocyanines).

The characteristic differences, especially in physical properties,
between E- and Z-isomers as well as the easy course and often considerable
quantum yields of photochemical isomerization enable us to regard many
classes of unsaturated compounds as potential photochromic systems. Here
the difference between the positions of the absorption bands of E- and
Z-isomers in certain groups of compounds is found to be in excess of
100 nm. Generally, because of the longer transition moment, the E-form
absorbs more intensely than the Z-form; the E-form is thermodynamically more
stable (see Chapters 2 and 3).

Stilbenes, polymethines, indigoids, azomethines, azo compounds, azines,
triazenes, and other compounds exhibit pronounced photochromism due to E-
and Z-isomerization. In some photochromic systems E-, Z-isomerizations
precede another photochromic process (for instance, stilbene photocyclization) or follow it (for instance, ring opening in spiropyrans).

Photoinduced valence tautomerism is another example of photoisomerization. It takes place when as a result of a rearranged electronic structure
(which may change the bond lengths and angles but which does not substantially affect the relative steric position of atoms and atomic groups)
equilibrium is reached between valence isomers [7]. This tautomerism includes: a) intramolecular ring opening (as, for instance, in spiropyrans;
2H-pyrans and -thiopyrans; 2H-chromenes and -thiochromenes, dihydroxan-
thenes; trans-10b,10c-dialkyldihydropyrenes); b) intramolecular cyclization
(for instance, in 1,3,5-hexatrienes, fulgides, stilbenes, bianthrones,
nitrones, 1,3-butadienes, and others). Intermolecular cyclodimerization is
described in the series of pyrimidine bases, coumarins, carbostyryls,
1,4-naphthoquinones and other quinones, and anthracene and acridinium derivatives.

In most cases light-induced reversible hydrogen transfer leads from
thermodynamically stable ortho-substituted aromatic structures to colored
compounds with a quinoid structure through a six-membered transition state.
Typical groups of such compounds include, in the case of keto—enol tautom-

erism, quinolones, alkylbenzophenones, arylimines, o-hydroxyaldehydes, o-hydroxyarylaldazines, -aryl azo compounds, benzophenones, and salicylates; in the case of acinitro tautomerism — o-nitrobenzene compounds; and in the case of hydrogen transfers in metal-complex compounds — metal dithizonates.

Photochromic reduction—oxidation systems are distinguished by a light-induced reversible electron transfer between the donor and the acceptor, which are found in at least two stable redox states. Examples of such reduction—oxidation systems include cell chlorophyll, the thiazine dye—Fe(II) or Sn(II), 3,6-dichlorofluorene—dye—disulfhydryl complex, as well as derivatives of p-phenylenediamine and triphenylamine in combination with the corresponding matrices in which reversible photoionization occurs.

The aim of theoretical and quantum-chemical studies of organic photochromic systems is to elucidate the following points:*

1) to estimate the energy levels of the ground and electronically excited states of all the particles involved in the photochromic process;

2) to characterize the absorption spectra of all the particles, i.e., to calculate the positions of absorption maxima, to ascertain the nature of electron transitions, and to assess their intensity and polarization direction;

3) to quantify the photochemical activity of photochromic molecules;

4) to describe the mechanisms of photochromic reactions and to estimate thermodynamic parameters characterizing the reaction (activation energies, activation enthalpy, entropy effects);

5) to elucidate the chemical structure of the compounds participating in the photochromic process with the aid of thermodynamic values, spectroscopic data, and parameters of electronic structure;

6) to find a connection between the structure and chemical behavior of compounds with a view to developing photochromic systems which would have optimal parameters from the standpoint of their practical application.

No methodologies described in the literature assure equally successful solutions to the above-stated questions for all potential photochromic systems. This reflects the state of the art of contemporary theoretical chemistry, in general, and of photochemistry and quantum chemistry, in particular. Therefore, workers in the field have been looking at individual problems in accordance with their importance for each specific photochromic system. The quality of the results of these efforts varies since they employ different theoretical approaches to the problems of photochemistry [8-10].

1. The use of qualitative models and correlation graphs:

a) the classification and description of synchronous photochemical reactions with the help of the Woodward—Hoffmann rules [11], the Fukui concept of frontier orbitals [12,13], the Hückel—Mobius concept as modified by Zimmerman [14], Dewar's ideas on the aromatic and anti-aromatic transition states [15,16], graph theory [17];

---

*Naturally, attempts at solving the scientific tasks in the field of photochromes mentioned here and elsewhere can produce good results, given a favorable combination of theoretical approaches with the experimental techniques of photochemistry.

b) qualitative description of synchronous polar cycloaddition with the help of diagrams of the donor—acceptor interaction of cycloadducts [18,19];

c) estimating molecular interaction energies with the help of perturbation theory to determine their capacity for entering into photocycloaddition [16];

d) qualitative assessment of the shape of potential curves along the coordinate of the reaction for certain photoreactions with the help of correlation diagrams of state; systematization of types of photochemical processes [20-22];

e) estimating the geometry and energy of photochemical intermediates in various electronically excited states to develop hypotheses about the reaction course and to identify its special features such as stereospecificity [8,23].

2. Calculating spectrophysical characteristics of photochromic compounds: their absorption, luminescence, radiationless deactivation, the efficiency of intramolecular energy transfer, etc.

3. The use of statistical and dynamic indices of the reacting components for describing the electronic structure of electronically excited Franck—Condon states and for developing ideas on the reactivity of molecules in photochemical reactions [9].

4. Calculating potential energy curves and surfaces or parts thereof for the ground and electronically excited states of molecules. These data, along with information about the geometry of various excited states, often provide more detailed information on the course of the reaction, the thermodynamic characteristics of the starting materials, intermediates, and reaction products, as well as on spectrophysical parameters [24-27].

In view of the great variety of theoretical concepts found in the literature, the varying standard of research, and in the absence of summarizing, comprehensive methodologies we have grouped the main theoretical works on organic photochromism by research method and photochrome group.

While the use of qualitative models for the photochromes has yielded relatively modest results, the quantum-chemical calculations of the spectra of photochromic system components and the indices of their reactivity have made it possible to find important correlations between structure and properties for practically all groups of organic photochromic systems and to assess the structure of photoinduced forms. The results of the calculations of potential energy surfaces are presented in a special section.

THE USE OF QUALITATIVE MODELS AND THE WOODWARD—HOFFMANN RULES
FOR DESCRIBING PHOTOCHROMIC SYSTEMS

The ideas of Woodward and Hoffmann [11] are normally drawn upon in classifying and describing in qualitative terms synchronous photochemical reactions such as cycloaddition and ring opening, electrocyclic and chelatropic reactions, group transfer, sigmatropic regroupings, and elimination. Of these only cycloaddition, ring opening, and electrocyclic reactions have by now been realized in photochromic systems.

The Woodward—Hoffmann rules provide the first basic idea of the steric course of photochemical reactions. [4 + 4]-Cycloaddition occurs during the cyclization of anthracene 1 and acridizinium derivatives and certain pyridine bases. In the case of coumarin compounds and 1,4-naphthoquinone 3, [2 + 2]-cycloaddition is realized:

In accordance with the Woodward–Hoffmann rules, these reactions are photochemically allowed if cycloaddition takes place from the same side of the reacting parts of both molecules (supra, supra-cycloaddition). Qualitative conclusions regarding the possibility of a reaction occurring are drawn from an analysis of correlation diagrams. For instance, one worker ([11], see Figure 10) presents a simplified correlation diagram of electronic states for a model of [2 + 2]-cycloaddition: 2 Ethylene $\rightleftharpoons$ Cyclobutane. The lowest excited state of both ethylene molecules corresponds to the electronic configuration $(SS)^2(SA)^1(AS)^1$; the bonding and antibonding molecular orbitals are symmetric with respect to two surfaces (σ_1 divides the C=C bond, σ_2 is parallel to both molecular surfaces and is equidistant from both). This excited state has the same symmetry as the first excited state of the reaction product — cyclobutane. This means that the symmetry of these states does not change as a result of the reaction and both states directly correlate and the reaction is symmetry-allowed.

It is impossible to obtain more detailed information from such correlation diagrams on potential curves or potential surfaces, the less so since this approach does not distinguish the spin multiplicity of reacting excited states.

Electrocyclic reactions, in keeping with the Woodward–Hoffmann rules, follow a strictly stereospecific course, either disrotatory in the excited state of diene structural fragments or conrotatory in the case of trienic structures. As a result, the substituents at the extreme ends of the dienic systems in the cyclobutane reaction product are in a cis-position relative to each other, while in the cyclic products derived from hexatrienic structures they are in a trans-position. Accordingly, hydrogen atoms in the product of stilbene cyclization — 4a,4b-dihydrophenanthrene — must be in the trans-position [28]. So far, no experimental proof of this fact has been obtained. In the case of 1,2-diphenylcyclopentene 5 the Z-structure was originally recorded since its molecule is a convenient model for studying reversible photocyclization uncomplicated by E/Z double-bond isomerization.

conrotatory
ring closure

trans-isomer

5   6

Compounds 7 and 8 form a relatively stable system among the photochromes of this class; colorless 1,1',3,3',5,5'-hexamethylstilbene 7 on ultraviolet irradiation is converted into red-colored hexamethyldihydrophenanthrene 8. The latter is not oxidized into a corresponding phenanthrene derivative, but it easily opens the ring in the back reaction [29].

Analogous behavior is exhibited by valence tautomers **9** and **10** [30].

For photochemical reactions which proceed through singlet states, analysis of secondary orbital interactions gives an idea of the stereochemical
results. On the other hand, experimental data on the structure of photoproducts make it possible to judge the multiplicity of the excited state.
Thus such secondary effects explain the preferable formation, in the case of
coumarins, of endo-adducts of the "head-to-head" type of cycloaddition, and
not of exo-adducts.

lowest unoccupied   highest occupied
molecular orbital   molecular orbital

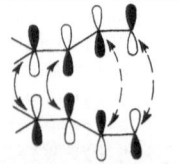   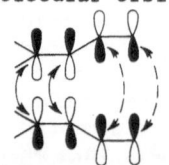

lowest unoccupied   highest occupied
molecular orbital   molecular orbital

"head-to-head"
endo-adduct

11   **12**

Of the binding interactions, the stronger ones (⌣) lead to the formation of a covalent bond while the weaker ones (⌣) determine the regio-
selectivity of the reaction. It then becomes clear why coumarin dimerization through the singlet state results in the formation of an endo-adduct of
the "head-to-head" type exclusively. If the reaction proceeds through the
triplet state, both endo- and exo-products of "head-to-head" cycloaddition
are formed along with traces of the "head-to-tail" type of product [31].
According to other workers [8,23], these differences in reaction stereospecificity depending on the multiplicity of reacting excited states may be
due to differences in the geometry of the reacting particles, which, in
turn, is attributable to a different electronic structure.

Assessment of the geometry of singlet and triplet intermediates with
due allowance for spin and electron—electron repulsion shows that in biradi-
caloid triplet states the electronic cloud is loose; this accounts for the
tendency of such systems towards the separation of the resultant radical
pairs, geometrical isomerization, and the loss of reaction stereospecificity.

Photodimerization of 9-substituted anthracenes **13** leads in most cases
to "head-to-head" dimers.

2

R=H, CH₃, ClCOOH, CHO

13   **14**   **15**   **16**

Conversely, in the analogous reaction of acridizinium **15** and pyridine
17 compounds, cycloadducts of the "head-to-tail" type are formed for the
most part — **16** and **18**, respectively.

17   **18**

On the basis of perturbation theory we can easily draw reliable conclusions regarding the nature of the excited state involved in intermolecular cycloaddition [12,13,16,32-34].

Salem [34] has proposed a useful method fairly readily applicable to reactions involving large π-electronic systems which, quickly and without complex calculations, can help to evaluate in qualitative terms the structure of the potential surface in the π-electronic approximation as well as the reaction path for conjugated reagents that are not too strongly polar. Unfortunately, no data have yet been reported on the application of this method to photochromic systems.

QUANTUM-CHEMICAL CALCULATIONS OF THE SPECTRA OF PHOTOCHROMIC SYSTEM COMPONENTS AND STATISTICAL AND DYNAMIC INDICES FOR ESTIMATING THEIR REACTIVITY

Knowledge of the energetic position of the ground and reactive electronically excited states (S_1, T_1) as well as of spectroscopic data (the position of the absorption maximum in the electronic spectrum, the absorption intensity and the direction of electronic transition polarization), and the connection between light absorption, the type of corresponding electronic transfers, and the molecular structure of photochromes is a fundamental factor in work on problems of photochromism, including for optimization of such systems for the purpose of their practical application. Quantum-chemical techniques have found useful applications for the solution of these problems along with experimental measurement techniques which have been greatly improved over the past two decades (emission spectroscopy, spectroscopic techniques for the study of fast reactions).

Given certain prerequisites (for instance, during consideration of narrow groups of compounds, general directions of conversion), it is possible to draw simple qualitative conclusions about the spectra of photoinduced forms even using the most elementary quantum-chemical approaches (for instance, the Hückel method). A good semiempirical method for calculating the spectral data of electronic systems (such as the SCF PPP CI method) makes it possible (given optimum parametrization) to obtain results which are in agreement with experimental ones within the measurement error (for instance, during the determination of the position of the absorption maximum); however, the band intensities in the spectra are, in most cases, overstated. By means of a special parametrization [35] it is possible to estimate with a high degree of accuracy the S → T transfer energies as well; bearing in mind the existence of additional centers of special parametrization we can also describe the n → π*-electron transfers [36]. To describe the electronic spectra of nonconjugated compounds and systems with delocalized π-bonds, a modification of the CNDO method with special parametrization (the CNDO/S variant) [37,38] is used. However, the lengthy procedure involved, as compared with the method in the π-electronic approximation, does not offer substantial advantages for calculating π → π*-transitions. Some of the electron transfers (π → π*, n → π*) can be characterized with the aid of wavefunctions of the electronic states between which the electronic transition occurs.

7

In many cases certain conclusions can be drawn from an analysis of the configuration interactions regarding the photochemical reactivity of electronic states as well as about its connection with structure [39,40].

In many photochromic systems, the lifetimes of the photoinduced form are brief and cannot be determined, and they can only be identified by electronic spectra. The latter are calculated for a number of possible structures. The most probable structure (or structures) can be chosen on the basis of the calculations, and its (their) parameters are correlated with the experimental data.

Modern quantum-chemical methods involving different levels of expenditure and varying degrees of accuracy can be used to calculate the most stable structure and geometry of molecules in the ground state (the geometry optimization technique) [41,42] and in the excited state [43,44]. By comparison, experimental techniques for determining the geometry of molecules in excited states are extremely complex and are applicable to only some of the systems [45].

Knowledge of the structure of the ground and excited states of photochromic system components provides useful information on the paths of reversible photoreactions, on the mechanisms of side processes, on the influence of environments differing in nature (viscosity and other parameters), and on the state of photochromic equilibria. For instance, general rules have been created for cyclic alternant and nonalternant hydrocarbons [46] which help draw conclusions regarding the geometry of energetically favorable structures in the ground and excited S_1- and T_1-states.

Static and dynamic indices of reactivity for electronically excited states can be calculated by a similar technique used in the case of the corresponding values for the ground state. The static values used include charge density, bond orders [47], electronic populations of bonds [48], and free valences [47,49]. With the help of these values one can characterize in crude terms the electronic structure of a molecule in its electronically excited state. The dynamic indices of reactivity include relative changes in bond energy on electronic excitation [50] and in localization energy [28, 51]. The use of reactivity indices assumes that a conclusion regarding the preference of the molecule for a given photochemical reaction can only be drawn from an analysis of the parameters characterizing its electronic structure in the Franck–Condon state (static indices) or that this tendency can be evaluated on the basis of changes in electronic energy during the molecule's crossing from the Franck–Condon state to the transition state. A note of caution should be sounded here against an uncritical use of such values.

Above all, the numerical values for such estimates are arrived at within the framework of simple one-electron models (for instance, by the simple Hückel method), which give no satisfactory description of either the character of excited singlet states or the differences between the singlet and triplet states. Strictly speaking, such indices of reactivity can only be used for reactions which proceed adiabatically or when the rate is limited by the primary photochemical processes.

In many cases such an analysis fails to produce absolute values, and workers often confine themselves to qualitative or semiqualitative conclusions regarding the relative change in reactivity of large series of structurally related compounds. A description of the major approximate operations and of their application to the solution of problems of reactivity is contained in the papers [37,52], while a detailed exposition of photochemical reactivity is given in the paper [9].

The photolysis of practically colorless spiropyrans involves the cleavage of the carbon–oxygen bond of the pyran ring and the formation of a merocyanine dye. Spiropyran molecules are composed of two topologically orthogonal parts linked by a sp^3-hybridized carbon spiroatom. For indoline
spiropyrans [53], intramolecular energy transfer from the indoline to the
pyran portion (or in the reverse direction, depending on structure–energy
correlations) is highly probable. Indirect evidence of this is supplied by
luminescent measurements [54]. It has been theoretically proven [55-57]
that because of the slight overlap of p-atomic orbitals next to the spirocarbon ones, along with an electrostatic interaction, an exchange interaction takes place. This effect, denoted as spiroconjugation, is reflected
in the changes of the electronic spectra of formally isolated molecular
halves of the spiro compounds. For spiro compounds possessing a relatively
high symmetry (such as spirodifluorenes with D_{2d} symmetry), interaction is
symmetry-forbidden. Compounds with a lower symmetry do not obey such selection rules, and this is what we should expect from the majority of photochromic spiropyrans.

The absorption and luminescence spectra of the spiropyrans were first
studied using one group of compounds as a prototype [54,58]. The spectra of
indoline spiropyrans were compared with those of substituted chromenes and
indolines, and the influence of the substituents on electron transfers was
analyzed to establish a series of dependences between structure and spectral
characteristics. The electron transfers relating in the spectrum to particular bands are localized in different parts of the molecule. The first
(a band of some 320 nm) and third (about 260 nm) electron transfers are
localized in the chromene part. Luminescent measurements showed that the
lowest singlet or triplet states have a $\pi,\pi*$-character. The lowest triplet
state has an n,$\pi*$-character only, given the presence of nitro groups in the
chromene part. The second (290-300 nm) and fourth (about 240 nm) spiropyran
transfers were located in the indoline part. Quantum-chemical calculations
by the CNDO/S method confirm, at least qualitatively, this relationship of
the bands in the spectrum [59].

To assess the intramolecular energy transfer [60] in the case of spirodifluorenes, Förster's theoretical concepts [61] were used.

The PPP method was used for numerical calculations of the wavefunctions
of the partially isolated chromophores of the a and b parts of the molecule
(Figure 1) [62]. The configuration interaction (CI) coefficients calculated
by this method are used to evaluate the interaction integrals. For asym-
metrically substituted spirodifluorenes, values of the interaction integral

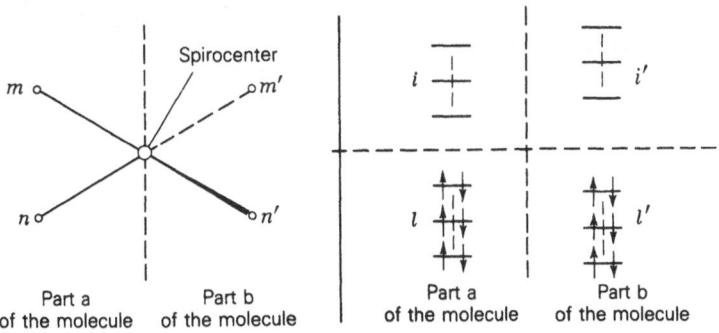

Fig. 1. The plot of the a and b portions of the spirodifluorene molecule
interacting through the MO spiroatom.

9

U other than zero are obtained, which explains theoretically the intramolecular energy transfer. No such assessments for spiropyrans are available.

    It is entirely possible that under certain conditions intermolecular
energy transfer and distant transmission of electronic excitation energy
within the lifetime of the electronically excited state (exciton movements,
as has been proved in the case of the molecular complexes of cyanine dyes;
see review [63]) also play a role. In concentrated solutions (in excess of
10^{-1} mole/liter) of trimethylindolinobenzopyran in polar solvents, the preferred formation of dimer particles and the monomer—solvent complex of the
merocyanine form [64] was observed. Complex formation leads to changes in
the energy position of the electronically excited states on account of
intermolecular interaction and may influence ring closure. The sandwich
compound spiropyran 19 and cyanine dye 20 have recently been obtained, and
fluorescence has been induced in the dye as a result of energy transfer from
the spiropyran to the cyanine dye [65]. The effect of this type of association with a cyanine dye on the photochromic properties of spiropyrans is yet
to be studied.

19   20

    An extensive body of experimental and theoretical material on the
photochromism of spiropyrans is contained in the papers submitted by Minkin
[66-68], Guglielmetti [59,69,70], Hirshberg and Fischer [53,71-74], and
their co-workers.

    The first quantum-chemical studies were carried out in 1954 [75].
To obtain data on the paths of ring opening in spiropyran 21 and on ring
closure in merocyanine 22, their electronic structure in the ground and
excited states was analyzed.

21   22

    The bond orders and electron densities of the merocyanine form 22 were
calculated by the simple MO method. Charge transfer on the heteroatom upon
electronic excitation (it was estimated by comparing the calculated π-elec-
tron densities in the ground and excited states) provides a qualitative
explanation for the negative solvatochromism. Qualitative arguments lead to
the conclusion [75,76] that of the four possible isomers 23-26, isomer 23
with the least steric hindrances should be the most stable, owing to the
shortest distance between the oppositely charged N^+ and O^- atoms.

23   24   25   26

    It follows from a comparison of the calculated π-bond orders that
Z/E-isomerization around the central bond which precedes cyclization
of the merocyanine form 22 should proceed at a higher rate in the excited
state than in the ground state (Figure 2).

10

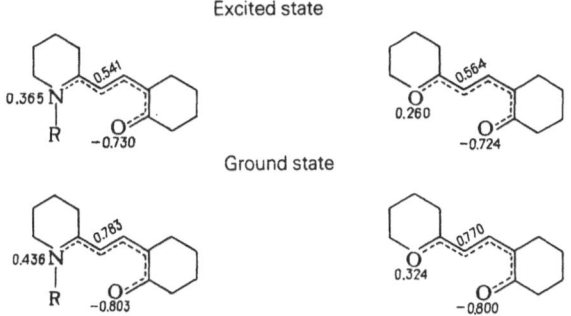

Excited state

0.541
0.365 N
R   -0.730

0.564
O
0.260
O
-0.724

Ground state

0.783
0.436 N
R   -0.803

0.770
O
0.324
O
-0.800

Fig. 2.   The electron density distribution and bond orders in the molecules
of merocyanine dyes according to the data of quantum-chemical
calculations.

As subsequent detailed studies showed, the photoreactions involving
spiropyrans should be examined from two angles [72]:

1. Redistribution of the electron density upon electron excitation; one
consequence of this is the opening of the pyran ring. The change in the
electronic structure can be attributed to a formal increase in the number of
π-electrons by two − σ-electrons of the broken bond assume the character of
π-electrons due to rehybridization of the AOs of carbon and oxygen. Ring
opening is immediately followed by the formation of a relatively high-energy
s-cis-isomer of merocyanine with a spiropyran configuration.

2. A change of configuration. In thermal or photochemical reactions
which follow the photochemical ring opening in spiropyrans, equilibrium is
reached between different cis- and trans-isomers of the merocyanine form.

There are no reliable data in the literature on the multiplicity of the
reacting electronically excited state. The spiropyran → merocyanine transition is facilitated by a triplet sensitizer with a suitable energy (for
instance, benzophenone). The study of phosphorescence, its quenching, and
photochemical ring opening also shows the triplet character of the reactive
state. Apparently, the reaction follows a fairly usual course [54,58,70,
77,78]:

$$S_0 \xrightarrow{h\nu} S_1 (\pi, \pi^*) \rightsquigarrow S_1 (n, \pi^*) \rightsquigarrow T (n, \pi^*) \longrightarrow \quad \text{ring opening}$$

Attempts were made to explain the photochemical opening of the pyran
ring within the framework of the analysis of the electronic structures of
various electronically excited states. The densities of valence electrons
on individual atoms were calculated by the CNDO/S method with Del Bene and
Jaffe parameters [38] for the S_1- and T_1-states. The electronic excitation
$S_0 \rightarrow S_1$ (π, π^*) is associated with the transition of 0.09 of the electron
density of the spirocarbon—oxygen bond to the cyclic core. The increase in
the positive charge resulting from this, especially on the oxygen atom,
causes polarization of the pair of binding σ-electrons on the oxygen atom.
This weakens the spirocarbon—oxygen bond in the excited state a good deal as
compared with the ground state. Concurrent changes occur in the electronic
structure of the entire pyran portion of the molecule which weaken the C—O
bond still further. Its breaking results in a merocyanine dye with delocalized bonds.

11

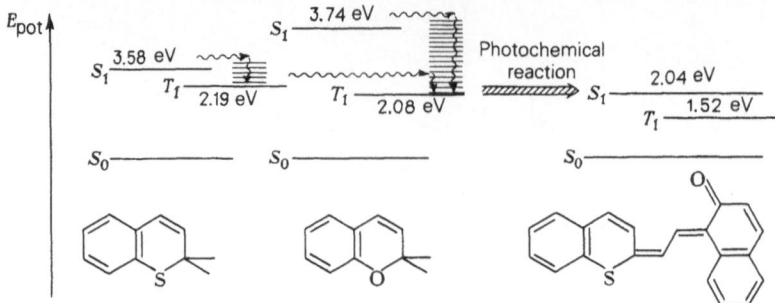

Fig. 3.   The energy levels of the structural fragments of spirothiopyranopyran 27 and the corresponding merocyanine.

    The connection between structure and the behavior of photochromic
compounds is important for practical applications of spiropyrans. Minkin,
Simkin et al. [66-68] have obtained important results in their work on the
problem.

    Relative energy levels of the ground and excited states of different
multiplicity [66] (Figure 3) were calculated for spirothiopyranopyran 27 by
the PPP method with β-variation and different parametrization for the singlet and triplet [35] states. These calculations helped to explain the
experimentally discovered pyran ring in the photoreaction and not a thiopyran ring. In accordance with the ideas of other workers it was believed
that ring opening proceeded in a triplet state [53,73].

    The triplet of the chromene fragment of spirothiopyranopyran in the
deactivation cascade has minimal energy and should have the longest lifetime. At the same time, the pyran ring opens. When the triplet state
of the thiopyran part of the molecule lies below the triplet state of the
pyran part, the thiopyran ring may open photochemically. Using such quantum-chemical calculations it is possible to predict structures in which the
thiopyran ring must open. They include compounds of type 28 or 29.

27

28

29

30 (R = H, CH$_3$, C$_6$H$_5$)

    Spiro(2H)-1-benzopyran-2,2'-benzo-1,3-dithiolanes 30 usually exist in
the form of a more stable spiropyran form [67]. It is stabilized by substituents in position 3 (see the numeration in 31). On the contrary, a
dimethylamino group in position 7 and a condensed benzene ring stabilize the
colored merocyanine form.

    On the basis of benzothiazolespiropyrans the dependence of the stabil-
ity of the merocyanine form on the nature of substituents was established
[69].

31   32

Decoloration rate constants in toluene (i.e., the rates of thermal cyclization into spiran 31 of the merocyanine form) increase with an increase in the volume of alkyl groups X in the series: methyl < ethyl < isopropyl < cyclohexyl. Electron-donating substituents in position 8 stabilize open form 32 [substituents with the oxygen atom (OCH_3, OC_6H_5) to a greater extent than the substituents with the sulfur atom (SCH_3, SC_6H_5); in this case both electronic and steric effects occur]. Naphthyl and para-substituted aryl residues ($X = C_6H_4Y$-p, $Y = H$, Cl, F, CH_3, OH; α-naphthyl, β-naphthyl), despite the steric hindrances, exert a stabilizing influence. The linear dependence of the ratio of the rate constant of the cyclization of substituted merocyanines to the rate constant of the unsubstituted compound on σ_p-substituents is in agreement with the stability of the spiropyran structure and the electron-donating properties of the substituents: $\log k/k_0 = 1.21\sigma_p$. Conversely, the effects of the substituents in position 6 correlate with the Brown–Okamoto substituent constants: $\log k/k_0 = 1.42\sigma_p^+$. A change in the nature of the heterocyclic residue associated with the pyran ring does not alter the relationship in any important way.

The authors of one of the more extensive recent papers [68] calculated quantum-chemical parameters for the spiropyrans possessing a series of heterocyclic rings linked with the pyran ring via the spiroatom. These parameters make it possible to assess the relative stability of the spiropyran and merocyanine forms. Attempts were made on the basis of these data to predict the behavior of unknown spiropyrans in photochromic equilibrium. In particular, localization energies calculated by the SCF PPP method were used. Thermal cyclization was regarded as an intramolecular nucleophilic addition in the α-position of the heterocyclic cation; the localization energy L_α^- of the heterocyclic cation was taken as a measure of the relative stability of the spiran. π-Electron densities in the α-position of the heterocyclic residue (q_α) were also used as a measure of merocyanine reactivity.

A clear connection exists between the stabilities of spiropyrans and merocyanine and the calculated values of L_α^- and q_α. Boundary values of L_α^- and q_α were found for the areas of existence of each of the valence tautomers. With few exceptions it has been possible to find with a sufficient degree of accuracy the areas of stability for the spiropyran and merocyanine forms, depending on compound type, which is of help in estimating the relative stability of unknown substances. For instance, the introduction of heterocyclic residue 33 into the pyran residue via the spiroatom should produce especially stable spiropyrans. In the series of compounds 34-43 with X = NR, O, S the stability of the spiropyran form increases as compared with the stability of the merocyanine form (the dot marks the spiroconjugation site).

33   34   35   36

37   38   39   40

13

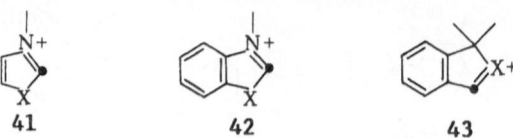

41	42	43

Deviations from the usually good correlation of calculated and experimental data are observed in the case of sulfur-containing compounds (X = S).

These conclusions relate to unsubstituted compounds only. Substituents in the pyran ring which exert an electronic or structural influence may have a decisive effect on the stability of both valence tautomers.

By calculating the relative energies of the valence tautomers in the equilibria **44** ⇌ **45** and **46** ⇌ **47** we may draw conclusions regarding the influence of annelation and the nature of heteroatom Z and substituents R, R', and R" on the stability of the forms **44** and **45** or **46** and **47**.

$$Z=O, S, NR$$

44	45	46	47

The transition from benzopyran to pyran raises the stability of merocyanine **45** so much that attempts to synthesize spiro(2H)pyran proper have so far been unsuccessful. The transition from benzopyran to dibenzo derivatives **48** and **49** or annelation of an additional heterocyclic nucleus improves the stability of the merocyanine form.

48	49

Donor or acceptor substituents in the benzopyran fragment do not have as strong an influence on the photochromic equilibrium as annelation or heteroatom substitution. Donor substituents in positions 7 and 8 and acceptor substituents in positions 6 and 8 should shift equilibrium towards the merocyamines. On the contrary, donor substituents in position 6 and acceptor substituents in position 7 enhance the stability of the spiropyran form and should shift equilibrium towards it. These conclusions deserve experimental verification.

Photochemical ring opening in the spiropyrans produces a high-energy isomer s-cis-merocyanine in which the spiropyran structure is still clearly present [72]. Although the structure of this isomer is not exactly known, its existence has been proven spectroscopically [71,72]. In the equilibria which follow the ring opening this compound is thermally or photochemically converted into various cis-trans-isomers. Which structures correspond to the equilibrium colored form is not yet precisely known despite all the intensive experimental [53,71-74] and theoretical research. Theoretical work on this problem is focused on the following criteria for the determination of merocyanine structure: analysis of experimentally and quantum-chemically calculated absorption spectra of the equilibrium photoinduced merocyanine mixture; development of a series of relative stability for isomer merocyanines with due regard for their electronic energies, steric effects, and the effect of solvent; the establishment of more likely pathways of cis-trans-isomerization proceeding from the structure of the high-energy s-cis-merocyanine as mentioned above.

As pointed out above, the first qualitative evaluation referred to the four most probable configurations 23-26. It is recognized that isomer 23 is more stable on account of the least distance between the terminal atoms N and O of the merocyanine chain. Some workers have tried to solve this problem by means of quantum-chemical calculations of electron energies in isomer merocyanines generated from benzothiazolinospiropyran [70,79,80]. Full energies have been calculated for compounds analogous to isomers 23-26. Four isomers 50-53 which possess a cis-configuration relative to the central bond are considered.

50   51   52   53

Interaction energies between free atoms are determined empirically from the correlation* [46,79-81] $E_{fi} = -(B/r^6) + Ae^{-Cr}$, where A, B (kJ/mole), and C (nm) are constants. The structures 50-53 are extremely unfavorable on account of severe steric hindrances; for the structures 23-26, with a trans configuration relative to the central bond, the following series of stabilities has been found: 26 > 23 > 25 > 24. The quantum-chemical calculations by a variety of approximate methods, e.g., SCF PPP and EHM,** and an allowance for (apart from electron energies) the energies of interaction between free atoms showed the preferred character of the cis-configuration with a slight rotation around the central bond; the most stable of the calculated structures corresponds to configuration 26 with an angle of rotation α about the central bond of 30°. The rotated configuration 23 is somewhat less stable.

Energies of interaction between free atoms have been determined for compounds with different substituents in position 3 (for the numeration see 31) − H, OCH$_3$, SCH$_3$, CH$_3$, C$_2$H$_5$, and C$_3$H$_7$; agreement was also established between these values and the rates of the thermal decoloration of the merocyanines. Analogous results were obtained for merocyanines from benzodithiolinospiropyrans 54. In these compounds steric effects are less important than in the case of benzothiazoline merocyanines. The conjugation effects of the phenyl group in position 3 stabilize the merocyanine form.

R = H, CH$_3$, C$_2$H$_5$, C$_6$H$_4$X, α-naphthyl

54

Structures 23-26 formed the basis for a mechanistic consideration of thermal or photochemical equilibria which follow the photochemical ring opening in spiropyrans [82]. It is impossible to give preference to any structures on the basis of excitation energy correspondence, calculated for different group structures by the SCF PPP method, and to the experimental spectra because of the too slight differences between these energies (a maximum of 0.07 eV -- 14 nm in the absorption spectrum of the merocyanine form) [66,68,82].

The π-bond orders and π-electron densities calculated for the ground and S$_1$- and T$_1$-excited states of merocyanines (Figure 4) make it

---

*fi − free interaction.
**The extended Hückel method.

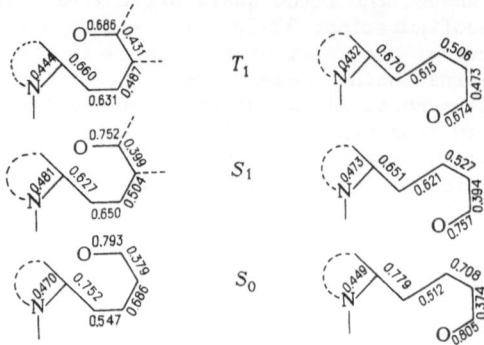

Fig. 4. Bond orders of the ground and S_1- and T_1-states of isomeric merocyanines.

possible to assume that the path described below for thermal and/or photochemical stages in the establishment of isomerization equilibrium that follow the photochemical ring opening in spiropyrans is quite likely:

The relatively low orders of π-bonds in the ground state of s-cis structures 55 or 56 indicate the possibility of the thermal stage of isomerization 55 → 57 or 56 → 60. Reverse transitions 57 → 55 or 60 → 56 must proceed thermally relatively readily as well, which is a prerequisite for merocyanine cyclization into spiropyran. Diminution of the order of the 4–5 π-bond upon electronic excitation into the state S_1 and especially into T_1 corresponds to the photochemical stage of isomerization 57 → 58 through rotation about the 4–5 bond in these states.

The data obtained can be interpreted as follows. Depending on how ring opening in spiropyran occurs, two different equilibrium processes are established:

formation of the "anti"-isomer 55 (the nitrogen and oxygen atoms are far apart) and the establishment of equilibria 55 ⇌ 57 ⇌ 58;
formation of the "syn"-isomer 56 (the nitrogen and oxygen atoms are next to each other) and the establishment of equilibria 56 ⇌ 60 ⇌ 61.

Direct isomerization of one of the compounds 55 ⇌ 57 ⇌ 58 into one of the isomers 56 ⇌ 60 ⇌ 61 is severely hindered as a result of the increased order of the corresponding π-bonds; this corresponds to the relatively high energies of activation of isomerization. The transition between two equilibria series can consequently be realized through the general cyclic spiropyran form 63 only.

However, Tyutyulkov et al. [82] point out, with every justification, that these quantum-chemical calculations apply to the gas-phase conditions. Before making final conclusions an attempt should be made to take into consideration solvent effects, something that has not yet been done.

The merocyanine spectra have been studied in greater detail than those of spiropyrans [66-68,82,83]. The SCF PPP method was employed to calculate excitation energies for the s-cis-configuration of the merocyanines 55 and 56 which form right after the photochemical opening of the pyran ring. They were compared with the experimental spectrum of the equilibria mixture after photolysis [82]. A planar structure with valence angles of 128° was taken as a basis for purposes of simplification. Since the geometry of the real molecule is not precisely known while the parameters of the atoms and bonds are less than optimal, only qualitative conclusions can be drawn from the results obtained. On the basis of these data the electronic structure of merocyanine can perhaps be better described with the help of the quinoid rather than the bipolar boundary form (however, see the paper [66], which gives preference to the bipolar boundary form). The excitation energies and oscillator forces calculated for four open merocyanine configurations, which are the most likely in the equilibrium of the colored form from the standpoint of steric requirements (23-26), differ so little from each other that it is impossible to establish a structure predominantly present in a state of equilibrium [66,82].

The connection between the color and structure of merocyanines was studied systematically. General laws governing the alteration of spiropyran structure through substitution or benzannelation to obtain a pre-set spectral shift of the longest wavelength merocyanine absorption in either direction [66-68] were established for indolinospiropyrans, spirodipyrans, and spirothiopyranopyrans. The energetically lowest singlet state whose population determines the photochromic process was located in the chromene part of the spiropyran. The calculated distribution of π-electron density in the ground and excited states in the merocyanines of indoline spiropyrans indicate a severe diminution of the dipole moment upon electron excitation, and consequently the negative solvatochromism of these compounds finds a good qualitative explanation. The absence of solvatochromism in the case of spirodipyrans and spirothiopyrans can be explained by the roughly equal dipole moments of the ground and excited states. The replacement of oxygen by sulfur in the benzo- or naphthospiran nucleus of spirodipyrans has no appreciable influence on the position of absorption maximum in the spectrum since the transition is localized in another fragment of the molecule. If the C–S bond opens, thiomerocyanine is formed, and its absorption as compared with oxygen-containing merocyanine is strongly bathochromically shifted (approximately by 0.35 eV, which in this spectral region corresponds to 120 nm). Benzannelation in position 3,4 of the pyran or thiopyran nucleus causes a hypsochromic shift, and a bathochromic shift if it occurs in position 5,6. In order to achieve a strong shift of the absorption maximum in the merocyanine spectrum, structural changes should be effected in that portion of the spirodipyran in which the C–X bond is broken in the course of the photoreaction.

In merocyanines from spiro(2H-benzopyran)-2,2'-benzodithiols, 5,6- or 7,8-benzannelation causes a hypsochromic shift, while 6,7-benzannelation causes a bathochromic shift of about 100 nm. This kind of influence is to be expected of powerful electron-accepting groups in position 4 (e.g., the nitro group). The electronic excitation of the merocyanine in this class of compounds is due to a considerable transfer of electron density from the benzo-1,3-dithiol ring onto the α- and γ-atoms of the carbon of the poly-methine chain, which is further enhanced by the electron-acceptor substitu-ents, which is what leads to the stabilization of the excited state. This corresponds to a decrease in the energy of the electronic excitation.

Calculations by the SCF PPP method for compounds 64-71 led to the following conclusions regarding the correlation between the structure and spectra of merocyanines [68]: a) condensed benzene and naphthalene nuclei have but an insignificant hypsochromic influence on the long-wavelength absorption band of the merocyanine form; b) azo substitution in the poly-methine chain (compound 66) does not lead to appreciable spectral shifts (for more on the effect of polymethines on the spectrum by means of azo sub-stitution see [84,85]); c) substitution of the vinyl group in the ring, condensed with the pyran ring by the sulfur atom or the C=O group, causes a strong hypsochromic shift of the long-wavelength band of the corresponding merocyanine 65; d) benzannelation or substitution in the indoline residue has no influence on the merocyanine spectrum. Methoxy and nitro groups in position 6 or 8 causes a bathochromic shift.

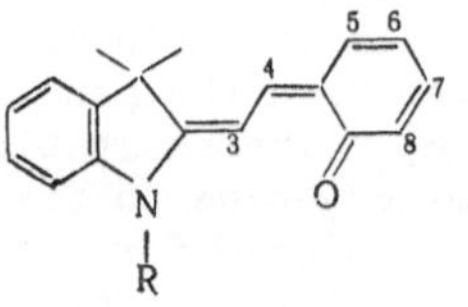

64 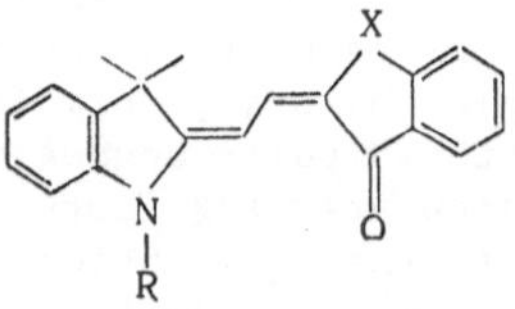   65

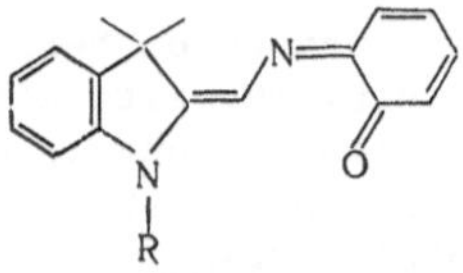

66 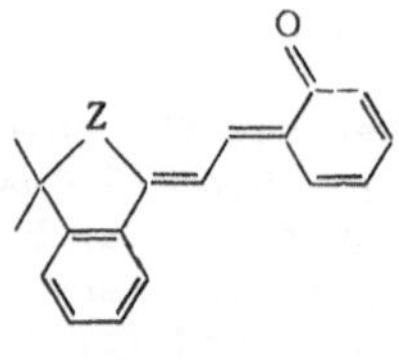   67

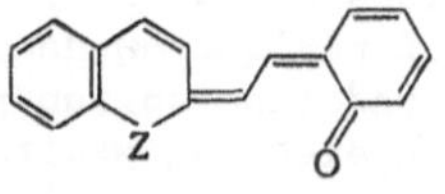

68 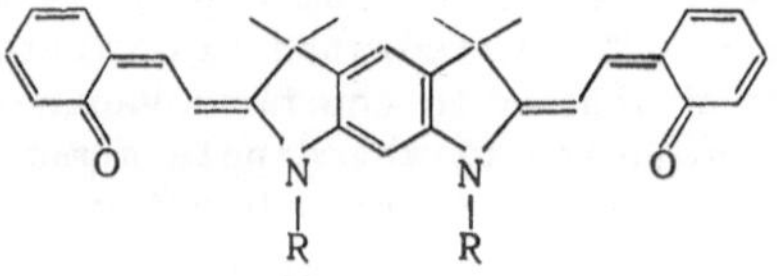   69

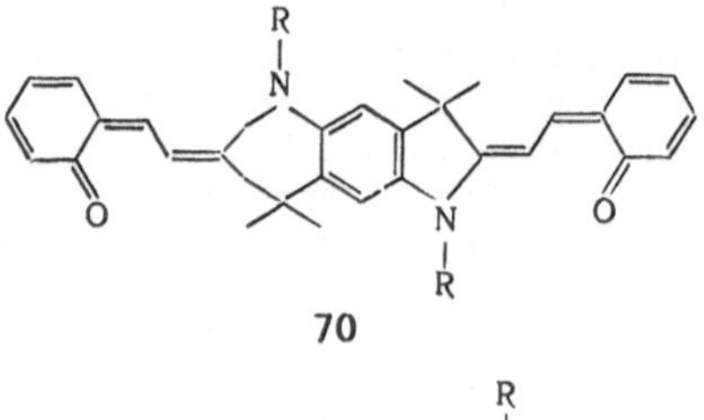

70

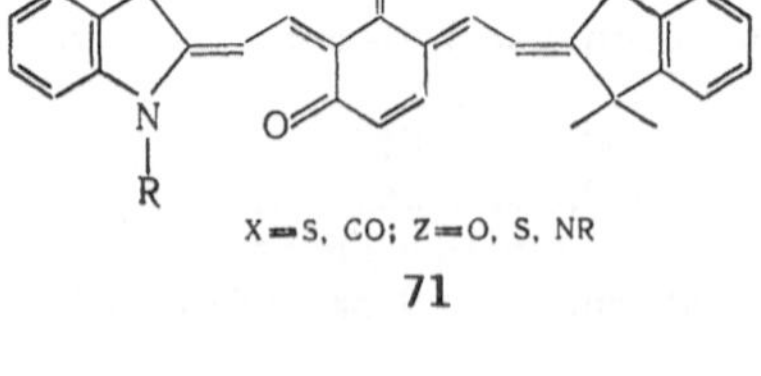

X=S, CO; Z=O, S, NR

71

Two quanta of light are required for both pyran rings in the bisspiro-pyrans to open. This does not offer any special advantages for the usual standard application of spiropyrans. Owing to large polarization, merocya-nines exhibit pronounced solvatochromism. No quantum-chemical work devoted to the effect of solvent on the light absorption of merocyanines has been reported. (For the theoretical approach to solvatochromic effects see paper [86].)

So far no experimental data are available on transition energies for S → T. Calculations performed by the SCF PPP method with parametrization for the triplet states taken from the paper [35] show that merocyanines from indolinospiropyrans, spirodipyrans, and spirothiopyranopyrans have triplets whose energies approach the energy of the ground state. Needless to say, this makes experimental observation of these transitions difficult.

Chromenes and Related Compounds

In photochromic chromenes, thiochromenes, selenochromenes, dihydro-quinolines, 1,2-dihydronaphthalenes, and fulgides the C–X bonds (X = O, S, Se, NR, CR_2) [87,88] split as they do in the case of spirans. Coloration of the open form here is attributed to the formation of the ortho-quinoid structure. Benzannelation of positions 5,6 or 7,8 as well as substituents in position 2 (which influence the conjugation) enhance the stability of the colored ortho-quinoid system, whereas benzannelation of positions 6,7 or sterically large groups in position 4 make the chromene more stable.

X = O, S, Se, NR, CR_2

72

X = O, S, Se, NR, CR_2

73

Ring opening occurs in the vibrationally excited first singlet state. This feature of the reaction can be explained with the help of the following arguments [87]. The potential curve of the vibrational states has a very flat minimum in the S_1-state so that the dissociation of the C–O bond occurs during vibrational excitation and not on excitation of the 0 → 0 transition; or the potential curve of the dissociative state intersects the curve of the S_1-state near or at the first vibrational level; or valence isomerization proceeds through the T-state which can be populated from the S_1-state only through absorption of at least one vibrational quantum.

Lukyanov et al. [88] found that selenochromenes are photochromic com-pounds. Experiments involving sensitization and quenching do not confirm the participation of triplet states in the reaction. Systematic quantum-chemical calculations for the interpretation of the reactivity of photo-chromic systems 72 [88] were performed in different variants of the π-elec-tronic PPP approximation and the CNDO/S valence approximation [38]. To evaluate the reactivity it was necessary to use the values of electron densities, bond orders, and Wiberg bond indices (the value used in the CNDO/S approximation to characterize bond strength) in the ground and S_1- and T-states. In its ground state the pyran ring contains more or less "pure" ordinary and double bonds. Upon electronic excitation to the S_1- or T-states the bonds even out. This change in the electronic structure is favorable to electrocyclic ring opening. The construction of correlation diagrams for this electrocyclic reaction already explains the exclusively photochromic and not thermochromic behavior of this class of compounds: occupied molecular orbitals of the cyclic and open forms of chromenes do not correlate with each other.

There are contradictions in the descriptions of excited states of 1,2-dihydronaphthalene. While Minkin et al. [89] ascribe to the S_1-state σ,π^*-character, Tinland and Decoret [90] consider states from S_1 to S_4 as π,π^*-excited; only the transition at 6.11 eV, judging by the calculations in the CNDO/S approximation, has σ,π^*-character. In the paper [90] photocleavage of the carbon—carbon σ-bond is attributed to polarization and the weakening brought about by the movement of electron density from saturated carbon atoms to the unsaturated residue.

In the groups of compounds under consideration, information on the triplet state is important. The lowest triplets of thiochromene and chromene have π,π^*-character, 1,2-dihydronaphthalene in the T_2 state has a noticeable σ,π^*-contribution. The T_1-state of thiochromene has a greater energy than the chromene triplet.

Correlation diagrams for electrocyclic opening of the nucleus of 1,8a-dihydronaphthalene show that this reaction should proceed both under the influence of light and heat. The slight changes that populated molecular orbitals undergo during ring opening made it possible to assume that triplet states participate in the reaction.

To evaluate the effect of substituents on quantum yields of the photochemical cyclizations of fulgides into 1,8a-dihydronaphthalenes [91], the sums of free valences for the excited state of the atoms involved in the cyclization were calculated on the basis of the wavefunctions computed in the PPP approximation.

$$\sum F_r^* = F_1^* + F_2^* \sim \phi \,(74 \rightarrow 75)$$

74   **75**

The quantum yields of cyclization (ring formation) were found to be in direct proportion to the sums of the free valences. A comparison of various reaction indices as they applied to photochemical cyclization of the fulgides showed that the free valences were best suited for the excited state; electron populations, localization energies, and bond orders sometimes yield inadequate results, especially so when describing the effects of substituents [92].

A quantum-chemical description of the electronic spectra of chromenes, 1,2-dihydroquinolines, 1,2-dihydronaphthalenes, and their colored isomers can easily be done by conventional methods. The lowest excited states of cyclic isomers have π,π^*-character [89]. Noncyclic ortho-quinoid structures are noted for a wide structured absorption in the 400-500 nm range. The broadening of the absorption band may be attributable to the existence of several isomers in a state of equilibrium (76-79).

76   **77**   **78**   **79**

The excitation energies of the four isomers 76-79 have been calculated in the PPP approximation [66]. According to the calculations, the longest-wavelength bands of different isomers, with the exception of the absorption of isomer 76, are closely grouped together. The absorption band calculated for isomer 76 lies roughly 0.3 eV in the longer-wavelength region than that for isomers **77-79**. It follows from steric considerations that a rapid s-cis-trans-isomerization should be expected for isomer **76**.

The lowest excited states have a π,π^*-character; in the case of the ortho-quinoid acyclic structures formed from 1,2-dihydronaphthalenes, the forbidden transition σ,π^* corresponds to the S_1-state (calculation by the CNDO/S method). The form photoinduced from the thiochromenes absorbs 100-120 nm in a longer-wavelength region than the same form from chromenes. Benzannelation shifts the absorption bathochromically. Slight changes in the dipole moment upon electronic excitation ($\Delta\mu \approx 1$ D) account for the absence of solvatochromism in these systems.

The papers [91,93] examine in some detail the electronic spectra of fulgides and the corresponding dihydronaphthalenes. Using structures **80** and **81** in a state of equilibrium as an example in an analysis of natural orbitals, the authors demonstrate that the S_1-state is characterized basically (in 47% of the cases) by a local excitation in the B part of molecule **81**, while the CT* transitions (B → A) contribute only some 10% to this state ($CT_{B \to A}$); the S_1-state of compound **80** is described well by a local excitation in the A fragment (~50%), and the contribution of the CT transition from B to A is approximately 25%.

Useful qualitative conclusions can be drawn from this description of electronic states regarding the effect of substituents on light absorption. The contribution of atomic orbitals (AOs) of the phenyl rings to the full wavefunction of the B portion of the molecule of the colored photoinduced form is small. Therefore, the effect of substituents in the phenyl residues on the bands in the absorption spectra should also be insignificant. The substituents in the A fragment of compound **81** can hardly enhance this effect either since, as mentioned earlier, the CT contribution to the transition energy is rather small. The results of numerical calculations of the substituted fulgides [91] confirm these regularities. By substituting sulfur for the carbonyl oxygen a long-wavelength broadening of the absorption region can be achieved.

80   **81**

The spectrum of the dihydronaphthalene isomer can be varied over a wide range by replacing the heteroatom in the A portion of the molecule or, in view of the greater CT contribution to the electron transfers, with the aid of various substituents in the condensed phenyl ring of the B fragment of the molecule.

Reversible Photocyclizations and Cycloadditions

Reversible photodimerization of aromatic hydrocarbons and their derivatives is responsible for considerable changes in electronic spectrum and refractive index. Changes in the refractive index are sufficiently great for them to be used in creating and reading a phase hologram. Especially effective by these criteria are such photochromes as anthracene derivatives and acridizinium ions.

Qualitative discussions based on the properties of the symmetry of molecular wavefunctions give a general approximate picture of dimerization

---

*CT — charge transfer.

and make it possible to reflect the influence of structural features of the components on the ease of dimer formation. What is more, the calculated quantum-chemical reaction indices give an idea of the reactivity of the individual representatives of these classes of compounds as well. Calculations for the changes in the electron densities of the substituted anthracenes of type **82** using the Longuet-Higgins ideas show that considerable asymmetry of the electron distribution in meso-positions favors dimer formation [94].

R, R', R'', R''' = H, CH$_3$, Cl, CN

82

The results of the calculations of localization energy as the difference between the energy of π-electrons of the reacting starting system (the energy of π-electrons of the hydrocarbon in the ground and excited states) and the transition state (the delocalized model for the transition state of the reaction) are in agreement with the experimentally determined dependence of photodimerization efficiency on the structure in the series of anthracene derivatives [95].

The electron populations by Mulliken [97], calculated [96] by the extended Hückel method, were found to be a suitable measure of reactivity for cyclization and photodimerization [48]. In this process, electron populations are calculated for the k and l atomic pairs between which a new bond is formed during the reaction, or the difference between these populations $\Delta n(k,l)$ for the electronically excited and ground states.

This reaction index was successfully used to describe the photodimerization of anthracene and 2-methoxynaphthalene [98]. Two interacting molecules are treated as a supermolecule, and electron populations for the ground and excited states are calculated for different intermolecular distances R. The ground state of the supermolecule is destabilized by all R's under consideration. In the excited state the n(k,l) values for positions 9 and 10 indicate a strong binding in all the values of R considered. The electron energies of the first excited state calculated by the extended Hückel method are minimal with distances of 0.5-0.3 nm. Analogous studies of 2-methoxynaphthalene **83** revealed the strongest binding interaction in the excited state [$\Delta n(1,4) = 0.022$ at 0.3 nm] of the centrosymmetric dimer structure in which only one ring of each of the components interacts [98].

The n(k,l) or $\Delta n(k,l)$ values were used also when interpreting the photocyclization of stilbenes, diarylethylenes **84**, azastilbenes, furyl- and thienylethylenes, 1,4-diarylbutadienes, and their oxygen- and sulfur-containing analogues [98,99].

83

Ar   Ar
(hetaryl) (hetaryl)

84

The photocyclization of stilbenes and related compounds [28] is a photochromic process which until recently did not find applications because of the ready oxidizability (e.g., by atmospheric oxygen) of the colored intermediate compound. The substitution of methyl groups for the hydrogen atoms helps to stabilize the colored photoproduct towards oxidation. In view of the great number of potential photochromic compounds of this type and the exceptional cyclization selectivity [28,100,101], more workers are expected to turn their attention to this area.

The following values were successfully used in calculating reactivity indices in photocyclizations:

a. The sum of free valences in the excited state $\Sigma F^* = F_r^* + F_s^*$ for positions in which cyclization occurs [28,100].

$$\left(F_\alpha^* + F_{\alpha'}^*\right) > \left(F_\alpha^* + F_\beta^*\right) > \left(F_\beta^* + F_{\beta'}^*\right)$$

85

b. Localization energies in the excited state [28,100] $L^* = E_\pi^*$ (parent molecule) $- E_\pi$ (transition state).

86   **87**   **88**

The models of the transition state **86-88**: $L_{\alpha\alpha'}^* < L_{\alpha\beta}^* < L_{\beta\beta'}^*$.

c. Electronic populations of Mulliken bonds $n(k,l)$ or population difference $\Delta n(k,l)$ [48,99].

It is assumed that the reactivity of a compound is determined by electronic factors exclusively, while steric factors apparently have no role to play. This is evidenced by the strict selectivity of the photocyclization of 1,2-diarylethylenes or distyrylbenzenes with products often forming as a result of the sterically least favorable reaction path.

The question of the structure of light-induced colored compounds from bianthrone and its structural analogues, which seemed solved in the 1960s, has again become open today [102]. To this day it is not clear whether the photochromism of dehydrobianthrones is due to an electrocyclic reaction or to bond rotation. The photochromism of compounds **89** is described in analogy with the photochemical cyclization of stilbene via the equilibrium of the cyclic and open forms [103,104].

X, Y = CO, O, S, NR

89   **90**

By substituting methyl groups for the hydrogen atoms in positions 4,5' or 4',5, stabilization of the colored photoinduced forms can be achieved. The arguments in favor of the proposed structure of the colored form were obtained as a result of the quantum-chemical calculations by the PPP method. The calculated energies of electron excitation of various possible structures of colored compounds were compared with the experimental spectrum with a view to obtaining indirect evidence for the structure of the colored compound [105]. Structures **91-95** were discussed: the low-energy triplet state of the form **91**, conformation **92**, rotated with respect to the central double bond; biradical **93**, composed of two planar anthrone molecules rotated through 90°; a planar biradical with the meso-naphthoquinone structure **94**;

cyclization product **95**, which is a valence isomer of the parent compound; the nonplanar conformation in the form of a double chair **96** of the ground state.

| 91 | 92 θ 30°; 93 θ 90° | 94 | 95 |

For these structures, the π-electron densities and π-bond orders were calculated besides the spectral data. It is hard to see what the authors proceeded from in assuming such a strong bathochromic shift involving severe disruption of geometry in the ground state as a result of the transition from the planar structure **91** to the double-chair conformation **96**. The calculated S $\rightarrow$ S* and T $\rightarrow$ T$_n$ absorption spectra of the parent compound **91** within the framework of the chosen approximation (the quantum-chemical method, parameters, and geometry) were in good agreement with the experimental data. Calculations do not favor a low-energy triplet state as a photo-induced compound. Structures 92-94 and **96** do not account for the absorption of the green-colored photoinduced form. Only for structure **95** was a complete correspondence between the calculated spectral data and the experimental spectrum obtained.

Further experimental studies of the chemical and spectroscopic properties of photoinduced forms allow a series of cogent arguments against structure **95** of colored particles to be marshalled [102,106]. Thus, colored structures including those of unsubstituted compounds in positions 4,5′ and 4′,5 are absolutely resistant to thermal and photochemical oxidation. In the case of structure **95** (especially in the unsubstituted compound) one could expect a rapid oxidation accompanied by aromatization, which makes equilibrium irreversible. Then again 3,4,5′,6′-dibenzodehydrobianthrones display photochromic properties, although the formation of a structure analogous to **95** seems rather unlikely on account of steric hindrances. Finally, the carbonyl band apparently disappears from the infrared spectrum of the photoinduced form, and a new as yet unidentified band appears. As a result, the problem remains unresolved.

The Photochromism of Indigoids

The quantum-chemical approach proved very useful in the search for the primary chromophore of indigoids. Calculations of the energies of electron excitation of systematically diminishing elements of indigoid structure showed that the transverse arrangements of electron-donor and acceptor groups in pairs over the central C=C double bond (**97**) corresponds to the main spectral characteristics of indigoids (the primary chromophore) [107] (cf. the data cited in papers [108,109]). In the series of heteroatoms X = NH, Se, S, O the energy of the electron excitation of the indigoids grows.

97

It should be noted that only thioindigoid derivatives (X = S) and compounds with X = Se, O display photochromism; compounds with X = NH are not

photochromic. The reason for this is believed to be the relatively strong
H...O H-bonds, which increase the activation energy of photochemical isomerization so as to make it unobservable. However, the problem appears to be
more complex than that (see Chapter 2). In N,N'-diacetylindigo, photoisomerization is achieved upon irradiation in the long-wavelength absorption
band [110]. Due to the relatively considerable difference between the
absorption maxima of the isomers (see Chapter 2), this compound may be regarded as a promising photochromic system from the standpoint of absorption
parameters. Analogous properties were discovered in N,N'-dimethyl- and
N,N'-diethylindigo [111]. The great polarization of indigoids in the excited state compared with the ground state accounts for their positive
solvatochromism.

The reversible light-induced color change in photochromic indigoids is
due to E → Z-isomerization about the C=C double bond [112]; the Z-isomer
absorbs in the shorter wavelength region than the E-isomer ("negative photochromism"). Previously, spectral changes were attributed to the nonplan-
arity of the Z-isomer on account of the repulsion of the oxygen atoms of the
carbonyl groups [113,114]. To establish dependences between the absorption
spectra and structure of the thioindigoid compounds [115,116], quantum-
chemical calculations by the PPP method were performed. The spectra of the
Z- and E-forms were successfully interpreted with the help of ideas on the
planar arrangement of the molecule's framework. As a result, the deviation
from the planar arrangement was ruled out as the reason for isomer spectral
differences.

The real reason for the different absorption spectra of the E- and
Z-forms is believed to be interactions between unbound S and O atoms (located in different nuclei) which have different magnitudes. The separation
of the calculated energy of electron excitation into one- and two-center
contributions leads to the conclusion that the interactions between unbound
S...O, O...O in the E- or Z-structure have a varying effect on absorption.
The primary contribution to the spectral shift comes from the S...O interaction; its numerical value may vary with the change in the distance between
S and O; the increased S...O interaction due to the shorter distance between
S and O corresponds to the bathochromic shift. The distance between S and O
can be reduced by reducing the S—C—C—C—O angles. Because of the different
S...O distances, compounds 98 and 99 absorb at different wavelengths, despite the fact that the π-electron systems in them are comparatively great
and the spectra should not be sensitive to differences within these structures.

98; λ_{max} 624 nm

99; λ_{max} 562 nm

Interesting conclusions regarding the dependence of light absorption on
the structure and character of interaction between unbound S and O atoms
were drawn from a theoretical study of structures 102-104.

100

101

102

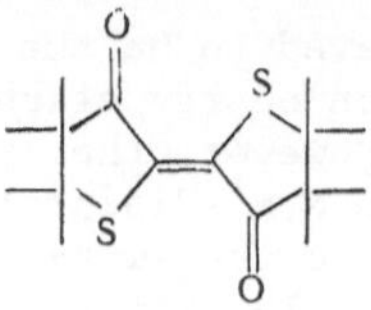

103 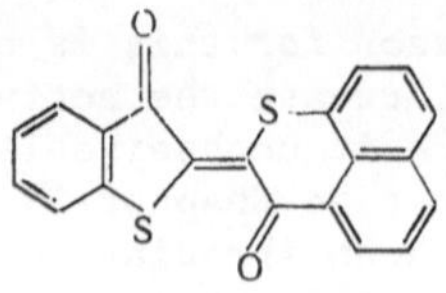   104

Judging by the calculation performed within the framework of the unbound interactions method, deviations from the planar arrangement by 15, 45,
and 45° in the E-isomers of compounds 101, 102, and 99 and by 30, 20, and
45° in the case of the Z-isomers of compounds 102, 98, and 99 have little
influence on the longest-wavelength absorption (a maximum of 10 nm at a
rotation through 30°) and are unnecessary for photochromism. The smallest
spectral shifts during photoisomerization are exhibited by thioindigoids
with only a single S...O structural unit (101, 102); the largest spectral
changes should be expected of compounds with the least S...O distance in the
E-isomer (104, 98).

Photoisomerization of Azines

Certain aromatic azines display photochromism due to E → Z-isomerization about the C=N bond [117]; this class of compounds exhibits negative
photochromism, i.e., the photoinduced form absorbs at longer wavelengths
than the starting form.

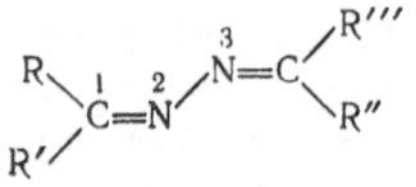

105

Two main types of photochromic reaction are discussed.

a. A $\underset{h\nu'\Delta}{\overset{h\nu}{\rightleftarrows}}$ B (105, R = R' = Ph, R'' = H, R''' = 9-anthryl).   The
s-trans-1E-3E-configuration is proposed for structure A.   The latter is
photochemically converted into B — the s-trans-1E-3Z-configuration.   In it
the anthryl residue no longer lies in the plane of the azine skeleton because of steric hindrances.   One sign of this is the hypsochromic shift of
the longest-wavelength band and the appearance of anthracene spectrum bands
that display a characteristic vibrational structure.

b. A $\underset{h\nu_3, \Delta}{\overset{h\nu_1}{\rightleftarrows}}$ B $\underset{h\nu_4, \Delta}{\overset{h\nu_2}{\rightleftarrows}}$ C (105, R = R'' = H, R' = R''' = 9-anthryl).
Structure A possesses the s-trans-1E-3E-configuration, B the s-trans-1E-3Z-
configuration, and C the s-trans-1Z-3Z-configuration.   The first quantum-
chemical calculations [117] by the simple Hückel method showed that the C=N
bond order in the excited state is lower than in the ground state by 25%,
while the N—N bond order is 70-80% higher.   This means that the barrier of
rotation around the C—N bond in the ground state is three times this value
for rotation about the N—N bond, while in the excited state it is some 50%
lower and in terms of order of magnitude is comparable with the increased
barrier of rotation about the N—N bond.

Calculations in the PPP approximation [118] for structures rotated at different angles on the C—C (9-anthryl) or N—N bond show that a great many conformers possess electronic spectra of photoinduced forms. This means that the search for the most stable conformation should also involve, apart from the electronic spectra, structural stability criteria. Calculations of the relative energies of different structures by the SCF PPP method with allowance for the interactions of unbound atoms [119] and a comparison of theoretical spectra with the experimental ones indicate that the most stable E-configuration is a structure with $\theta_{N-N} = 0°$ and $\theta_{C-C(9-anthryl)} = 30°$; the Z-isomer is at its most stable at $\theta_{N-N} \sim 0°$, $\theta_{C-C(9-anthryl)} = 60°$ or $\theta_{N-N} = 0°$ and $\theta_{C-C(phenyl)} = 40°$ [120,121]. The same conclusion is drawn from a comparison of the calculations of the spectra with the experimental data.

The negative photochromism of anthraldazines is attributed to the bathochromic shift of the longest-wavelength band which occurs as a result of the E/Z-isomerization about the C=N bond being offset by the hypsochromic shift because of the rotation around the N—N bond or the C—C (9-anthryl) and C—C (9-anthryl') bonds [121].

The possible influence of n-electron pairs of nitrogen atoms in the azine chain on structure is poorly reflected by the standard SCF method in the CNDO/2 and INDO* approximations widely used at present. Of the existing semiempirical methods employed, the self-consistent field technique in the NDDO** [122] valence approximation, which takes into account the anisotropy of two-center coulomb interactions, best meets the requirements. 1-(α-Naphthyl)formaldazine 106, 107 [123] was used as a model compound for such studies. Depending on the angle of rotation θ around the C—C (α-naphthyl) bond, total energies were calculated and the energy minimum found at $\theta_{C-C(\alpha-naphthyl)}$ equal to 30 and 150°. It was found that the results of this calculation lead to structures that are optimal by the data obtained by the SCF PPP method with the interactions of unbound atoms taken into account. Therefore, the influence of the unshared electron pair of nitrogen is insignificant and other interactions of unbound atoms could play a role. An analysis of the energy of the electrostatic two-center interaction with respect to the calculated total energy indicates that with rotamer configuration 106 the decisive influence on stability is exerted by interactions between N...H and N...C, while in the case of rotamer 107 – between H...H [123,124]. Rotamer 106 is slightly more stable than rotamer 107.

106   **107**

Hydrogen Migration in Photochromic Systems

Photochromism due to hydrogen transfer is known for several types of structure [1]. Despite the existence of special quantum-chemical concepts of the pathways of hydrogen transfer [125], specific quantum-chemical calculations have been performed only for the anils of salicylaldehyde [126], 2- and 4-(o-nitrobenzyl)pyridines, and o-nitrotoluene. For the anils of aromatic o-hydroxyaldehydes the following conversion sequence was proposed

---

*INDO — intermediate neglect of differential overlap.
**NDDO — neglect of diatomic differential overlap.

on the basis of the results of low-temperature photolysis in rigid paraffin matrices [127-129]

Irradiation of the colorless E-enol **108** at a low temperature produces the keto form **109** which, with an increase in temperature, turns into the Z-keto isomer **110**, stabilized by an intramolecular hydrogen bond. At the same time, the photo-E → Z-isomerization of the enol around the C=N bond is observed, which leads to Z-enol **111**. The quinoid tautomer does not participate in the thermal reverse Z → E-isomerization. The Z-keto isomer **110** is in a state of thermal equilibrium with E-enol **108**, and its concentration is solvent dependent; in alcoholic solutions it is quite considerable. According to available data [127], keto form **109** in the ground state arises from the excited singlet of the Z-keto isomer, which in turn arises adiabatically from the n,π*-state of the E-enol along with the intersystem crossing of the latter into the n,π*-triplet and subsequent deactivation.

$$108 \xrightarrow{h\nu} \pi, \pi^*\,(108) \longrightarrow n, \pi^*\,(108) -\begin{cases} \longrightarrow (110)^* \longrightarrow 109 \\ \longrightarrow {}^3n, \pi(\text{enol}) \longrightarrow \text{enol} + h\nu \end{cases}$$

$$109 \longrightarrow 110 \longrightarrow 108$$

Thus, the photochemical reaction **108** → **109** while irreversible, per se, becomes reversible only due to the bypass route through the thermal reactions **109** → **110** → **108**. Potashnik and Ottolenghi [130] established that photoisomerization **110** → **109** as well as **108** → **109** occurs in fluid solutions as opposed to liquid paraffin media. In systems having a low polarity, the dark-fading relaxation of E-keto isomer **109** into enol **108** proceeds through an intermediate dimer; the latter's excitation results in dissociation to enol monomers. Calculations of the π-electronic structures of salicylidene-anil in a variety of electronically excited states were carried out by the semiempirical SCF PPP CI method with a variable β [126]. It was shown that upon electronic excitation of the enol form into the S_1-state the C–O bond should be shortened, which follows from the calculated increased order of this bond; similarly, by following changes in the bond orders we may establish that the C=N bond will become extended while the C–C=N and N–(C₆H₅) bonds become shorter at the same time. This involves some structural changes — an increase in the angles C–C=N and C=N–(C₆H₅), and increased distance between oxygen and nitrogen. The electron density on the oxygen atom decreases by 0.1 π-electron because of the gain in the C–O bond; the π-electron density on the nitrogen atom increases the alkalinity. All this means that changes in the electronic structure on S_0 → S_1 excitation correspond to the formation of a quinoid structure.

Parallel to photoenolization, E → Z-isomerization proceeds possibly through an electronically excited state; added evidence of bond weakening, and thus an easier isomerization course, comes from the π-bond order, which has a lower value in the excited state than in the ground state.

The photochromic reaction of o-nitrobenzylaryls or -hetaryls apparently has a similar mechanism (see Chapter 4).

113	112	114 [131]

The aim of Tinland's paper [132] was to establish which of the two possible hydrogen-transfer reactions 112 → 113 or 112 → 114 is theoretically more favorable. To this end, quantum-chemical calculations were performed in the PPP π-electron approximation for the ground state and for the S_1- and T_1-states of compounds 112-114. Judging by the results upon electron excitation of o-nitrobenzene compounds from S_0 into S_1, the electron density on the pyridine nitrogen atom decreases (a drop in basicity) while the electron density on the oxygen atom in the nitro group jumps (an increase in basicity). The change in electronic structure upon excitation is evi-

dence in favor of the mechanism 112 $\overset{h\nu}{\to}$ 112* → 114* (or 114) (statistical examination) which, however, fails to be completely in agreement with the experimental data available. No reliable conclusion can be drawn on the basis of the calculated data regarding the multiplicity of electronically excited states in which hydrogen transfer would have the least obstructed passage.

DISCUSSION OF THE MECHANISMS OF PHOTOCHROMISM USING CURRENT IDEAS ON THE POTENTIAL ENERGY SURFACES

Potential energy surfaces are a function of the potential energy of a system from all nuclear coordinates. To characterize and describe a chemical reaction it is enough to examine individual parts of the potential energy surfaces (potential surface). In most cases, workers confine themselves to the cross section of the potential surface along the reaction coordinate (the potential surface profile). This energy profile [133] gives an idea of the mechanistic, structural, and energetic aspects of chemical conversions generally characterized by the minimum energy path [134,135].

Apart from the fact that potential surfaces serve as sources of information on the geometry of various excited states, approaches to the experimental definition of which are rather limited, they also provide essential primary data for a discussion of the course of photochemical reactions as well as for estimating equilibrium constants and rates of conversion. Unfortunately, work on this problem is still in the embryonic stage.

Potential surfaces are characterized by local minima and barriers. Minima on the potential surface of electronically excited states are designated as "spectroscopic minima" because of the possibility of studying them by means of spectroscopic techniques due to the relatively long lifetime of a system in this state. Starting with this minimum, by means of changing the geometry of a molecule (chemical reaction) another "nonspectroscopic minimum" can be attained upon overcoming the barrier; this corresponds to the completion of a definite stage of the photochemical reaction. The height of the barrier between both minima has a decisive effect on the molecule's properties: high barriers inhibit the photochemical reaction, while unduly low barriers account for a process which runs a nonuniform course (e.g., the reaction may prove to have a low stereospecificity).

The more or less considerable convergence of potential surfaces of the same multiplicity creates areas of deviation from the intersections. In such an area the molecule effects a very rapid radiationless transition from a higher energy surface to a lower energy one (internal conversion). If potential surfaces of different multiplicity closely resemble each other, we speak of areas of "non-avoided crossing" in which transition between potential surfaces of different multiplicity is minimally forbidden and the molecule can undergo an intersystem crossing as a result of spin—orbital interaction.

The conversion between the potential surface of an excited state in the region of a nonspectroscopic minimum and the maximum on the potential surface of the ground state results in the formation of a "funnel" through which the activation onto the potential surface of the ground state quickly takes place. The distance between two potential surfaces in the "funnel" region has a substantial effect on the stereochemical results of a photochemical reaction. The short distance accounts for a brief presence in the "funnel" region and thus for the stereospecific coordinated course of the reaction when the potential surfaces of the ground and excited states are energetically subject to a severe energy separation which may result in a metastable photointermediate which in most cases proceeds to yield products of a nonstereospecific, uncoordinated reaction [8,9,20-24].

Quantum-chemical techniques of varying degrees of approximation are suitable for calculating potential surfaces. The chosen areas of the surface close to the minima and saddle (stationary) points are calculated point by point. A more elegant approach consists of calculating stationary points by means of geometry optimization (see, e.g., [37]). Mehlhorn and Dietz's paper [9] contains a review of the application of some of the approximate quantum-chemical techniques used to solve these problems. At the moment, reliable quantum-chemical calculations of potential surfaces or potential curves can be performed only for small molecules and simple chemical conversions which are of no immediate interest in the context of photochromic processes and which can only serve as their models.

On the other hand, substantial headway has been made with the use of a combined technique. A well thought-out combination of calculated quantum-chemical parameters and experimental data for the initial material, product, or intermediate in a variety of electronically excited states with various qualitative considerations (e.g., the principle of conservation of symmetry during the course of certain chemical reactions or Salem's ideas on the nature of intermediate compounds or structures which correspond to the minima on a potential energy surface) now makes it possible, by means of qualitative and semiquantitative energy profiles, to form an idea of the course of many types of photochemical reactions. However, Turro et al. [136] caution against an uncritical use of such structure—energy diagrams. One should always bear in mind that reaction energy profiles so schematized are but an extremely simplified, if graphic, representation of potential surfaces from which it would be too much to hope to derive more than a very general qualitative picture of the reaction involved.

E/Z-Photoisomerizations

The study of ethylene isomerization has yielded fundamental data on the course of the reaction involved [137]. In this case the reaction coordinate can be graphically represented by means of the angle of rotation θ around the C—C bond. The point of departure in qualitative considerations is the question of the nature, structure, and energetics of the ground and excited states of the starting and end products as well as of the intermediate product. As the MO energy change diagram (Figure 5) indicates, the energy of the bonding π-MO grows with an increase in the angle of rotation, the bond is progressively weakened as a result of the diminishing overlap, and the energy of the antibonding π-MO declines. At θ equal to 90° between two $2p_z$-AO which form a π-bond in the planar molecule, no interaction exists, to all intents and purposes. Subsequent rotation leads to a further π-MO energy drop and the energy of the $\pi*$-MO goes up again.

A more clearly defined picture emerges as one moves from representing the process involved with the help of MOs towards a diagram of energy levels of electronic states, e.g., calculated quantum-chemically. The following considerations are useful for a relative evaluation of the energy levels of the ground and excited states and the intermediate. A molecule rotated through 90° can have a biradical D or zwitterionic Z structure with both electrons of the π-bond occupying one atomic orbital each (in the form rotated through 90° both π-electrons are unpaired) or, correspondingly, both electrons occupy one AO with the other remaining unoccupied (Figure 6). The rotation causes the electron–electron interaction to diminish further. This signifies a diminishing energy of the electron states for which the π-bond has no role ($S_2 > S_1 > T_1$). The electron energy of the ground state, for whose stability the realization of the π-bond is essential, grows as the deflection angle increases. These qualitative data, placed in the context of the energy profile of the reaction, give the picture presented in Figure 7 (see [138]). Both possible isomerization mechanisms are clear from this representation.

Isomerization through the S_1 excited state (see Figure 7a). The non-spectroscopic minimum ("funnel") is achieved during movement across its potential surface along the reaction coordinate and the rotation about the

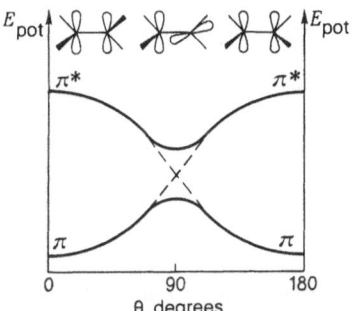

Fig. 5. Change in the energy of the π-MO of ethylene upon rotation about the C=C bond.

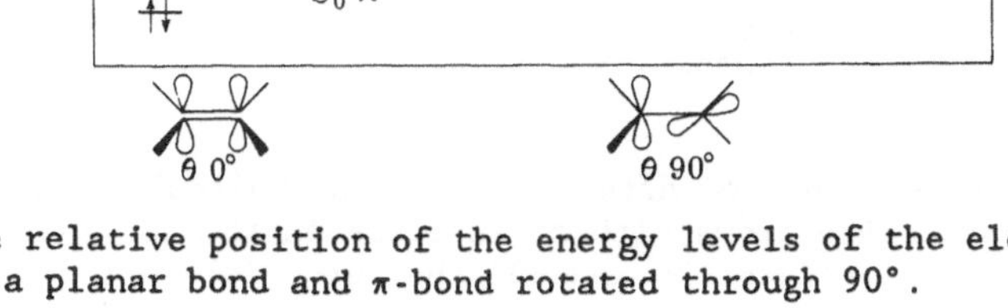

Fig. 6. The relative position of the energy levels of the electronic state of a planar bond and π-bond rotated through 90°.

31

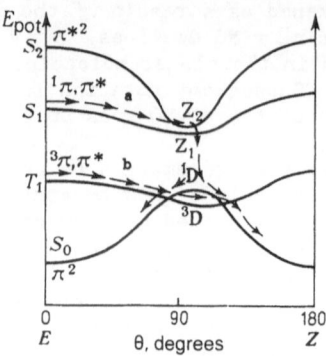

Fig. 7. Profiles of the potential surfaces of the ground and excited
states upon a qualitative consideration of E/Z-isomerization.

C–C bond (possibly through a low barrier). A very rapid radiationless deactivation results from this state onto the potential surface of the ground
state which exhibits a maximum in this region. In turn, a rapid geometric
relaxation results from this maximum either into the original E-configuration or into the Z-configuration as a photoproduct.

As the rotation angle about the C–C bond increases, the system moves
from the T_1 excited state (achieved either by direct excitation or by conversion from another excited state or through a T–T-energy transfer) to the
area of the intersection of the potential surfaces of the S_0- and T_1-states,
where an intersystem crossing to the potential surface of the ground state
appears very likely (see Figure 7b).

Stilbene calculations provide a typical example of a theoretical description of E/Z-photoisomerization. Quantum-chemical investigations cover
a very wide range of problems ranging from calculations of the adiabatic
potential energy curves of various electronically excited states [139,140],
to evaluations of the likelihood of the intersystem crossing [141], to a
description of the reaction in the nonstationary, dynamic mode [142,143].

A description of excited states within the framework of the SCF PPP
method taking into account only singly excited electronic configurations for
a solution to the problem of the states (S_1 or T_1) through which the reaction proceeds does not give unambiguous results [144]. Momicchioli et al.
[145] investigated this reaction by the SCF CI technique, allowing for
singly excited configurations. They found that the S_1- and T_2-states reach
the energy maximum at a 90° angle of rotation around the C–C bond; T_1, however, lacks this maximum (Figure 8a). Calculations showed that the T_1
potential surface intersects the saddle point (trough) which separates the
E- and Z-isomers. An evaluation of the likelihood of intersystem crossing
due to spin–orbital interaction indicates that it is fairly strong even at
small values of the rotation angle [146,147]. All these considerations
taken together seem to indicate a possible isomerization path through the
triplet state. On the contrary, results of later work [148,149] favor a
singlet mechanism. The importance of doubly and multiply excited electronic
configurations for describing reacting electronically excited states becomes
clear from these studies especially when we examine the nonplanar structures
of π-electron systems [144].

The S_1-state is populated after the stilbene molecule becomes excited;
the energy of this state grows with an increase in the angle of rotation.
At rotation angles of between 30 and 60° the potential surface of the

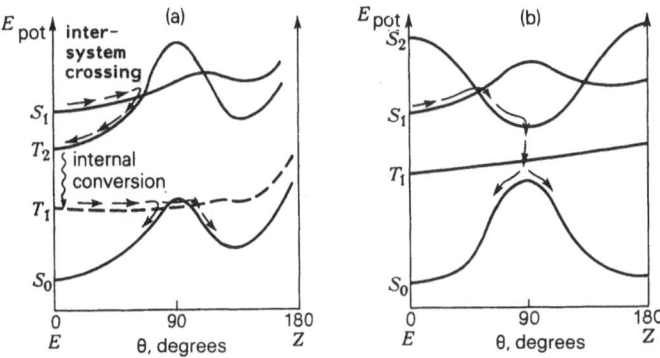

Fig. 8. Profiles of the potential surfaces upon the photoisomerization of stilbene: a) according to the data of Momicchioli et al. [145]; b) according to the data of Orlandi and Siebrand [148].

S_1-state intersects the potential surface of the second S_2 excited state. A radiationless transition from S_1 to this potential surface takes place in the area of the intersection which then leads to a nonspectroscopic minimum ("funnel"). From there the system very quickly moves to the maximum of the potential surface of the ground state, and finally a geometric relaxation occurs into an E- or Z-configuration (Figure 8b). The results of experimental studies can be successfully interpreted with the help of these ideas [150,151].

The foregoing arguments cannot be unconditionally applied to the isomerization of other structurally related photochromic compounds. Even relatively small structural changes can influence the character of such a reaction [151,153], and polar substituents or a heteroatom in the molecule may produce profound changes in the wavefunctions and contribute to the formation of zwitterionic intermediate structures (in such cases the influence of a solvent cannot be ignored). As a result of this, a quite different type of reaction can of the be initiated; e.g., upon reaching the zwitterionic singlet excited state a competition between isomerization and cyclization may occur.

Compounds with C=N and N=N bonds can isomerize not only by the rotation mechanism but by inversion as well. In this case it is convenient to use a a change in the angle α ($\overset{\alpha}{\underset{}{>}}C\overset{..\overset{R}{\diagup}}{=}N$) as the reaction coordinate. An energy barrier in the triplet state upon rotation around the N=N bond was discovered by quantum-chemical calculations of E/Z-isomerization (see Chapter 3).

Of great importance in the photochromism of spiropyrans are the equilibria of the cis-trans-isomerization of the merocyanine form which follow the photochemical ring opening. Because of the large cyclic terminal groups, steric effects have an important role to play in this instance. The pentamethinomerocyanine chain, largely responsible for the color of the compound, becomes stabilized (owing to steric factors) in the s-trans-structure form through the central, formally ordinary bond. Calculations performed by the extended Hückel method for concrete compounds give curves [70] of the dependence of potential energy on the angle of rotation around the central, formally ordinary bond. It becomes clear from their examination that the form with a rotation of some 30° about this bond is the most stable of all.

There are ideas on various cis-trans-equilibria and the structure of
the final product of ring opening in spiropyrans (the most stable isomer)
which are summarized in the scheme on p. 16, which takes into account stability data. The quantum-chemical calculations of the various equilibria in
the isomerization models of the pentamethinomerocyanine chain (without the
heterocyclic end groups) largely confirm the validity of these ideas.
According to calculations by the NDDO method [123], which is suitable for
calculating the barrier of rotation around the ordinary bond, the trans-
isomer proves the most stable (in a polar solvent). Calculated values of
the solvation energy do not change the stability series of different isomers
nor the relative barrier heights for thermal isomerization in the ground
state.

Calculation in the PPP approximation, allowing for doubly excited electronic configurations [154], of the potential energy at certain points along
the possible reaction path on the potential surface of the singlet state
gives a picture qualitatively similar to that which represented the isomerization of ethylene. The special features of the electronic structure of
polymethines with charge alternation along the polymethine chain and the
possibility of the formation of dipolar structures and states with charge
transfer on rotation about the bond at the isomerization stage suggest a
considerable solvent influence. Solvation energy was estimated by the
relatively simple Hoijtink scheme [155], with the values of atomic charges
calculated by the SCF PPP method, allowing for doubly excited electronic
configurations. The results confirmed the earlier assumption that the
interaction with a polar solvent has a particularly strong stabilizing
effect on the states involving charge transfer and dipolar structures with a
90° angle of rotation of the methylene groups.

Figure 9 presents potential curves for the isomerization of trans-
pentamethinomerocyanine around the formally 4–5 double bond and the ordinary
5–6 bond. These results show that one should exercise caution in attaching

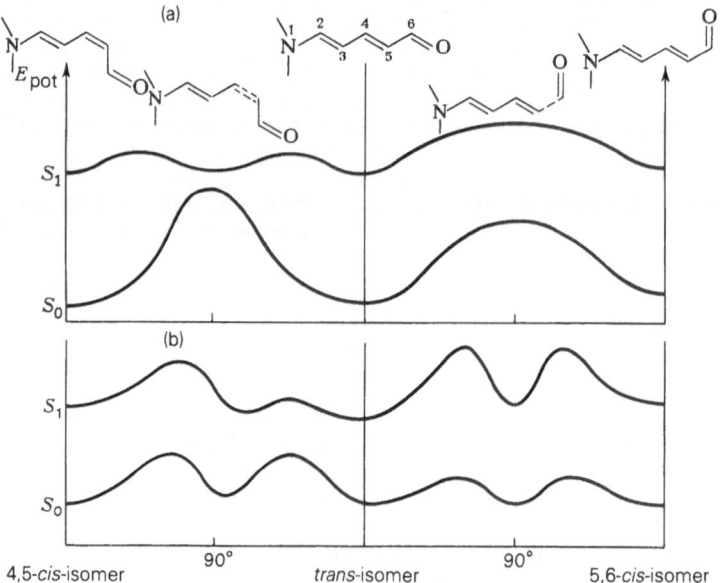

Fig. 9. Profiles of the potential surfaces upon the photo-trans-cis-
isomerization of pentamethinomerocyanine (computed by the
SCF PPP CI method, allowing for doubly excited configuration):
a) without allowance for solvent influence; b) with allowance
for the solvation by polar solvents.

unduly great importance to the numerical values involved. Thermal isomerization at the ordinary bond represents a lower barrier than in the case of
rotation around the 4—5 double bond; the reaction in the ground state proceeds predominantly through the lower barrier to form the 5,6-cis-isomer.
In the S_1-state the correlations are reversed: from the spectroscopic minimum the reaction path leads through the lower barrier to the nonspectroscopic minimum ("funnel"), and the distance on the energy scale between the
potential surfaces (curves) of the ground and excited states is not considerable. However, the relatively high barrier blocks the reaction path
towards the nonspectroscopic minimum upon rotation around the 5—6 bond. Because of the considerable distance on the energy scale between the potential
surface of the ground state and the minimum on the potential surface of the
S_1-state upon rotation around the 5—6 bond, the likelihood of a radiationless transition from the higher surface to the lower must be compared to the
situation which arises upon rotation around the 4—5 double bond. Only a
consideration of all the possible analogous equilibria made it possible to
interpret the series of isomerizations which followed the photochromic
process proper.

There are few papers dealing with other important photochromic reactions at the time of writing, notably those concerning analysis of the
potential curve surfaces of the ground and excited states. This is true, in
particular, of ring opening (closure) in spiropyrans. One possible reason
for this state of affairs is the rather considerable size of the molecules
and the relatively large number of geometrical parameters which undergo
changes along the hypothetical reaction coordinate and which should be taken
into account in the calculation.

Reversible Photocyclizations

The studies of the mechanisms of cyclization—recyclization reactions
through an analysis of potential surfaces were all confined to the model
systems containing only a small number of atoms (e.g., the photocyclization
of butadiene into cyclobutene) and, to a large extent, to a qualitative examination of the cyclization of hexatriene systems of diarylethylene and
bianthrone.

Calculations by the valence-bond method show [156] that the conclusions
made on the basis of the Woodward—Hoffmann rules are grossly simplified and
that the actual reaction course is far more complex. The energy profile
calculated for the electrocyclic reaction of butadiene indicates that both
con- and disrotatory cyclizations which proceed through the first singlet
state are unfavorable on account of the relatively high barrier. Analogous
to the case of E/Z-isomerization of stilbene, the disrotatory rotation of
the molecule's ends results in the potential curves of the S_1- and S_2-
states drawing apart (S_2 with a substantial contribution from the doubly excited electronic configuration). The radiationless transition of the system
to a lower potential surface leads to a nonspectroscopic minimum lying close
to the maximum on the potential surface of the ground state. Through this
"funnel" the disrotatorily deformed butadiene gets stereospecifically onto
the maximum of the ground state of the cyclobutene (Figure 10). Later
nonempirical quantum-chemical calculations [157] corroborate these results.

This topology of potential surfaces corresponding to the stereospecific
course of the reaction is highly desirable for photochemical conversions in
the above-mentioned example since this guarantees the homogeneous course of
the reaction without any side processes. For the photochromic system this
means a greater number of ring conversions.

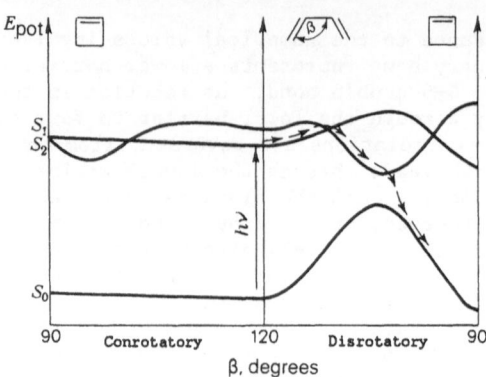

Fig. 10. Profiles of potential surfaces upon the photocyclization
of butadiene.

Even relatively slight changes in molecular structure can make a photochemical reaction impossible. The reasons for this can be understood in
certain circumstances through a simple analysis of the change in the relative position of the MOs of different compounds along the reaction coordinate. A change in the sequence for both the highest occupied and lowest
unoccupied MOs in the transition from dibenzophenanthrene 115 to hydrocarbon
116 results in the latter compound failing to undergo photocyclization
analogous to the photocyclization of stilbene to 4a,4b-dihydrophenanthrene
[99].

115   116   117

The photochromism of bianthrones was discussed with the help of potential curves built on the basis of available experimental data [158]. A
molecule slightly rotated with respect to the central double bond (angle θ)
was taken as the most stable structure. The rotation angle θ depends on the
nature of the substituents in positions 4,4' and 5,5'.

When X = Y = CO, the planar structure (θ 0°) was discovered experimentally for 2,7'-dimethylbianthrone in the colored cyclic form 90 whose energy
minimum of the ground state lies approximately 15 kJ/mole above the energy
minimum of the corresponding bianthrone (Figure 11). The potential surfaces
of the ground state of the starting bianthrone and the photoproduct inter-
sect some 60 kJ/mole above the minimum of the ground state of the photoinduced form (these values were obtained from thermal equilibrium data).
This value corresponds to the value of the barrier for the reverse thermal
reaction of the photoinduced form. The first excited singlet state of the
photoinduced form lies approximately 140 kJ/mole above the ground state
minimum, while the S_1-state of the 2,7'-dimethylbianthrone lies 260 kJ/mole
above the ground state minimum (spectroscopic values). Electron excitation
is followed by a rapid vibrational relaxation, as a result of which the
system gets into the area of intersection of the potential surface of the
S_1-state of bianthrone with the potential surface of the photoinduced form
inherent in which is the greatest likelihood of intersystem crossing between
both potential surfaces. The radiationless deactivation from the S_1 surface

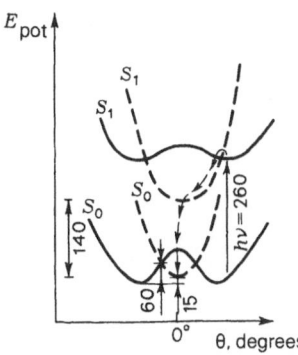

Fig. 11. Profiles of potential surfaces upon the photocyclization of
bianthrone: ———) bianthrone 90; ----) photoproduct 91.
Energies are given in kJ/mole.

of the colored form ends on the potential surface of the ground state. The
relatively high barrier between the minima of the bianthrone and cyclic
photoinduced form on the potential surfaces of the ground states makes a
rapid thermal back reaction impossible.

Following a change in experimental conditions (e.g., a switch from a
solvent to a rigid matrix), the shape of the potential surface of the S_1-
state of the bianthrone form changes so drastically that an intersection
with the potential surface of the photoinduced form practically does not
exist. As a result, the rigid matrix may not display photochromism.

An alternative explanation exists for the case involving large substantial geometric changes in the excited state when the minima on the
bianthrone potential surfaces in the ground and excited states no longer lie
above each other.

Hydrogen Phototransfer

A rapid transfer of the hydrogen atom may occur during the lifetime of
electronically excited states [125]. Ideas on the paths of such reactions
have been worked out using the photochromic anils of salicylaldehyde as an
example [127-129]. Hydrogen transfer between the X—H group and heteroatom Y
($r_{X-H...Y}$) is regarded as the reaction coordinate in this examination.
Energy considerations favor the likelihood that the potential curves of the
ground and excited states along the reaction coordinate are arranged above
each other in such a way that minimum lies above minimum and maximum above
another maximum. This is evidence of the adiabatic course of reaction;
i.e., the photochromic reaction proceeds entirely on the potential surface
of the excited state — from the spectroscopic minimum of the starting material (the parent substance) to the spectroscopic minimum of the photoproduct. The deactivation of the excited state follows either a radiationless
course or proceeds with the emission of a quantum of light, depending on the
product's molecular structure and the distance, on the energy scale, to the
potential surface of the ground state.

A relatively low barrier on the potential surface of the ground state
(Figure 12) is a prerequisite for a fairly rapid hydrogen atom transfer in
the electronically excited state. Because of the formation of primarily
orthoquinoid structures which determine the color and which are less stable
compared with the starting compound (in most cases aromatic structures), the
reverse reaction proceeds through relatively low barriers in the ground
state (see [130]).

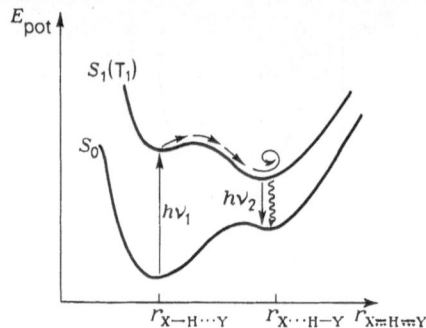

Fig. 12.   Profiles of potential surfaces upon rapid photochemical
proton transfer.

THE INFLUENCE OF SOLVENT ON PHOTOCHROMISM

Photochromic reactions involve changes in molecular structures. The
properties of the medium and the nature of solvents may have completely
different effects on the ground and excited states of the particles involved
in the photochromic process. The varying sensitivity of electronically
excited states to the effect of environmental factors may lead to changes in
the relative series of their stabilities. If such differences have already
been observed in the "spectroscopic" state, the chances of the solvent
having a deciding influence on the possibility of a choice of photoreactions
are all the greater. Because of the different stabilization of intermediates or structures to which the minima on the potential surfaces correspond,
a change in the sequence of the surfaces along the reaction coordinate may
occur. Although in this case the photoreaction may begin in the usual way,
changes in direction may occur during its subsequent course, which will
eventually lead to a quite different set of results. Finally, the solvent
may substantially influence the height of the various energy barriers.

The effect of solvents in thermal reactions, primarily from an experimental standpoint, is examined in Reihardt's survey [159], while in a theoretical context — in the paper by Abronin et al. [160].

The paper [161] presents a wide-ranging review of the possibilities
of describing the influence of the nature of solvent on the electronic spectrum with the help of the classical methods and techniques of quantum chemistry. There has been no shortage of attempts to describe the influence of
solvents on the ground state of the molecule and its reactivity. Work in
this area followed two basic directions.

1. Creation of the so-called continuum model. In this process the
solvent is regarded as a structureless dielectric with no particular specific molecular structure (continuum) which surrounds the solute molecule.
Solvent effects are evaluated by means of well-known microscopic constants
such as dielectric permeability, refractive index, and surface tension factor. To estimate the solvation energy, the concept of the energy of cavity
formation (cavitation energy) in the solvent is employed. The energy of
interaction of the solute molecule with the surrounding solvent can be
formally divided into the van der Waals and electrostatic components. To
calculate the solvation energy within the framework of the continuum model,
classical and quantum-chemical approaches are normally used. A basic flaw
of these approaches is that they cover only nondirectional, nonspecific
interactions, while specific interactions such as hydrogen bonds, charge-
transfer complexes, etc. are left out of account.

2. The development of a discrete "solvent—solute molecule" model [162]. Taken into consideration here are the molecules of the solvent which surround the solute molecules by means of calculating the cluster composed of the solute molecule and a definite number of solvent molecules. It is essential to find a cluster structure with the least energy. This requires optimizing the geometry of supermolecules (solute molecule and solvent molecules) by special methods.

The results of this approach and the calculations by semiempirical quantum-chemical methods do not as yet justify the heavy outlay in obtaining them. The so-called solvaton models are far more simple [163] and make it possible to calculate the electrostatic energies of interaction between the solute molecule and the solvent. It is assumed here that the solute molecule, in keeping with its electronic structure, induces charges in its surroundings. The solute molecule is surrounded by fictitious charges (solvatons) which interact with the point charges of the solute molecule's atoms. Methods have recently been developed whereby the electrostatic energy of solvation in the light of classical models is calculated directly by a quantum-chemical technique (the quantum-chemical model of the reaction field [164], the quantum-chemical model of solvatons [165]).

A certain amount of experience has been gained in the use of these models, but only as applied to the properties of molecules in the ground state. Research has been done inter alia on the influence of solvents on the rotation barriers during rotation around ordinary bonds (equilibrium of conformers); on the basicity of amines; on the equilibrium of tautomers; on the electron paramagnetic resonance (EPR) and electronic spectra. So far no systematic verification has been conducted to see if these models are also suitable for interpreting the effect of solvents on the properties of electronically excited states, notably on photochemical reactivity.

The numerical calculation of solvation effects in photochemical reactions is still in the incipient stage. However, simplified approaches are available which go some way towards explaining the special features of the reaction course, depending on the polarity of the solvent used. Salem et al. [20] attempted, with the help of their concept of correlation diagrams for photochemical reactions, to solve the question of changes in the relative positions of the potential surfaces of biradical and zwitterionic intermediates as a result of medium effects. While examining the correlation diagrams of state, which undergo stabilization or destabilization, one may make qualitative assumptions about the effect of solvent on the photochemical reactivity. Thus, reduced reactivity following hydrogen abstraction can be attributed to a greater decrease in the energy level of the $\pi,\pi*$-triplet state compared to the level of the $n,\pi*$-triplet under the influence of the polar solvent. Because of this, the $\pi,\pi*$-triplet state could have become reactive, but it did not since it correlates with the unfavorable (in energy terms) position of the product's state (from the energy standpoint an "uphill" process), as a result of which the reaction does not occur.

Analogous examinations proved essential for E/Z-isomerizations of the double bond of olefins. In polar media the energy of the zwitterionic Z-state (see Figure 6) is below the energy of the biradical state ^1D. As a result, an additional barrier arises on the potential surface of the ground state along with a second minimum upon rotation through 90° (stabilization of a strongly polar structure rotated through 90°). The potential surface of the $\pi,\pi*$-excited singlet state has two minima at angles of rotation of slightly under and over 90° which are separated by a low barrier upon rotation through 90° (Figure 13). Deactivation of the excited singlet state occurs before the rotation angle of 90° (hypothetical transition state

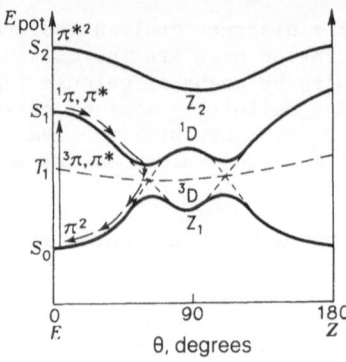

Fig. 13. The effect of polar solvents on the profiles of potential
surfaces of E/Z-photoisomerization.

for Z/E-isomerization). As a result, the isomerization efficiency in polar
solvents declines.

In this fashion we may obtain qualitative results on the influence of
polar solvents on photochemical reactivity during ring opening (e.g., upon
α-ring opening in 2,4-cyclohexadienones) and on photochemical electron
transfer (e.g., in exciplex formation).

REFERENCES

1.   G. H. Brown (ed)., Photochromism, Vol. 4, Wiley, New York (1971).
2.   S. Dähne, Z. Wiss. Phot., **62**:183 (1968).
3.   V. A. Barachevskii, G. I. Lashkov, and V. A. Tsekhomskii, Photochromism
     and Its Uses, Khimiya, Moscow (1977).
4.   A. V. El'tsov (ed.), Photochemical Processes in Layers, Khimiya, Lenin-
     grad (1978).
5.   G. H. Dorion and A. F. Wiebe, in: Photochromism − Optical and Photo-
     graphic Application, Focal Press, New York (1970), p. 120.
6.   J. Epperlein, B. Hofmann, and K. Stopperka, J. Signal AM, 4:155 (1976).
7.   G. Schreder, G.F.M. Ott, and R. Mereni, Usp. Khim., **36**:993 (1967).
8.   J. Michl, Mol. Photochem., 4:243 (1972).
9.   A. Mehlhorn and F. Dietz, J. Signal AM, 5:389 (1977).
10. F. Dietz and A. Mehlhorn, J. Signal AM, 6:5 (1978).
11. R. B. Woodward and R. Hoffmann, Conservation of Orbital Symmetry.
     Academic Press.
12. K. Fukui, in: Molecular Orbitals in Chemistry, Physics, and Biology,
     P.-O. Löwdin and B. Pullman (eds)., Academic Press, New York (1964),
     p. 513.
13. I. Fleming, in: Frontier Orbitals and Organic Chemical Reactions,
     Wiley, New York (1976), p. 111.
14. H. E. Zimmerman, J. Am. Chem. Soc., **88**:1564 (1966); Angew. Chem. Int.
     Ed., **8**:1 (1969).
15. M.J.S. Dewar, Tetrahedron, Suppl. 8, Part 1:75 (1966).
16. M.J.S. Dewar and R. C. Dougherty, The PMO Theory of Organic Chemistry,
     Plenum, New York (1975).
17. A. Rassat, Tetrahedron Lett., 4081 (1975).
18. N. D. Epiotis, Angew. Chem., **86**:825 (1974); J. Am. Chem. Soc., **94**:1941,
     1946 (1972).
19. S. Inagaki, H. Fujimoto, and K. Fukui, J. Am. Chem. Soc., **97**:6108
     (1975).
20. W. G. Dauben, L. Salem, and N. J. Turro, Acc. Chem. Res., **8**:41 (1975).
21. R. C. Dougherty, J. Am. Chem. Soc., **93**:7187 (1971).

22. A. Devaquet, Pure Appl. Chem., 41:455 (1975).
23. J. Michl, Mol. Photochem., 4:257 (1972).
24. A. Devaquet, Top. Curr. Chem., 54:1 (1975).
25. S. D. Peyerimhoff, Jerusalem Symp. Quant. Chem. Biochem., 6:359 (1974).
26. H. Basch, Jerusalem Symp. Quant. Chem. Biochem., 6:183 (1974).
27. S. D. Peyerimhoff and R. J. Buenker, NATO Adv. Study Inst. Ser., 8:257 (1973-1974).
28. M. Scholz, F. Dietz, and M. Mülstedt, Usp. Khim., 38:93 (1969).
29. K. A. Muszkat, D. Gegiou, and E. Fischer, Chem. Comm., 447 (1965).
30. H. R. Blattmann, D. Meuche, E. Heilbronner, et al., J. Am. Chem. Soc., 87:130 (1965).
31. R. Hoffmann, P. Wells, and H. Morrison, J. Org. Chem., 36:102 (1971).
32. G. Klopman (ed)., Chemical Reactivity and Reaction Paths, Wiley, New York (1974).
33. W. C. Herndon, Chem. Rev., 72:157 (1972).
34. L. Salem, J. Am. Chem. Soc., 90:543, 553 (1968).
35. R. Zahradnik, I. Tesarova, and J. Pancir, Coll. Czech. Chem. Commun., 36:2867 (1971).
36. G. Hohlneicher and W. Säenger, Jerusalem Symp. Quant. Chem. Biochem., 2:193 (1974).
37. P. Birner, H.-J. Hofmann, and C. Weiss, MO-theoretische Methoden in der Organischen Chemie, Akademie-Verlag, Berlin (1979).
38. J. Del Bene and H. H. Jaffe, J. Chem. Phys., 48:1807 (1968).
39. J. Fabian, J. Prakt. Chem., 320:361 (1978).
40. J. Fabian, J. Signal AM, 7:67 (1979).
41. J. W. McIver and A. Komornicki, Chem. Phys. Lett., 10:303 (1971).
42. M. Scholz and H. Köhler, J. Quantenchemie — Ein Lehrgang. Vol. IV, Berlin, VEB Deutscher Verlag der Wissenschaften (1980).
43. J. Michl, Top. Curr. Chem., 46:1 (1974).
44. R. Hoffmann, Pure Appl. Chem., 24:567 (1970).
45. K. K. Innes, Excited States, 2:1 (1975).
46. A. Imamura and R. Hoffmann, J. Am. Chem. Soc., 90:5379 (1968).
47. A. Streitwieser, Molecular Orbital Theory for Organic Chemists, Wiley (1961).
48. K. A. Muszkat and S. Sharafi-Ozeri, Chem. Phys. Lett., 20:397 (1973).
49. O. E. Polansky, Z. Naturforsch., 29a:529 (1974).
50. A. Mehlhorn and J. Stumpe, Z. Chem., 16:36 (1976).
51. M. Scholz, M. Mühlstädt, and F. Dietz, Tetrahedron Lett., 665 (1967).
52. M. V. Bazilevsky, The Molecular Orbital Method and the Reactivity of Organic Molecules, Khimiya, Moscow (1969).
53. T. Bercovici, R. Heiligman-Rim, and E. Fischer, Mol. Photochem., 1:23 (1969).
54. N. W. Tyer and R. S. Becker, J. Am. Chem. Soc., 92:1295 (1970).
55. H. E. Simmons and T. Fukunaga, J. Am. Chem. Soc., 89:5208 (1967).
56. R. Hoffman, A. Imamura, and G. D. Zeiss, J. Am. Chem. Soc., 89:5215 (1967).
57. R. Boschi, A. S. Dreiding, and E. Heilbronner, J. Am. Chem. Soc., 92:123 (1970); H. Labhart and G. Wagniere, Helv. Chim. Acta, 42:2219 (1959).
58. N. W. Tyer and R. S. Becker, J. Am. Chem. Soc., 92:1289, 1295 (1970).
59. B. Tinland, R. Guglielmetti, and O. Chalvet, Tetrahedron, 29:665 (1973).
60. H. Labhart, E. R. Pantke, and K. Seibold, Helv. Chim. Acta, 55:658 (1972).
61. Th. Forster, in: Modern Quantum Chemistry, ed. by O. Sinanoglu, Vol. 3, Academic Press (1965).
62. G. Parr, Quantum Theory of Molecular Electronic Structure, Benjamin, New York (1963).
63. F. Dietz, J. Signal AM, 1:157, 237 (1973).
64. J. B. Flannery, J. Am. Chem. Soc., 90:5660 (1968).

65. E. E. Polymeropoulos and D. Mobius, Ber. Bunsenges. Phys. Chem., **83**:1215 (1979).
66. B. Ya Simkin, V. I. Minkin, and L. E. Nivorozhkin, Khim. Geterotsikl. Soedin., 76 (1974).
67. V. I. Minkin, L. E. Nivorozhkin, I. S. Trofimova, et al., Zh. Org. Khim., **11**:828 (1975).
68. B. Ya. Simkin, V. I. Minkin, and L. E. Nivorozhkin, Khim. Geterotsikl. Soedin., 1180 (1978).
69. R. Guglielmetti, M. Mosse, and J. C. Metras, J. Chem. Phys. Physico-chim. Biol., **65**:454 (1968).
70. R. Guglielmetti, J. Photogr. Sci., **22**:77 (1974).
71. R. Heiligman-Rim, Y. Hirshberg, and E. Fischer, J. Phys. Chem., **66**:2470 (1962).
72. R. Heiligman-Rim, Y. Hirshberg, and E. Fischer, J. Phys. Chem., **66**:2465 (1962).
73. T. Bercovici and E. Fischer, J. Am. Chem. Soc., **86**:5687 (1964).
74. E. Fischer, Fortschr. Chem. Forsch., **7**:605 (1967).
75. J. L. Masse, Comp. Rend., **238**:1320 (1954).
76. O. Bloch-Chaude and J. L. Masse, Bull. Soc. Chim. France, 625 (1955).
77. R. S. Becker and K. J. Roy, J. Phys. Chem., **69**:1435 (1965).
78. V. A. Kuz'min, Ya. N. Malkin, V. P. Martynova, et al., Dokl. Akad. Nauk SSSR, **228**:127 (1976).
79. A. Samat, R. Guglielmetti, Y. Ferre, et al., J. Chem. Phys. Physico-chim. Biol., **69**:1202 (1972).
80. Y. Ferre, E.-I. Vincent, J. Metzger, et al., Tetrahedron, **30**:787 (1974).
81. J. B. Hendrickson, J. Am. Chem. Soc., **89**:7036 (1967).
82. N. Tyutyulkov, S. Stojnov, M. Taseva, et al., J. Signal AM, **3**:435 (1975).
83. B. Tinland and C. Decoret, Gazz. Chim. Ital., 792 (1971).
84. H. Hartmann, J. Signal AM, **7**:101 (1979).
85. H. Hartmann, J. Signal AM, **7**:181 (1979).
86. W. Liptay, in: Optische Anregung Organischer Systeme. II. Internat. Farbensymposium 1964, Verlag Chemie, Weinheim (1966).
87. R. S. Becker and J. Michl, J. Am. Chem. Soc., **88**:5931 (1966).
88. B. S. Lukjanov, M. I. Knjazschanski, L. E. Niworoschkin, et al., Tetra-hedron Lett., 2007 (1973).
89. V. I. Minkin, B. Ya. Simkin, L. E. Nivorozhkin, et al., Khim. Getero-tsikl. Soedin., 67 (1974).
90. B. Tinland and C. Decoret, Tetrahedron Lett., 3019 (1971).
91. H.-D. Ilge, J. Sühnel, and R. Paetzold, J. Signal AM, **5**:177 (1977).
92. K. Gustav, J. Sühnel, W. Weinrich, et al., Z. Chem., **19**:219 (1979).
93. K. Gustav, J. Sühnel, W. Weinrich, et al., Z. Chem., **18**:387 (1978).
94. H. Bouas-Laurent and C. Leibovici, Bull. Soc. Chim. France, 1847 (1967).
95. J. Bertran and G. H. Schmid, Tetrahedron, **27**:5191 (1971).
96. R. Hoffmann, J. Chem. Phys., **39**:1397 (1963).
97. R. S. Mulliken, J. Chem. Phys., **23**:1833, 1841 (1955).
98. K. A. Muszkat, G. Seger, and S. Sharafi-Ozeri, J. Chem. Soc., Faraday Trans., II, **71**:1529 (1975).
99. A. H. A. Tinnemans, W. H. Laarhoven, S. Sharafi-Ozeri, et al., Recl. Trav. Chim. Pays-Bas, **94**:239 (1975).
100. M. Scholz, F. Dietz, and M. Mühlstädt, Z. Chem., **7**:329 (1967).
101. W. H. Laarhoven, Th. J. R. M. Cuppen, and R. J. F. Nivard, Recl. Trav. Chim. Pays-Bas, **87**:687 (1968).
102. G. Kortum, Ber. Bunsenges. Phys. Chem., **78**:391 (1974).
103. A. Schönberg and K. Junghans, Chem. Ber., **98**:2539 (1965).
104. K. A. Muszkat and E. Fischer, J. Chem. Soc., B., 662 (1967).
105. R. Lorenz, U. Wild, and J. R. Huber, Photochem. Photobiol., **10**:233 (1969).

106. T. Bercovici, R. Korenstein, K. A. Muszkat, et al., Pure Appl. Chem., **24**:531 (1970).
107. M. Klessinger and W. Lüttke, Tetrahedron, **19**(2):315 (1963).
108. J. Fabian and G. Tröger-Naake, J. Prakt. Chem., **318**:801 (1976).
109. D. Leupold and S. Dähne, Theor. Chim. Acta, **3**:1 (1965).
110. W. R. Brode, E. G. Pearson, and G. M. Wyman, J. Am. Chem. Soc., **76**:1034 (1954).
111. R. Pummerer and G. Marondel, Lieb. Ann. Chem., **602**:228 (1957).
112. A. D. Kirsch and G. M. Wyman, J. Phys. Chem., **79**:543 (1975).
113. G. M. Oxengendler and E. P. Gendrikov, Zh. Fiz. Khim., **33**:2791 (1959).
114. W. Lüttke and M. Klessinger, Chem. Ber., **97**:2342 (1964).
115. J. Sühnel, K. Gustav, R. Paetzold, et al., J. Prakt. Chem., **259**:17 (1978).
116. J. Sühnel and K. Gustav, J. Prakt. Chem., **320**:917 (1978).
117. R. Paetzold, M. Reichenbacher, and K. Gustav, ICPS Dresden, 1974.
118. K. Gustav and S. Vettermann, Z. Chem., **18**:26 (1978).
119. E. L. Eliel et al. (eds.), Conformational Analysis, Wiley (1965).
120. K. Gustav and S. Vettermann, Z. Chem., **18**:456 (1978).
121. K. Gustav, S. Vettermann, and I. Rose, J. Prakt. Chem., **321**:395 (1979).
122. J. A. Pople, D. P. Santry, and G. A. Segal, J. Chem. Phys., **43**:5129, 5136 (1965).
123. K. Gustav, S. Vetterman, and P. Birner, Z. Chem., **19**:259 (1979).
124. K. Gustav, S. Vetterman, and P. Birner, Wiss. Z. Friedrich-Schiller-Univ., Jena, Math.-Naturwiss. Reihe.
125. M. A. Vorotyntsev, R. R. Dogonadze, and A. M. Kuznetsov, Elektrokhim-iya, **10**:687, 867 (1974); R. R. Dogonadze et al., ibid., **3**:739 (1967).
126. B. Tinland, Tetrahedron, **26**:4795 (1970).
127. W. F. Richey and R. S. Becker, J. Chem. Phys., **49**:2092 (1968); J. Am. Chem. Soc., **89**:1298 (1967).
128. M. Ottolenghi and D. S. McClure, J. Chem. Phys., **46**:4613, 4620 (1967).
129. M. D. Cohen and S. Flavian, J. Chem. Soc., B, 317, 321, 329, 334 (1967).
130. R. Potashnik and M. Ottolenghi, J. Chem. Phys., **51**:3671 (1969).
131. J. D. Margerum, L. J. Miller, E. Saito, et al., J. Phys. Chem., **66**:2324 (1962).
132. B. Tinland and C. Decoret, Tetrahedron Lett., 2467 (1971).
133. K. Muller, Angew. Chem., **92**:1 (1980).
134. H. Eyring and M. Polany, Z. Phys. Chem., **12**:279 (1931).
135. M. V. Basilevsky, Chem. Phys., **24**:81 (1977).
136. N. J. Turro, J. McVey, V. Ramamurthy, et al., Angew. Chem., **19**:597 (1979).
137. N. J. Turro, Modern Molecular Photochemistry, Benjamin, New York (1978).
138. A. J. Merer and R. S. Mulliken, Chem. Rev., **69**:639 (1969).
139. P. Borell and H. H. Greenwood, Proc. R. Soc., Ser. A, **298**:453 (1967).
140. A. Warshel and E. Huler, Chem. Phys., **6**:463 (1974).
141. H. H. Jaffe, D. M. Hayes, and K. Morokuma, J. Chem. Phys., **60**:5108 (1974).
142. K. G. Kay and S. A. Rice, J. Chem. Phys., **57**:3041 (1972).
143. A. Warshel and M. Karplus, Chem. Phys. Lett., **32**:11 (1975).
144. K. Schulten, I. Ohmine, and M. Karplus, J. Chem. Phys., **64**:4422 (1976).
145. F. Momicchioli, M. C. Bruni, I. Baraldi, et al., J. Chem. Soc., Faraday Trans., II, **70**:1325 (1974); **71**:215 (1975).
146. R. L. Ellis, R. Squire, and H. H. Jaffe, J. Chem. Phys., **55**:3499 (1971).
147. G. Lancelot, Mol. Phys., **29**:1099 (1975).
148. G. Orlandi and W. Siebrand, Chem. Phys. Lett., **30**:352 (1975).
149. P. Tavan and K. Schulten, Chem. Phys. Lett., **56**:200 (1978).
150. J. Saltiel and J. T. D'Agostino, J. Am. Chem. Soc., **94**:6445 (1972).
151. M. Sumitani, N. Nakashima, and K. Yoshihara, Chem. Phys. Lett., **68**:255 (1979).

152. W. T. Borden, J. Am. Chem. Soc., **97**:5968 (1975).
153. P. M. Crosby and K. Salisbury, Chem. Commun., 477 (1975).
154. F. Dietz, W. Föster, C. Weiss, et al., J. Signal AM, **9**:177 (1981).
155. G. J. Hoijtink, E. De Boer, V. Van der Meij, et al., Recl. Trav. Chim. Pays-Bas, **75**:487 (1956).
156. W.Th.A.M. Van der Lugt and L. J. Oosterhoff, J. Am. Chem. Soc., **91**:6042 (1969).
157. D. Grimbert, G. Segal, and A. Devaquet, J. Am. Chem. Soc., **97**:6629 (1975).
158. R. S. Becker and C. E. Earhart, J. Am. Chem. Soc., **92**:5049 (1970).
159. C. Reihardt, Solvent Effects in Organic Chemistry, Verlag Chemie, Weinheim (1979).
160. I. A. Abronin, K. Ya. Burstein, and G. M. Zhidomirov, Zh. Strukt. Khim., **21**(2):145 (1980).
161. Optische Anregung organischer Systeme, Berichtsband des 2. Internationalen Farbensymposiums, 1964, Schloß Elmau, Verlag Chemie, Weinheim (1966); see: works by W. Liptay, E. Lippert, J. Schlag, G. Briegleb, G. Kortum, and M. Pestemer.
162. B. Pullman (ed)., Environmental Effects on Molecular Structure and Properties, D. Reidel, Dordrecht (1976).
163. G. Klopman, Chem. Phys. Lett., **1**:200 (1967).
164. J. H. McGreery, R. E. Christoffersen, and G. G. Hall, Chem. Phys. Lett., **26**:501 (1974); J. Am. Chem. Soc., **98**:7191 (1976).
165. H. A. Germer, Theor. Chim. Acta, **34**:145, **35**:273 (1974).

Chapter 2

PHOTOCHROMIC THIOINDIGOID DYES

M. A. Mostoslavskii

Dye Chemistry Research Institute
Rubezhnoe Branch of the Scientific-Research Institute
of Organic Intermediates and Dyes, USSR

In this chapter we examine photochromic indigoid dyes of structures I-V and some others. The indigoids include compounds V (X = NH); the thioindigoids, compounds I (X = S). The indigoid dyes include all compounds I-V, regardless of the nature of the heteroatom (X, X').

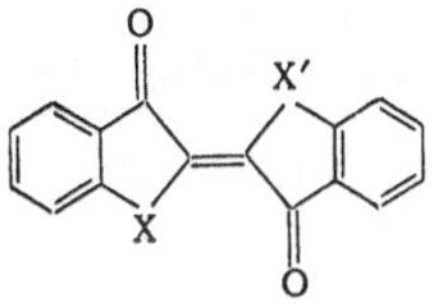

E-I 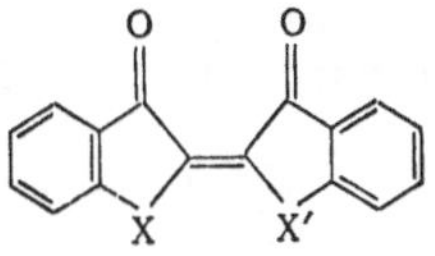   Z-I

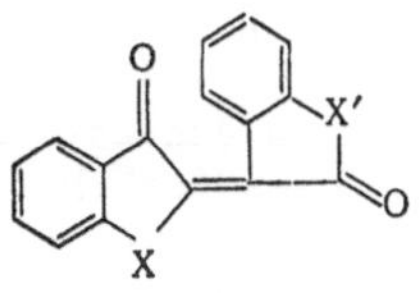

Z-II 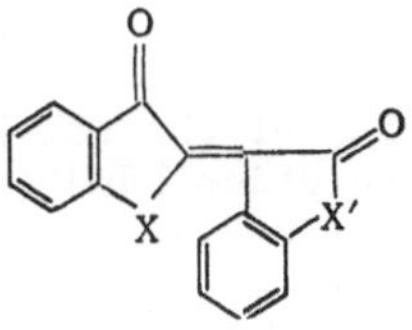   E-II

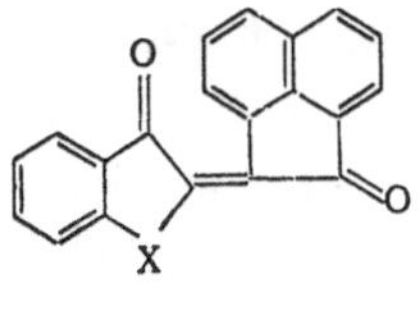

Z-III 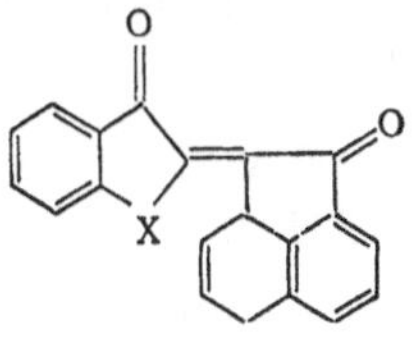   E-III

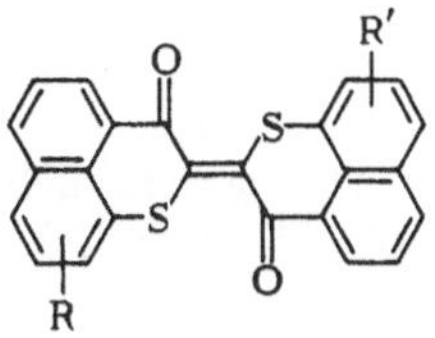

E-IV 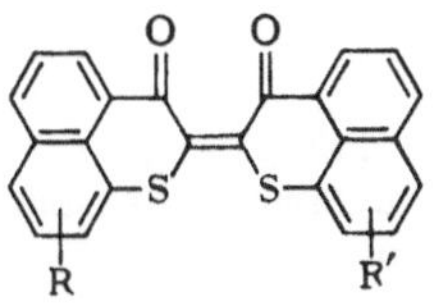   Z-IV

E-V   Z-V

(numeration according to Table 9; for ordinary numeration, see p. 75).

The indigoid dyes are crystalline, high-melting, intensely colored substances which are not readily soluble in organic solvents or water and are capable of sublimating in a vacuum at a high temperature [1,2,165].

The photochromism of indigoid dyes is due to photochemical E → Z- or Z → E-isomerization. The isomer whose carbonyls are in the trans-position* is thermodynamically stable.

Photochromic indigoid dyes can be used for actinometric purposes, as the working substances of photochromic materials for use in optical data storage devices, and as substances which accumulate the energy of sunlight (solar energy).

The emphasis in the study of photochromism is on exploring the possibility of controlling the properties of photochromes that are important for practical purposes (by changing the structure of dyes, by varying the medium, temperature, pressure, force fields) which include the following:

λ_{max} and ϵ_{max} of the thermodynamically stable initial form (the E-form in the case of thioindigoids), which must be adequate to the wavelength and capacity of the laser used for recording and reading the data;

λ_{max} and ϵ_{max} of the labile form (the Z-form in the case of thioindigoids), which must be adequate to the wavelength and the capacity of the laser used for data erasure;

color contrast, which determines the possibility of a visual read-out of data recorded on a photochromic material (the color contrast is measured by the distance between λ_{max} of the initial and photoinduced forms);

the photosensitivity of the stable form to excitation in the region of its long-wavelength absorption band (it improves for thioindigoids upon an increase** in $\phi_{E\to Z}$, increase** in D_E, decrease** in D_Z, and decrease** in $\phi_{Z\to E}$); the photosensitivity of the photoinduced form in the region of its long-wavelength absorption band (it improves in the case of thioindigoids with an increase** in $\phi_{Z\to E}$, increase** in D_Z, decrease** in D_E, and decrease** in $\phi_{E\to Z}$; stability to photobleaching (it improves with a decrease in ϕ of side photoreactions);

the solubility of the dye in the materials' matrix (it should ensure, at a given thickness of the photochromic layer, a sufficiently high

---

*In accordance with Z/E-nomenclature a "senior" substituent (heteroatom > C=O > aryl > alkyl > H) is isolated in each sp²-carbon atom. Then the mutual arrangement of the senior substituents about the "ethylene" bond is specified: for I, IV, and V compounds — the arrangements of the heteroatoms; for II and III compounds — that of heteroatom and carbonyl. E (from entgegen, i.e., "opposite") signifies that the senior substituents are in a trans-configuration; Z (from zusammen, i.e., "together") means that they are in a cis-configuration.
**At λ excitation.

optical density both at the excitation wavelength and at the reading wavelength);

stability of the photoinduced form (in the case of the thioindigoids it
is dependent on the rate constant of the thermal Z → E-isomerization which
must be as low as possible, and on the sensitivity of the Z-form to admixtures that slip into the matrix by accident or that form within it).

Given below is the upper limit of the positive properties of photochromic indigoid dyes which were exhibited by representatives of this
class. The conversion induced by visible light in the forward and reverse
directions is 95-100% [3-6]. The number of cycles in photoisomerization
reaches 20,000 [7,8]. The distance between λ_{max} of the Z- and E-isomers is
as much as 150 nm [9]. The value of ϵ_{max} for the E-isomer approaches
5×10^4. The rate constant of the thermal Z → E-isomerization at room temperature is roughly 10^{-7} sec^{-1} [10,11]. The value of $\phi_{E \to Z}$ may reach 0.3
[7,8], and $\phi_{Z \to E}$ – 0.96 [12]. These parameters stimulate theoretical [13-17]
and applied [7,8,18-21] research into the photochromism of thioindigoid
dyes.

THE STRUCTURE AND SPECTRA OF THIOINDIGOIDS

The establishment of correlations between the structure and absorption
spectra of dyes is usually determined either by relying on a single formula
of the dye or by examining the relations of the contributions made by the
resonance boundary structures in the ground and excited states. A more
correct procedure is to examine the usual [22,23] molecular diagrams obtained from calculations by the MO LCAO PPP CI method as "the formulae" of
the ground and excited states. Let us examine, for instance, diagrams of
the indigo molecule [22]. Figure 1 shows π-bond orders, charges on the
atoms, and their signs. Both in the ground and in the first excited states
the indigoid molecule is strongly polarized.

In a discussion of electronic spectra, in the formation of which two
states participate, it is logical to examine the diagram of the redistribution of electron density and change in π-bond orders, the PEPIPS* diagram
for short (Figure 2), rather than the diagrams shown in Figure 1. While
constructing the PEPIPS diagram from the atomic charges (with their signs)
in diagram (b), the charges of these atoms were subtracted (along with their
signs) in diagram (a) (see Figure 1), while from the π-bond order in diagram
(b), the π-orders of corresponding bonds in diagram (a) were subtracted.
The increase in the negative charge on the atom (or the π-bond order) is
marked with unfilled circles (rectangles). The efflux of electrons is
marked with filled circles. The area of the circle is proportional to the
change of the charge, that of the rectangle — to the change of the π-bond
order.

Clearly, the $S_0 \to S_1$ transitions in indigo are accompanied by a charge
transfer primarily from the heteroatom onto the carbonyl carbon, and the
oxygen of the carbonyl and the carbon atoms in positions 5, 6, 7, 7a (intramolecular charge transfer) (cf., however, [24,25]) participate in the
redistribution of the charge. The greatest change in π-order (decrease) is
observed in the central double bond. The bonds of the five-membered ring,
the carbonyl, and the 7–7a bond change order markedly as a result of the
transition.

Wide use of the PEPIPS diagrams for quantitative comparisons will be

---

*PEPIPS is the Russian acronym for "redistribution of electron density and
change in π-bond orders."

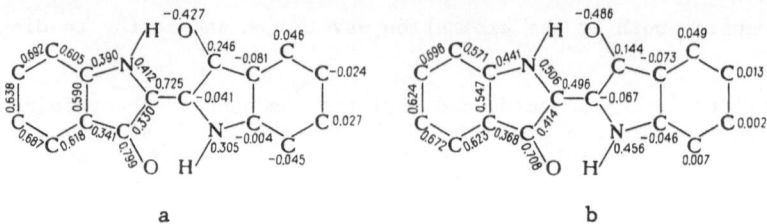

Fig. 1. The molecular diagram of indigo: (a) the S_0-state; (b) the S_1-state
[22].

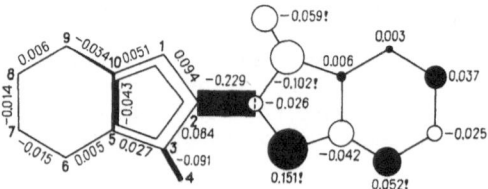

Fig. 2. The diagram of the electron density redistribution and change in
the π-bond orders (PEPIPS diagram) for the $S_0 \rightarrow S_1$ transition in
indigo (according to the data of [22]).

facilitated by the steady growth in the bank of results of quantum-chemical
calculations of the ground and excited states of organic substances.* The
use of the PEPIPS diagram is also possible for qualitative predictions especially if Murrell's hypothesis, which has been found to be valid in many
cases, is accepted [27]. This hypothesis assumes that changes in molecular
structure that bring the structure of the ground state closer to that of the
excited state (through the introduction of substituents and bond deformation, etc.) reduce the energy quantum of the electron transfer.

The most pronounced charge redistributions and changes in bond order
are important for qualitative forecasting. This makes it possible to use
the PEPIPS indigo diagram for establishing connections between structure and
spectra for the entire group of dyes having formula I. A good many useful
consequences follow from the PEPIPS indigo diagram and Murrell's hypothesis
[27]:

1. The diagram indicates that the $S_0 \rightarrow S_1$ transition causes an efflux
of electron density from the carbon atoms. The heteroatom's enhanced ability to donate electrons is expected to reduce transfer energy. Indeed, a
bathochromic shift occurs which increases antibatically with the heteroatom's electronegativity**: oxindigo < thioindigo < selenoindigo < indigo <
N,N-dimethylindigo (Table 1, Nos. 1, 4, 8, 10, and 12).

---

*A change in electron charges on atoms and bond order (but not in the form
of graphic representation or diagram) was used earlier for discussing
intramolecular interactions (for example, see [26]).
**Because of the absence of a series of electronegativities including values
not only for O, S, and Se but also for NH and NCH_3, the relative values of
the electronegativities of the "heteroatoms" were estimated on the basis of
the relative arrangement of straight lines in Figure 3. The figure is based
on the data [28] on the ionization potential (IP) of simple molecules containing the above "heteroatoms." Point 3_7 of the nitrogen atom in Figure 3
is above the straight line which contains the ionization potentials of substances with the NH group.

Table 1. Maxima of the long-wavelength absorption band of the Z- and E-isomers of some indigoid dyes in benzene, their shift as compared with the maximum of indigo, and the color contrast of the Z- and E-isomers

Indigo dye	E-isomer λ, nm ($\epsilon \cdot 10^{-4}$)	$\Delta\lambda$, nm	Z-isomer λ, nm ($\epsilon \cdot 10^{-4}$)	Color contrast $\Delta\lambda_{E/Z}$, ($\Delta\epsilon \cdot 10^{-4}$)	ϵ_Z/ϵ_E
1. Oxindigo [29]	452(1.27)	-149	397(1.34)	55(-0.07)	1.05
2. $\Delta^{2,2'}$-Bis(4,4'-dimethylthiolan-3-one) [30]	454(4.1)	-147	400(4.1)	54(0)	1.0
3. Thiophenindigo [31]	505	-96	476	29	–
4. Thioindigo [17,32]	543(1.61)	-58	484(1.46)	59(0.15)	0.91
5. N,N'-Bis(3,5-dinitrobenzoyl)indigo [33]	552(0.62)	-49	436(0.28)	116(0.34)	0.45
6. N,N'-Diacetylindigo [34,35]	562(0.72)	-39	435(0.45)	127(0.27)	0.62
7. N,N'-Diheptadecanoylindigo [33]	567(0.71)	-34	437(0.39)	130(0.32)	0.55
8. Selenoindigo* [36]	570(1.12)	-31	495(0.69)	75(0.43)	0.62
9. N,N'-Dibenzoylindigo [33]	574(0.77)	-27	460(0.39)	114(0.38)	0.51
10. Indigo	601	0	–	–	–
11. Perinaphthothioindigo [5]	625(2.6)	24	502(1.8)	123(0.8)	0.69
12. N,N'-Dimethylindigo [35]	644(2.31)	43	588(0.83)	56(1.48)	0.36

*In chloroform.

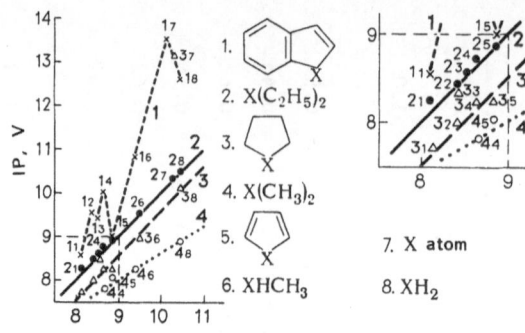

IP (in water) of compounds with X = S

Fig. 3. Ionization potentials (IPs) [28] of substances with analogous structure containing heteroatoms X = O, S, NH (or NCH$_3$). Ionization
potentials of substances with X = S are plotted on the x-axis. The
points belonging to substances with one and the same heteroatom are
linked together with straight lines. The number of the point com-
prises the number of the line (the first digit) and the number of
the substance's general formula, containing an X heteroatom (second
digit). Numbers of lines: 1) X = O; 2) X = S; 3) X = NH; 4) X =
NCH$_3$. The formula and the numbers of substances are given in the
figure.

2. The efflux of electron density from the nitrogen atoms is made
easier by the introduction of donor substituents; therefore, the latter
should cause a bathochromic shift. This is confirmed by the data given
below on the effect of substituents on the nitrogen atoms on the position
of λ_{max} of the absorption band of the solutions of N,N'-substituted indigo
in sym-tetrachloroethane (the values of $\Delta\lambda_{max}$ are given in comparison with
λ_{max} of indigo in sym-tetrachloroethane – 620 nm):

Substituents on N and N'	$\Delta\lambda$, nm
C$_6$H$_5$, C$_6$H$_5$ [37,41]	25
CH$_3$, H [31]	26
C$_2$H$_5$ [38]	35.5*
CH$_3$, CH$_3$ [31]	26
C$_2$H$_5$, C$_2$H$_5$ [39]	50**
CH$_2$C$_6$H$_5$, CH$_2$C$_6$H$_5$ [37,41]	55

*In hexane.   **In chloroform.

3. The efflux of electron density from the nitrogen atoms is made more
difficult by the presence of acceptor substituents which should cause a
hypsochromic shift. Indeed, this is observed in the case of acyl groups
(see Table 1, Nos. 5-7 and 9).

4. The efflux of electron density from positions 4, 5, and 7 presup-
poses a bathochromic shift upon the introduction of donor substituents. For
instance, the 5-methoxy group causes a substantial bathochromic shift (Table
2). The halogen atoms act as donors in the vast majority of cases.

5. The introduction of an acceptor into positions 4, 5, and 7 should
make the efflux of electron density from them more difficult. Thus, the
nitro group in position 5 (Tables 2 and 3) causes a hypsochromic shift, but
no experimental data on the effect of acceptors in positions 4 and 7 are
available.

Table 2. The influence of substituents on the shift of λ_{max} (nm) of the absorption band in solutions of symmetrical derivatives of indigo and thioindigo as compared with indigo and thioindigo, respectively

Substituent	Substituted indigo forms in tetrachloroethane	Substituted thioindigo forms in benzene
4,4'-di-Cl	-4[43], 5[37,40-41]	
4,4'-di-Br	5[40,41]	
4,4'-di-I	15[40,41]	
4,4'-di-OCH$_3$		15[47]*
5,5'-di-NO$_2$	-25[37,40,41]	-32[45]
5,5'-di-F	10[37,40-42]	
5,5'-di-Cl	0[43],15[37,40-42,44]	14[46]
5,5'-di-Br	0[43],15[37,40-42]	14[46]
5,5'-di-I	5[37,40-42]	17[46]
5,5'-di-CH$_3$	15[37,40,41]	13[46]
5,5'-di-OCH$_3$	40[37,40,41]	63[47]*
6,6'-di-NO$_2$	30[37,40,41]	22[45]
6,6'-di-F	-35[37,40,41]	
6,6'-di-Cl	-30[43], 15[37,40,41]	-9[4], -8[48]
6,6'-di-Br	-15[37,40,41]	-9[4]
6,6'-di-I	-15[37,40,41]	-3[47]**
6,6'-di-CH$_3$	-10[37,40,41]	
6,6'-di-OCH$_3$	-35[37,40,41]	-51[47]
7,7'-di-NO$_2$		-21[45]
7,7'-di-F	-45[40], 0[42]	
7,7'-di-Cl	-14[43], -5[36,40-42]	0[46]
7,7'-di-Br	0[40,42]	1[46]
7,7'-di-I	0[40,42]	3[46]
7,7'-di-CH$_3$	9[49]***	6[46]
7,7'-di-OCH$_3$		49[47]*

*In xylene.
**In pyridine.
***In chloroform.

6. If the magnitude of charge redistribution is any guide, the shift of the maximum of the absorption band upon the introduction of a concrete substituent should increase in the series: 4-substituents < 5-substituents < 7-substituents < N-substituents. In reality the shift upon the introduction of substituents in position 5 is greater than in position 7.

In indigoids only the substitution in position 5 or 6 is not accompanied by steric hindrances for obvious reasons and can be used in evaluating uncomplicated electronic influences exerted by the substituent on the spectrum. Substitution in positions 4 and 7 can be accompanied by upsetting the coplanarity of the substituent and the framework of the dye, as well as by steric proximity effects of the "peri-effect" type [5], the "ortho-effect" type, and the like [25,54]. These complications account for the impossibility of a strict fulfillment of consequence 6.

7. The influx of electron density in position 6 will be hindered by the presence of a donor substituent. Indeed, donors (including halogens) cause hypsochromic shifts upon introduction into position 6 (see Tables 2 and 3).

8. The influx of electron density into position 6 will be facilitated by the presence of an acceptor substituent; it is shown that acceptors in position 6 causes bathochromic shifts (see Tables 2 and 3).

Table 3. Maxima of the longwave absorption band of solutions of mono- and disubstituted thioindigo [45,46,50-53]

Sustituent	λ, nm		$\epsilon \cdot 10^{-4}$	Δλ, nm	
	in benzene	in pyridine	in benzene	in benzene	in pyridine
Without substituents	545	546	1.6	0	0
5-SO₂CH₃	542	543	1.7	-3	-3
5-NO₂	540	541	1.6	-5	-5
5-OC₂H₅	561	563	1.1	16	17
5-NH₂		550*			4
6-SO₂CH₃	552**	553	1.8	7	7
6-NO₂	561	562	1.8	16	16
6-OC₂H₅	531	531	1.0	-14	-15
6-NH₂		530*			-16
5,5'-di-SO₂CH₃		512			-34
5,5'-di-NO₂	513	512		-33	-34
5-SO₂CH₃, 5'-OC₂H₅	562**	564	1.0	17	18
5-NO₂, 5'-OC₂H₅	562	562	1.0	17	16
5-NO₂, 5'-NH₂		557*			11
5,6'-di-NO₂		545*			-1
5-SO₂CH₃, 6'-OC₂H₅	527**	527	1.3	-19	-19
5-NO₂, 6'-OC₂H₅	527	528	1.1	-19	-18
5-NO₂, 6'-NH₂		535*			-11
5,5'-di-OC₂H₅	588	587		43	41
5,5'-di-NH₂		583*			37
5-OC₂H₅, 6'-SO₂CH₃	573**	573	1.1	28	27
5-OC₂H₅, 6'-NO₂	578	578		33	32
5-NH₂, 6'-NO₂		565*			19
5-NH₂, 6'-NH₂		545*			-1
6,6'-di-SO₂CH₃	557			11	
6,6'-di-NO₂	567	567		22	21
6-SO₂CH₃, 6'-OC₂H₅	539**	538	1.6	-7	-8
6-NO₂, 6'-OC₂H₅	549	548	1.4	4	2
6-NO₂, 6-NH₂	540			-6	
6,6'-di-OC₂H₅	515	517	1.0	-30	-29
6,6'-di-NH₂	526			-20	
5,5'-di-CH₃	558		1.4	13	
7,7'-di-CH₃	551		1.5	6	
5,5'-di-Cl	559		1.6	14	
5,5'-di-Br	559		1.7	14	
5,5'-di-I	562		1.6	17	
7,7'-di-Cl	545		1.5	0	
7,7'-di-Br	546		1.5	1	
7,7'-di-I	548		1.3	3	

*No solvent indicated.
**In p-xylene.

9. Deformation of the central double bond under the influence of some structural factors should cause the coloration to deepen as the bond order declines upon excitation. Whether this effect can be observed in reality is still not quite certain. It is routine practice, during investigations of the effect of steric hindrances in the region of some bond on the molecule's spectrum, to introduce alkyl groups of various sizes into this bond or the adjacent atoms. Their electronic influence on the spectrum is slight [55]. Only indigo can yield derivatives containing alkyls attached to nitrogen atoms and having additional steric hindrances in the C=C bond. However,

concurrently with the possible appearance of rotation about the central double bond in N,N'-disubstitution in the indigo molecule can influence the spectrum as follows:

1) liquidation of the hydrogen bond; 2) changing the donor capacity of the heteroatom; 3) withdrawal of the carbonyl group from the plane of the molecule. It is a difficult task to separate these effects.

Summarizing the foregoing (consequences 1-5, 7, and 8), the PEPIPS indigo diagram (see Figure 2) can be regarded as a modern formula of indigo which explains the relation between the structure and spectral properties of formula I dyes. Its accuracy and predictive power are no greater than those of the method of quantum chemistry used for calculation purposes. However, this formula makes it possible to use more of the results of calculations than the use of the terminology of resonance theory (cf. [56]).

The PEPIPS diagram can help provide a modern definition of what a chromophore is. It seems logical to accept that during an electron transfer the redistribution of the charge density Δq_m on the atoms of the chromophores should be greater than the mean value in the conjugated system $|\Delta q_m| > (1/n)\Sigma|\Delta q_i|$, where m is the index of the chromophore's key atom; i are indices of the atoms entering into the conjugated system; n is the number of atoms in the system.

Upon the introduction of weak substituents into the molecules, charge redistribution in them will be slight and the chromophore will remain unchanged. On the other hand, if the substituent is very actively involved in charge redistribution during light absorption, it will be among the "central atoms," which will mean that the introduction of the substituent changed the chromophore. Different "chromophores" are readily identified for different electron transfers in one and the same substance.

Judging by the data available from the PEPIPS indigo diagram (see Figure 2), the key atoms of indigo's chromophores are: nitrogen, carbon, and carbonyl oxygen as well as the carbon atom in position 7. They are marked with a ! sign near the numbers, indicating charge redistribution. The chromophore is also represented by broad lines of the bond order redistribution that single out the C=C bond, the heterocycle, and one of the bonds of the annelated aromatic ring. On the whole, the atoms and bonds which form part of indigo's chromophore system by Kuhn's formula are singled out* [57]. The additional "separation" of atom 7 is not corroborated by experiments since the influence of the substituent is less from position 7 than from position 5. Apparently this discrepancy is a result of the imperfection of the quantum-chemical calculation.

Figure 4 gives some idea of the genesis of the absorption spectrum of indigoid dyes. Benzo[b]thiophen-3(2H)-one, the starting material for obtaining thioindigo, exists in the keto form (Figure 4a) and is a model of the half molecule of thioindigo, even though not quite adequate [61]. The substitution of the methyl group in position 4 does not alter the character of benzo[b]thiophen-3(2H)-one's spectrum (see Figure 4a and b). It is easy to notice the similarity of the spectra of 5-methylbenzo[b]thiophen-3(2H)-one and a ketone with an open ring — 2-methylthio-5-methylacetophenone

---

*V. A. Izmail'skii [38,58-60] explained indigo's coloration by the interaction of two contrapolarized systems superimposed upon each other and sharing the common C=C bond. Upon light absorption, charge transfer occurs from the two-donor system —NH—C=C—NH— to the two-acceptor system —CO—C=C—CO—.

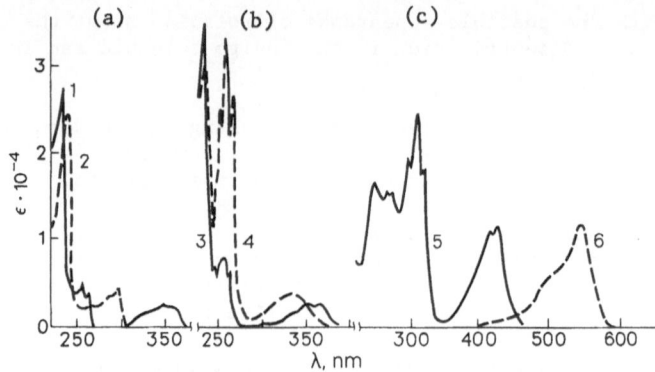

Fig. 4. The absorption spectra of solutions in hexane: 1) benzo[b]thio-
phen-3(2H)-one; 2) 3-methoxybenzo[b]thiophene; 3) 5-methylbenzo-
[b]thiophen-3(2H)-one; 4) 2-methylthio-5-methylacetophenone;
5) 2-benzylidene-5-methylbenzo[b]thiophen-3(2H)-one; 6) 5,5'-di-
methylthioindigo in cyclohexane.

(see Figure 4b). The close similarity of the vibrational structure of the
250-290 nm absorption band of 5-methylbenzo[b]thiophen-3(2H)-one and the
300-340 nm absorption band of 2-benzylidene-5-methylbenzo[b]thiophen-
3(2H)-one (see Figure 4b and c) makes it possible to assume that these bands
are similar in origin. A band in the 270-340 nm region is also exhibited by
the spectrum of 5,5'-dimethylthioindigo (see Figure 4c). The more complicated molecule fails to lead to the appearance of an additional long-wavelength absorption band, and a new band seems to appear as a result of the
disappearance of the old long-wavelength band.

benzo[b]-
thiophen-3(2H)-one,
R = H, CH₃

2-methylthio-5-
methylacetophenone

2-benzylidene-5-
methylbenzo[b]-
thiophen-3(2H)-one

The paper [53] shows that the chromophore of thioindogenides, aurones,
and selenoindogenides is similar to half of the indigoid chromophore (see
Figures 2 and 4b).

We could begin constructing the molecule of an indigoid dye in our
mind's eye with the central ethylene bond. The successive introduction of
donor (alkylthio) and acceptor (carbonyl) groups shifts the long-wavelength
absorption band towards longer wavelengths (Table 4). However, the realization [64] of "the chromophore of thioindigo" [57] as 3,4-bis(methylthio)-
3-hexene-2,5-dione revealed its inadequately deep color due to steric hindrances for the simultaneous arrangement in the plane of acetyl and alkylthio groups [64]. This was to be expected.

As a result of the realization of the thioindigoid chromophore in the
compounds given below, the following was found experimentally:

λ_{max} 318 nm [66]

λ_{max} 450 nm [67]

λ_{max} 405 nm [66]

Table 4. Maxima of intense long-wavelength absorption bands of compounds containing carbonyl, the alkylthio group, and the ethylene bond

Compound	Medium	λ, nm (log ϵ) E-isomer	λ, nm (log ϵ) Z-isomer
$CH_3-CH=CH-CH_3$ [62, pp. 32-37,100,104]	gas phase	177 (4.1)	175 (4.3)
$CH_2=CHSC_2H_5$ [63]	heptane	228,240 (3.84,3.66)	
$CH_3COCH=CHCH_3$	gas phase	220 (4.12)	221 (3.94)
$CH_3COCH=CHC_3H_7$		222 (4.09)	226 (3.83)
$C_2H_5SCH=CHCl$ [65]	cyclohexane	237,256 (3.69,3.77)	235,246 (3.92,3.76)
	methanol	236,254 (3.68,3.75)	233,245 (3.87,3.71)
$CH_3COCH=CHCOCH_3$		228 (4.10)	223 (3.82)
$C_6H_5COCH=CHCOC_6H_5$		268 (4.3)	260 (4.3)
$C_2H_5SCH=CHSC_2H_5$ [63]	heptane		254,285 (4.01,3.48)
$(CH_3)_2CHCOCH=CHSC_2H_5$ [66]	cyclohexane		292 (3.18)
3-(Methylthio)-3-hexene-2,4-dione [66]	ethanol, cyclohexane	316; 311 (3.88;3,84)	
3,4-Bis(methylthio)-3-hexene-2,5-dione [66]	ethanol, cyclohexane	321; 315 (3.91;3.93)	324; 318 (3.95;3.94)

The calculated value of λ_{max} (common to all these compounds) is 453 nm. Deviation from it is in excess of 100 nm.

A λ_{max} of 457 nm was precomputed for the E-s-E-s-E-conformation [67].

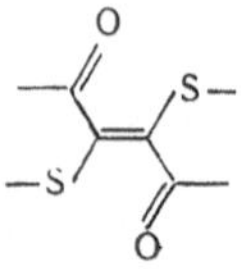

E-s-Z-s-Z
453 [67], 453 nm [66]

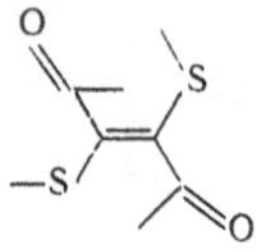

E-s-E-s-E
457 [67], 424 nm [66]

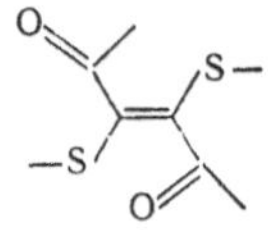

E-s-E-s-Z
485 nm [67]

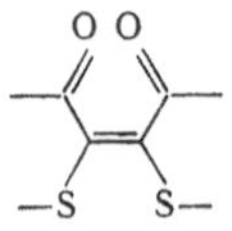

Z-s-Z-s-Z
378 [67], 398 nm [66]

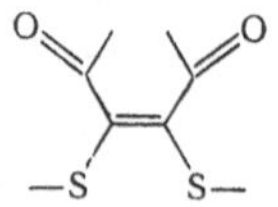

Z-s-E-s-E
436 [67], 415 nm [66]

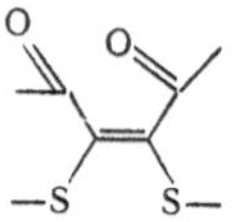

Z-s-E-s-Z
418 nm [67]

while a λ_{max} of 453 nm [67] for the E-s-Z-s-Z-conformation was realized in the thioindigo molecule. Although 2,3,5,6-tetrahydro-2,2',5,5'-tetramethyl-thieno[3,2-b]thiophene-3,6-dione has a λ_{max} of 467 nm, i.e., is more deeply colored than 2,2'-bis(4,4'-dimethylthiolan-3-one), the interesting hypothesis [67] regarding the greater suitability as a chromophore for deeply colored dyes of the E-s-E-s-E-conformation (rather than the E-s-Z-s-Z-conformation that is not realized in the thioindigo molecule) remains inadequately corroborated because of the smallness of the observed effects.

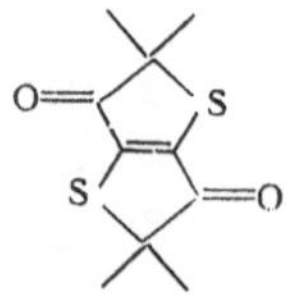

The data summarized in Table 3 [45,50-53] show that donor substituents in position 5 of thioindigo cause bathochromic shifts and in position 6 hypsochromic shifts; acceptor substituents in position 5 cause hypsochromic shifts and in position 6 bathochromic shifts, all of which is in agreement with the predictions (see pp. 50-53). Some additional effects are also observed.

In position 5 of the monosubstituted dye the substituent, irrespective of its own polarity, causes a smaller shift of λ_{max} (average absolute value of 7.25 nm) than half of the shift observed upon the introduction of two such substituents in positions 5,5' (average absolute value of the shift 38.5 nm, of half the shift — 19.25 nm).

Similar phenomena were observed in the cyanine series [55] and 4,4'-di-substituted diphenylamines [68,69]. They were interpreted as the "effect of symmetric structure" [68,69] or as the result of electron density on the atoms of the polymethine chain leveling out at the same electronegativities of the terminal links [55]. However, for thioindigoid dyes substituted in position 6 rather than in position 5 the "symmetric structure" effect is totally absent. On the contrary, a slight but undoubtedly opposite effect is observed for the amino and nitro groups.*

The thesis [71,72] regarding the "secondary role" of the annelated benzene rings in the colorability of the indigoid dyes is a matter of dispute. The effect of the introduction into a conjugated system of additional π-bonds diminishes with an increase in the size of the indigoid dye's molecule. This is reflected in the data cited in the order: initial structure λ_{max}, nm; ν_{max}, cm^{-1}; final structure λ_{max}, nm; ν_{max}, cm^{-1}; general shift of the maximum $\Delta\lambda$, nm; $\Delta\nu$, cm^{-1}; shift of the maximum by a single introduced π-bond, ν_{max}, cm^{-1}.

$\Delta^{2,2'}$-Bis(4,4'-dimethylthiolan-3-one) 450; 22222 (in cyclohexane) [67]; thiophenindigo 502; 19920 (in cyclohexane) [66]; 42; 2302; 1150. Thiophenindigo; thioindigo 546; 18315 (in benzene); 44; 1605; 401. Thioindigo; 5,6,5',6'-dibenzothioindigo 580; 17241; 44; 1074; 268. Thioindigo; 4,5,4',5'-dibenzothioindigo 562; 17794; 16; 521; 130. Thioindigo; 6,7,6',7'-dibenzothioindigo 555; 18018 (in trichlorobenzene); 9; 297; 74.

The transition from $\Delta^{2,2'}$-bis(4,4'-dimethylthiolan-3-one), i.e., the "principal chromophore," to thioindigo ensures the shift of the maximum of the absorption band by 100 nm. We should note the appreciably smaller bathochromic effect upon the introduction of benzene rings into positions 4,5 and 6,7 compared with the effect of their introduction into positions 5,6. As is shown in [43], to be discussed later on (see pp. 50-53), this shift of λ_{max} is directly associated with $\phi_{E/Z}$ of the thioindigoid dyes. The additional shift of the long-wavelength absorption band can be achieved through the introduction of substituents in the annelated ring. Some of them are capable of shifting λ_{max} of symmetrical disubstituted dyes by more than 40 nm (see Table 3).

Because the uniqueness of positions 5 and 6 is independent of the substituent's polarity, the effect is apparently associated with the different conductivity of the dye's hetero ring groups which are in the para position with respect to the substituents. The substituent in position 6 does not act on the half molecule with which it is not directly connected since a carbonyl does not convey electronic effects. Therefore, the introduction of a second substituent into position 6' causes rather the same shift of λ_{max} as the introduction of the first one. On the contrary, the substituent's donor or acceptor action in position 5 is partly offset by the shift of the electrons to the other half of the molecule (due to the ability of the sulfur atom to convey electronic effects). Upon the introduction of the second substituent into position 5' such a compensation becomes impossible, and the apparent influence of exactly the same second substituent turns out to be several times that of the first.

Let us assume [73] that the substituents introduced into the molecule of the indigoid dye affects, in the first instance, the light absorption of the half of the molecule of which it forms a part (the "a" effect). This

_____

*Interestingly, the introduction of a single nitro group into position 2 of the material causes a larger hypsochromic shift than the introduction of two nitro groups in positions 2 and 7 "because of the mutual neutralization of their action on the chromophore system" [70], viz.,

influence is modified by the interaction with the other half of the molecule (the "b" effect). Let us label the λ_{max} shift as a result of the first effect as a, and the shift as a result of the second effect as b, and let us assume that for dyes sharing a similar skeleton (e.g., a thioindigo struc-ture) for each substituent in a certain position a and b are constant. Upon the introduction of the substituents R and R' (see Table 3) the a shifts should be added in all cases; the b shifts are added only upon the introduc-tion of substituents of different kinds, donor and acceptor, into the op-posite halves of the molecule.

If we take a λ_{max} of 545 nm for thioindigo, we arrive at the following formula for calculating the wavelength of the absorption maximum of a disub-stituted dye:

$$\lambda_{max} = 545 + a_1 + a_2 + b_1 + b_2$$

Upon the introduction of two donors or two acceptors the absolutely lower value b is subtracted from the higher value: $\lambda_{max} = 545 + a_1 + a_2 + b_1 - b_2$. For the monosubstituted dye $a_2 = 0$, $b_2 = 0$, and, consequently, $\lambda_{max} = 545 + a_1 + b_1$. For the symmetrically disubstituted dye $a_1 = a_2$, $b_1 = b_2$, and then $\lambda_{max} = 545 + 2a_1$.

The values of a and b for different substituents were found from ex-perimental data (see Table 3) on λ_{max} of the monosubstituted and the cor-responding symmetrically disubstituted dye and were used in precomputing λ_{max} of the unsymmetric disubstituted forms.

The values of the a and b parameters for some of the substituents introduced into the thioindigo molecule [53] are given below in the order: substituent; a, nm, cm^{-1}; b, nm, cm^{-1}:

5-NO_2; −16.5, 589; 10.5, −386. 5-SO_2CH_3; −17, 608; 13, −473. 5-OC_2H_5; 21, −654; −6, 164. 5-NH_2; 18.5, −581; −14.5, −448. 6-NO_2; 10.5, −339; 4.5, −151. 6-SO_2CH_3; 5.5, −181; 0.5, −18. 6-OC_2H_5; −15.5, 551; 0.5, −34. 6-NH_2; −10, 348; −6, 205.

In 1962 the λ_{max} value for 5,6'-dinitrothioindigo was precomputed at 546 nm [73]. After the synthesis of this dye in 1973 λ_{max} was found to be 545 nm [53]. For 14 unsymmetrically substituted thioindigo derivatives the mean deviation of the λ_{max} value from the experimental one computed on the basis of the above-cited formula was found to be a mere 3.2 nm [53]. The maximum deviation was observed for materials which contained a donor in one half of the molecule and an acceptor in the other.

PRODUCTION, THERMAL STABILITY, AND SPECTRA OF Z-ISOMERS OF THE THIOINDIGOIDS

Solutions of oxindigo, thioindigo, selenoindigo, and N,N'-disubstituted indigo are all photochromic [32,35,74], which is generally attributed to E/Z-photoisomerization. Indeed, the thioindogenides

are photochromic only when the structure of their molecule allows geometri-cal isomerism (R ≠ R') and are not photochromic when R = R', which is in-direct evidence of the photochromism of these thioindigo-related substances being due to E/Z-photoisomerization [75].

Pending an X-ray crystallographic analysis of the Z-isomers of the indigoid dyes, we may formally doubt the existence of the Z-isomers [76] and may look for other explanations of the photochromism of thioindigo and its analogues. However, we propose to continue our exposition as though firm proof was available that the solution of the E-form upon irradiation produces the Z-isomer.

The formation of the Z-isomer of the indigoid dye in the course of synthesis or upon heating a solution of the E-isomer has not been reported. This shows that the Z-isomer is thermodynamically less stable. No quantitative data on the difference in enthalpies of the E- and Z-isomers of indigoid dyes are to be found in the literature. If this difference is not too great at high temperatures, there will be noticeable amounts of the Z-form in an equilibrium mixture. Indeed, some sterically hindered thioindigoid dyes exhibited [77] the condensation of the Z-isomer along with the E-form on a cooled quartz glass from sublimation vapors as did 4,4'-dimethyl-6,6'-dichlorothioindigo.

Photoisomerization of E-form solutions with subsequent (usually chromatographic) separation is the only practically applicable technique for obtaining Z-isomers. The purest samples of the Z-isomers for thioindigo and its substituted forms were found to contain as much as 10% of the E-form [32,79]. According to the data cited in the paper [4], a pure Z-isomer of perinaphthothioindigo was photochemically obtained and isolated by pouring a dioxane solution over water. On E → Z-photoisomerization of saturated solutions of E-N,N'-diacylindigoids, pure Z-isomers [33] crystallized and were isolated.

The ability of an indigoid dye to exist as geometrically isomeric forms is not always easy to register (see, e.g., observations on the discussion [35,79,80] which established that the lifetime of the Z-form of N,N'-dialkylindigoids did not exceed a few seconds). One obstacle here is not only the low thermal stability of the Z-form but also the low value of $\phi_{E \to Z}$.

Although in all the conditions studied the E → Z-photoisomerization of indigo* is not observed [82,83], during chemical reactions indigo is capable of forming Z-form derivatives [84,85]. The nature of the intermediate particles that are formed in the course of the synthesis is yet to be investigated along with the reasons for their ability to assume the Z-configuration. No attempt has been made to accelerate photochemically the dark (fading) reactions that result in the formation of Z-derivatives of indigo. The closeness between λ_{max} of N,N'-oxalylindigo, whose Z-structure is fixed by valence bonds, and the photoinduced form N,N'-diacetylindigo is considered to be an argument in favor of the Z-structure of the latter [84].

The appearance of a "vibrational structure" band in the electronic spectrum is attributed to the nonadiabatic nature of transitions [86,87]. During the $S_0 \to S_1$-transition the E-indigoid dyes display a strong change of order in the central bond and a noticeable alteration in the bonds of the five-membered ring (see Figure 2). This corresponds [88] to a change in the equilibrium distances between the corresponding atoms. As a rule, the long-wavelength absorption band of the Z-isomers of indigoid dyes lacks a vibrational structure. However, it does not always have the Gaussian form, while some of the dyes in aliphatic solvents (cf. [89]) display a discernible inflection (shoulder) on the short-wavelength branch. This is typical of

---

*Indirubin and 2-benzo[b]thiophen-3-indolindigo do not isomerize either [37,81].

the thioindigo and 2-benzo[b]thiophene-1-acenaphthene-indigo series which
apparently have minimum steric hindrances in the C=C bond. Therefore, the
search for fluorescence in solutions of Z-isomers of substances from these
series in aliphatic solvents does not seem hopeless. A characteristic
feature of the absorption band of the Z-isomer of thioindigo is believed
[90] to be a long "tail" which extends (at very low levels of intensity) all
along the absorption band of the E-form. Upon excitation by light a shift
of the photoequilibrium towards the E-form (see Figure 1 in the paper [90])
was observed in the long-wavelength continuum of the E-form due to the similarity of the absorption intensities of the E- and Z-forms and $\phi_{Z \to E}$ which is
greater than $\phi_{E \to Z}$.

No attempt has yet been reported to ascertain the relation between the
light absorption intensity of the substituted forms of thioindigo and the
character and location of the substituents; for thioindigo ϵ_{max} is 1.6×10^4
(in benzene). These regularities are not associated with the shift of λ_{max}
and are formulated as follows (see Table 3):

the introduction of halogen atoms and methyl groups does not appre-
ciably affect ϵ_{max} [values of ϵ_{max} amount to $(1.5-1.7) \times 10^4$]; only for
7,7'-diiodothioindigo and 5,5',7,7'-tetramethylthioindigo does ϵ_{max}
drop to 1.3×10^4; the introduction of at least one substituent with a
pronounced donor ability results in a sharp drop of ϵ_{max} to $(1.0-1.1) \times$
10^4 irrespective of the substitution site;

as far as one can judge by the monosubstituted forms* the introduction
of electron-acceptor substituents raises somewhat the molar extinction
up to $(1.7-1.8) \times 10^4$. If both donor and acceptor substituents are
introduced into the molecule, the acceptor prevents extinction from
falling, which is usually caused by the donor only for 6,6'- but not
for 5,5'- and 5,6'-substituted forms.

The electronic system of the thioindigoids containing at least one
naphtho[1,8-b,c]thiopyran fragment (Tables 5 and 6) absorbs light far more
intensely [ϵ_{max} $(2.5-4.0) \times 10^4$] than thioindigo and their substituted
forms. Because of this, changes of ϵ_{max} of less than 0.5 along with substituent changes are treated as too small to deserve detailed examination.

Normally a thermodynamically less stable isomer has a lower intensity
in the long-wavelength absorption band. Indeed, the ratio ϵ_Z/ϵ_E is less
than unity for the majority of indigoid dyes.** For unsymmetrical dyes V it
is 0.48-0.6 (see Table 6), for pyrenothiopyranone − 0.5-0.6 (Table 7), for
symmetrically substituted forms of perinaphthothioindigo − 0.58-0.7 (see
Table 5), for substituted forms of 2-naphtho[1,8-bc]thiopyran-1-acenaphth-
ene-indigo − 0.68-0.8 (see Table 6, formula VI), for thioindigo and its substituted forms − 0.75-0.9, for 2-benzo[b]thiophene-1-acenaphthene-indigo −
0.8-0.95 (see Table 7). A certain tendency is observed towards an antibatic
dependence between $\Delta\lambda_{E/Z}$ and ϵ_Z/ϵ_E.

The hypsochromic shift of λ_{max} and decrease in ϵ_{max} indicate increased
steric hindrances. One may assume that the steric hindrances during the
transition from the E- to Z-isomer diminish in the series: V ≈ pyrenothiopyranone > IV > VI > thioindigo and its substituted forms > 2-benzo[b]thio-
phene-1-acenaphthene-indigo.

---

*Substituted forms containing two electron-acceptor substituents dissolve
too poorly for their ϵ_{max} to be reliably measured.
**The exception is 2-benzo[b]thiophene-9-phenanthrene-indigo [91] (apparent-
ly because of the strong steric interaction between the phenanthrene hydrogen in position 10 and the carbonyl oxygen of the second half of the molecule) as well as oxindigo (Tables 1 and 7).

Table 5. Maxima of the long-wavelength absorption band and color contrast of solutions of geometrical isomers of perinapthothioindigo and its substituted forms in trichlorobenzene

Substituents in the nucleus of perinaphtho-thioindigo	λ, nm ($\epsilon \cdot 10^{-4}$)		Color contrast	
	E-isomer	Z-isomer	$\Delta\lambda_{E/Z}$, nm ($\Delta\epsilon \cdot 10^{-4}$)	ϵ_Z/ϵ_E
Perinaphthothioindigo [4]	637(2.9)	513(1.8)	124(1.1)	0.62
6,6'-di-Cl [4]	644(2.5)	518(1.8)	126(0.7)	0.72
7,7'-di-Cl [4]	650(3.0)	520(2.0)	130(1.0)	0.67
6,6'-di-OCH$_3$ [4]	632(2.8)	502(1.9)	130(0.9)	0.68
7,7'-di-OCH$_3$ [4]	693(2.6)	549(1.7)	144(0.9)	0.65
7,7'-di-Br	650(3.7)	519(2.2)	131(1.5)	0.59
7,7'-di-I	652(3.4)	520(2.0)	132(1.4)	0.59
6,6'-di-NO$_2$	629(4.3)	506(2.7)	123(1.6)	0.63
7,7'-di-NO$_2$	676(2.6)	544(1.7)	132(0.9)	0.65
7,7'-di-SO$_3$H	622	505	117	

The color contrast reflects the difference in the absorption spectra of the stereoisomers. To illustrate the complexity of the picture, let us examine the spectra of the thermodynamically unstable Z-isomers of thioindigo, stilbene, and their "hybrid" E-isomer.* The first two substances are less intensely colored while the third is colored more deeply than the corresponding thermodynamically stable isomer. The E-isomers of substances of the general formula VII

E-isomer   VII, X = O, S, Se, NH, NCH$_3$

are more deeply colored than the Z-isomers. When X = O, S, Se, or NH, the E-isomers are less thermodynamically stable, while when X = NCH$_3$ the Z-isomer [94-96] is less stable. Apparently, a methyl at the nitrogen creates sufficient steric hindrances within the Z-isomer for the E-isomer to become more thermodynamically advantageous. However, the E-isomer remains more deeply colored. Probably the position of λ_{max} is less subject to steric interactions than the character of the mutual polarization of adjacent groups, and yet it has not changed and should be even more pronounced because of the (supposed) increase in the donor ability of the heteroatom X = NCH$_3$ compared with X = NH.

The unquestionable reason for any differences (including spectral ones) between geometric isomers is the different arrangement in space of the molecule's fragments with respect to one another linked together by the π-bond. A change in the fragment's arrangement alters the molecule's symmetry,** the interactions of adjacent groups (not linked with chemical bonds) and, occasionally, the degree of noncoplanarity.

---

*See the footnote on p. 46.
**As a result of E/Z-isomerization, the symmetry of the molecule of a symmetrical indigoid dye changes from C_{2h} to C_{2v}. To the best of our knowledge no analysis of the effect this fact should have on the spectral properties of the indigoid dyes has been reported.

The theoretical elaboration of the problem of the relationship of the spectra of geometric isomers is difficult on account of the interaction between unrelated groups being extremely varied so that in the general case it cannot be broken down into the additive contributions of the various effects [54,97].

We do not have a PEPIPS diagram for Z-thioindigo. If it is qualitatively analogous to the PEPIPS diagram for E-indigo (see Figure 2), in the first singlet of the Z-form the charge on the carbonyls should grow upon light absorption, while the order of the central double bond should decrease. The latter is characteristic of many unsaturated compounds.

Oksengendler and Gendrikov [3] believed that the hypsochromic shift of the coloration of the Z-form was due to its noncoplanarity, compared to the E-form. If the noncoplanarity were the result of a "twisting" of the central double bond, the bond order would decrease to approach that observed in the excited state; in other words, a bathochromic shift would occur. The "stretching" of the bonds of the heterocyclic ring normally requires greater energy expenditure than the exit of individual atoms from the plane. If noncoplanarity arose as a result of the carbonyl's withdrawal from the plane of the heterocyclic ring, an increase in the order of the $C=O*$ bond would be likely. According to the PEPIPS diagram (see Figure 2), on transition into the first singlet the π-order of the $C=O$ bond increases. The withdrawal of the carbonyl from the plane should cause a bathochromic shift in the coloration. If noncoplanarity comes as a result of a distortion of the heterocycle's form (cf. [88]), a qualitative discussion of the shift in the long-wavelength absorption band is difficult.

The foregoing discussion shows that the steric hindrances arising upon transition from the E- to Z-form cannot induce a hypsochromic shift of the coloration (cf. [40]). The reason for the hypsochromic shift lies in the interaction between sterically adjacent unrelated groups (located in different nuclei), oppositely polarized in E-thioindigo and similarly polarized in Z-thioindigo (cf. [53]). In the E-thioindigo molecule oppositely charged carbonyl and sulfur interact through space. Due to mutual polarization in the ground state, the negative charge on the carbonyl and the positive charge on the sulfur increase, which, in accordance with the PEPIPS diagram and Murrell's principle, should make transition into the excited state easier and cause a bathochromic shift (compared with the Z-form).

In the Z-thioindigo molecule two carbonyls and two heteroatoms interact through space. This makes the increase of the positive charge on the heteroatoms and of the negative charge on the carbonyls more difficult and causes a hypsochromic shift (compared with the E-form). A considerable color contrast results.

Even in very simple substituted forms of ethylene the color contrast becomes substantial if sterically adjacent groups differ widely in electronegativity in the E-form and coincide in electronegativity in the Z-form, as is the case in the example below [98].

E-isomer

R = SCH$_3$; E-isomer, λ_{max} 364 nm, log ϵ 4.01; Z-isomer, λ_{max} 340 nm, log ϵ 4.10; $\Delta\lambda_{E/Z}$ 24 nm; R = SNa; E-isomer, λ_{max} 412 nm, log ϵ 4.10; Z-isomer, λ_{max} 368 nm, log ϵ 4.06; $\Delta\lambda_{E/Z}$ 44 nm

---

*Carbonyl's higher frequency in the Z- than in the E-isomer confirms this.

The CN group has an acceptor character, the SCH$_3$ group — a donor. Upon the replacement of the SCH$_3$ groups by SNa groups, sulfur's donor ability increases as does the color contrast.

Interactions of unbound atoms can be revealed by studying infrared spectra. Two different methods are employed for this purpose: the frequencies of E-isomer vibrations are compared with those of a model compound's vibrations (e.g., one containing only one half of the indigoid dye), or, alternatively, with the frequencies of Z-isomer vibrations.

Claims of the absence of the interaction through space between sulfur and carbonyl in the E-thioindigo, made with a reference to the infrared spectra of the compounds A-D [72], stem from an error.* In reality upon transition from ketone A to dione B the change in $\nu_{C=O}$ is 14 cm^{-1} less than upon the transition from ketone C to thioindigo D.

A, X = CH$_2$, $\nu_{C=O}$ 1717 cm^{-1} $\Delta\nu$ 39 cm^{-1} B, X = CH$_2$, $\nu_{C=O}$ 1678 cm^{-1}
C, X = S, $\nu_{C=O}$ 1709 cm^{-1} $\Delta\nu$ 53 cm^{-1} D, X = S, $\nu_{C=O}$ 1656 cm^{-1}

Since A differs from C and B differs from D only by the presence of the CH$_2$ group instead of the sulfur atom, the deviation $\Delta\Delta\nu_{C=O}$ 14 cm^{-1} reflects the effects of the sulfur of one half of the molecule on the carbonyl of the other. Whether this interaction proceeds through bonds or through space is impossible to establish here.

Using empirical data on the influence of conjugated double bonds and aromatic rings on the carbonyl's frequency [99],** we can estimate the interval of the $\nu_{C=O}$ values which may be due to a usual conjugation without the additional polarizing influences of the heteroatom. A and C should be regarded as five-membered ketones condensed with one aryl, and B and D as five-membered ketones condensed with an aryl and a double bond (Table 8).

In this table the region of expected (anticipated) values of $\nu_{C=O}$ 1673-1683 cm^{-1} in the five-membered aliphatic ring conjugated with an aryl and an α,β-double bond is found by the formula

$$\nu_{C=O} = \nu_{C=O}^{5\text{-memb.}} - \Delta\nu_{C=O}^{Ar} - \Delta\nu_{C=O}^{\alpha,\beta}$$

where $\nu_{C=O}$ is the expected value; $\nu_{C=O}^{5\text{-memb.}}$ is the carbonyl's frequency in the five-membered ring. $\Delta\nu_{C=O}^{Ar}$ and $\Delta\nu_{C=O}^{\alpha,\beta}$ are, respectively, the shift in the

---

*Ammon and Hermann [72], after the subtraction 1709 − 1656 cm^{-1} obtained 44 cm^{-1} instead of the correct value 53 cm^{-1}.
**The suggestion [31] to estimate the degree of affinity to indigo ("indigoid character") by the proximity of $\nu_{C=O}$ of the substance to $\nu_{C=O}$ of indigo (1626 cm^{-1}) does not hold, since it conflicts with the estimation of the indigoid character by the depth of the dye's coloration (for 2,2'-bisthiophenindigo $\nu_{C=O}$ is less than in thioindigo and selenoindigo, and yet its color has a large hypsochromic shift). In a wide series of indigoid dyes (Table 8) no similarity in the change of ν_{max} and $\nu_{C=O}$ (cf. [31]) is observed.

Table 6. Maxima of the long-wavelength absorption band of solutions of geometrical isomers of dyes V, Va-c, and VI

V, X = S   Va   Vb

Vc   VI

No. of formula; R; R'	Solvent	λ, nm ($\epsilon \cdot 10^{-4}$) E-isomer	λ, nm ($\epsilon \cdot 10^{-4}$) Z-isomer	Color contrast $\Delta\lambda_{E/Z}$, nm ($\Delta\epsilon \cdot 10^{-4}$)	ϵ_z/ϵ_E
V; H; 6-OC$_2$H$_5$	benzene	582(2.7)	470(1.2)	112(1.5)	0.44
	nitrobenzene	586.5(2.6)	478.5(1.4)	108(1.2)	0.54
V; H; 6-Cl	benzene	590(3.5)	480(1.7)	110(1.8)	0.49
	nitrobenzene	596(3.0)	493(1.5)	103(1.5)	0.50
V; H; 4-CH$_3$, 6-Cl	benzene	591(4.3)	479(2.2)	112(2.1)	0.51
V; H; H	benzene	592(3.1)	482(1.5)	110(1.6)	0.48
V; H; 5-NO$_2$	benzene	594(3.2)	486(1.8)	108(1.4)	0.56
V; H; 5-CH$_3$	benzene	597.5(3.2)	486.5(2.3)	111(0.9)	0.72
Va; H	benzene	600(3.2)	487(1.4)	113(1.8)	0.44
Vc; H	benzene	609(2.5)	504(1.4)	105(1.1)	0.56
V; Cl; 6-OC$_2$H$_5$	benzene	585(3.0)	469(1.4)	116(1.6)	0.47
V; Cl; 4-CH$_3$, 7-Cl	benzene	595.5(2.1)	481(1.1)	104(1.0)	0.48
	nitrobenzene	601.5(3.1)	495(1.7)	106.5(1.4)	0.55
V; Cl; 6-Cl	benzene	596(3.2)	485(1.6)	111(1.6)	0.5
	nitrobenzene	601.5(3.1)	497.5(1.7)	104(1.4)	0.55
V; Cl; H	benzene	598(2.9)	487(1.5)	111(1.4)	0.52
	nitrobenzene	603(3.0)	497.5(1.5)	105.5(1.5)	0.50
V; Cl; 5-NO$_2$	benzene	601(3.5)	492(1.9)	109(1.6)	0.54
	nitrobenzene	605.5(3.1)	504(1.7)	101.5(1.4)	0.55
Va; Cl	benzene	606.5(2.8)	490.5(1.3)	116(1.5)	0.46
	nitrobenzene	611(2.5)	502.5(1.4)	108.5(1.1)	0.56
Vc; Cl	benzene	614.5(2.5)	507(1.5)	107.5(1.0)	0.60
Vb; Cl	benzene	619(3.8)	509(2.0)	110(1.8)	0.53
V; Br; 6-OC$_2$H$_5$	benzene	587(3.0)	470(1.6)	117(1.4)	0.53
	nitrobenzene	591(3.2)	481(1.6)	110(1.6)	0.50
V; Br; 4-CH$_3$, 6-Cl	benzene	597(2.7)	482(1.4)	115(1.3)	0.52
	nitrobenzene	600(3.3)	492(1.6)	108(1.7)	0.48
V; Br; 6-Cl	benzene	597.5(2.8)	485.5(1.6)	112(1.2)	0.57
	nitrobenzene	599(3.2)	495(1.7)	104(1.5)	0.53
V; Br; H	benzene	598(2.9)	485.5(1.5)	112.5(1.4)	0.52
	nitrobenzene	605(2.8)	496(1.5)	109(1.3)	0.54
V; Br; 5-NO$_2$	nitrobenzene	603(3.7)	500(2.1)	103(1.6)	0.57
Vb; Br	benzene	605(3.3)	489.5(1.6)	115.5(1.7)	0.48
	nitrobenzene	611(2.8)	499(1.4)	112(1.4)	0.50
Vc; Br*	benzene	615	507	108	
Vc; Br	nitrobenzene	622(2.7)	514(1.5)	108(1.2)	0.56
V; I; H	benzene	600(3.0)	487(1.6)	113(1.4)	0.53
	nitrobenzene	603.5(1.8)	500(1.8)	103.5(1.2)	0.60

Table 6. (Continued)

No. of formula; R; R'	Solvent	λ, nm (ε·10⁻⁴) E-isomer	λ, nm (ε·10⁻⁴) Z-isomer	Color contrast Δλ_{E/Z}, nm (Δε·10⁻⁴)	ε_Z/ε_E
V; I; 5-NO₂	benzene	603.5(4.0)	494.5(2.2)	109(1.8)	0.55
	nitrobenzene	608(3.2)	507(1.8)	101(1.4)	0.56
Vb; I	benzene	606(3.4)	491(1.6)	115(1.8)	0.47
	nitrobenzene	611(2.8)	499(1.6)	112(1.2)	0.57
Vc; I	benzene	618(2.8)	506.5(1.4)	111.5(1.4)	0.50
Va; Br	benzene	622(2.9)	511(1.7)	111(1.2)	0.59
	nitrobenzene	625.5(3.3)	519(1.8)	106.5(1.5)	0.55
V; I; 6-OC₂H₅	benzene	589(3.3)	471.5(1.6)	117.5(1.7)	0.48
	nitrobenzene	593.5(3.1)	484(1.5)	109.5(1.6)	0.48
V; I; 4-CH₃, 6-Cl	benzene	597.5(3.5)	485(1.7)	112.5(1.8)	0.49
V; I; 6-Cl	benzene	598.5(3.3)	486.5(1.7)	112(1.6)	0.52
	nitrobenzene	603(3.2)	497.5(1.7)	105.5(1.5)	0.53
Va; I	benzene	621(3.3)	511(1.7)	110(1.6)	0.52
VI; H	benzene	521(1.5)	448.5(1.1)	72.5(0.4)	0.73
VI; NO₂	benzene	522.5(2.3)	453.5(1.7)	69(0.6)	0.74
VI; Br	benzene	527.5(1.6)	455(1.3)	72.5(0.3)	0.81
VI; I	benzene	531(1.9)	458.5(1.3)	72.5(0.6)	0.68

*Substance not readily soluble in benzene.

carbonyl's frequency as a result of conjugation with the aryl and with the α,β-C=C double bond. For the necessary numerical data see [99]. The region of the expected value of $\nu_{C=O}$ (1715-1725 cm⁻¹) in the five-membered aliphatic ring condensed with one aryl is found by the formula

$$\nu_{C=O} = \nu_{C=O}^{5\text{-memb.}} - \Delta\nu_{C=O}^{Ar}$$

The heteroatom's influence along the bonds is already fixed in benzo-[b]thiophen-3-(2H)-one (C; a deviation of $\nu_{C=O}$ from the expected value by 6 cm⁻¹) and is more strongly felt in the thioindigo (D; a deviation of $\nu_{C=O}$ of 17 cm⁻¹), where interactions through space are possible. For compounds A and B, containing no heteroatom, $\nu_{C=O}$ falls in the region of expected values (see Table 8).

In thioindogenide VIII, X = S, one carbonyl group is contiguous with the heteroatom, and thus analogous to the carbonyl group in the E-form of thioindigo. The frequency of this thioindogenide carbonyl ($\nu_{C=O}$ 1672 cm⁻¹) is lower than expected for α,β-unsaturated ketones, which is evidence of additional carbonyl polarizing influences. These can apparently be exerted only by the heteroatom of the ring. The second carbonyl group, contiguous with the carbonyl of the heterocycle, is analogous to the carbonyl in the Z-form of thioindigo and has a $\nu_{C=O}$ (1705 cm⁻¹) close to the frequency of the saturated ketones VIII, X = S.

VIII

Table 7. Maxima of the long-wavelength absorption band of solutions of Z- and E-isomers of thioindigoids

Thioindigoid	λ, nm (ε·10⁻⁴) E-isomer	λ, nm (ε·10⁻⁴) Z-isomer	Color contrast $\Delta\lambda_{E/Z}$, nm ($\Delta\epsilon$·10⁻⁴)	ϵ_Z/ϵ_E
2,2-Bis(pyreno[3,4-bc]thio- pyran)-indigo [9,92]	728(3.28)	584(1.67)	144(1.67)	0.51
2-Benzo[b]thiophene-2-pyreno-[3,4-bc]thiopyran-indigo [9,92]	657(2.40)	551(1.31)	106(1.09)	0.55
2-Naphtho[2,1-b]thiophene-2-pyreno[3,4-bc]thiopyran-indigo [9,92]	662(2.88)	542(0.92)	120(1.96)	0.32
2,2-Bis(pyreno[4,3-bc]-thiopyran)-indigo [9,92]	668(4.41)	543(2.81)	125(1.60)	0.64
2-Benzo[b]thiophene-2-pyreno-[4,3-bc]thiopyran-indigo [9,92]	609(2.79)	507(1.67)	102(1.12)	0.60
2-Naphtho[2,1-b]thiophene-2'-pyreno[4,3-bc]thiopyran-indigo [9,92]	615(2.52)	512(1.50)	103(1.02)	0.60
2-Benzo[b]thiophene-1-ace-naphthene-indigo [17,93]*	512(1.12)	467(1.05)	45(0.07)	0.94
2-Naphtho[1,8-bc]thiopyran-1'-acenaphthene-indigo*	519(1.46)	448(1.10)	71(0.36)	0.75
2-Benzo[b]thiophene-9-phenanthrene-indigo [17]*	556(0.82)	470(0.89)	86(-0.07)	1.08
2-Naphtho[1,8-bc]thiopyran-9-phenanthrene-indigo [17]*	578(0.38)	510	68	
2-Benzo[b]selenophene-2-naphtho-[1,8-bc]thiopyran-indigo [12]*	608(1.6)	484(0.84)	124(0.76)	0.52

*In benzene.

Table 8. Values of ν_{max} of long-wavelength bands in the electronic spectra of solutions of indigoid dyes and deviation of $\nu_{C=O}$ from the expected values of the corresponding ketones

Indigoid dye	ν_{max}, cm⁻¹ (λ_{max}, nm)	$\nu_{C=O}$, cm⁻¹ (in KBr)	Deviation of $\nu_{C=O}$ from the expected values low frequency	Deviation of $\nu_{C=O}$ from the expected values high frequency
Oxindigo	452	1 692	-19	-9
N,N'-Diacetylindigo	562	1 689	-16	-6
2,2'-Bis(indan)indigo		1 678	-5	5
Thioindigo	545	1 656	17	27
6,6'-Diazaindigo	15 980(629)	1 645	28	38
Selenoindigo	570	1 642	31	41
5,5'-Diazaindigo	18 220(549)	1 642	31	41
N,N'-Dimethylindigo	644	1 639	34	44
4,4'-Diazaindigo	16 660(600)	1 635	38	48
N-Methylindigo		1 634	39	49
7,7'-Diazaindigo	17 980(556)		40	50
Indigo	16 500(606)	1 626; 1 630	47;43	57;53
1-Indanone		1 717	-2	8
Benzo[b]thiophen-3-one		1 709	6	16

The difference in frequencies of the stretching vibrations of the carbonyl of the acetyl groups in the E- and Z-positions relative to the carbonyl of the ring is comparable (33 cm^{-1}) with the difference of $\nu_{C=O}$ of the E- and Z-forms of thioindigo (46 cm^{-1} [85]). In oxindigo this difference is 32 cm^{-1} [29].

Interactions between unbound atoms are also ascertained in the course of a theoretical examination of the fine structure of the molecule. Thus, in calculations performed by the MO LCAO PPP CI method a matrix of bond orders is printed out which are not only linked in a classical formula but are also separated by several simple bondings [25]. As an example we shall cite figures from such matrices for the states S_0 and S_1 of compound V (Table 9) obtained by Mostoslavskii, Georgieva, and Minkin (cf. [100]). It is easy to see that the bond order between unbound atoms of sulfur and oxygen is comparable with the π-order between atoms linked by σ-bonds (cf. [101]). (See also pp. 24-26.)

Unmistakable evidence of the interaction of unbound atoms is the violation of the vinylogy principle in indigoid dyes [71,102]. As the length of the conjugated system between the chromophore halves increases, there are hypsochromic rather than bathochromic shifts.

X = S,	n = 0,	λ_{max} 541 nm
X = S,	n = 1,	λ_{max} 520 nm
X = S,	n = 0,	λ_{max} 620 nm
X = NH,	n = 1,	λ_{max} 597 nm

In the absence of polarizing interactions between the heteroatom and the carbonyl in E-indigo and E-thioindigo molecules the transition to vinylogues would cause a bathochromic shift. If the interaction of unbound groups occurred in indigo and thioindigo molecules which differed widely in nature or intensity (e.g., with the intramolecular hydrogen bond playing the determining role for the indigo), the λ_{max} shift would be different upon transition to a vinylogue in the case of the thioindigo and indigo. In reality it is practically the same (approximately 20 nm).

Although in vinylogues heterocycles forming part of the dye's molecule are so far apart as to rule out intramolecular interaction between the carbonyl of one half of the molecule and the heteroatom of the other, indigo's vinylogue is bathochromically shifted compared to thioindigo's. This shows that the "interaction of the unbound atoms" is not the main reason for the differences in the coloration of the E-isomers of indigo and thioindigo. The primary reason, as mentioned above, is the differing electronegativities of the heteroatom.

For model compounds it has been shown [75,103] that indogenide (VIII, X = NH; λ_{max} 486 nm, ϵ_{max} 0.6 × 10^4) is bathochromically shifted compared to thioindogenide (VIII, X = S, λ_{max} 461 nm, ϵ_{max} 0.36 × 10^4); i.e., the NH group as a "heteroatom" gives a bathochromic shift by substituting it for sulfur.

Apparently, the structure of half of the E-form of the symmetric indigoid dye predetermines not only interactions transmitted through bonds but also interactions between the heteroatom of one half of the molecule and the carbonyl of the other. This is only possible given a purely electronic character of the interactions, undisturbed in any appreciable way by differences in steric factors (associated with differing geometrical dimensions of the heterocycles lying at the base of various dyes). To corroborate this

Table 9. Orders of π-bonds of the sulfur atoms with other atoms in the molecule E-V; X = S; R' = H (see p. 46). Enclosed in parentheses are the numbers of interacting atoms. Calculations were performed by the MO LCAO SCF CI method.

State	With carbon atoms		With oxygen atoms	
	of the same half of molecule	of the other half of molecule	of the same half of molecule	of the other half of molecule
S_0^E	0.345(14-13) 0.34(14-1) 0.240(16-17) 0.311(16-15)	-0.053(14-2) -0.143(14-23) -0.095(16-23) -0.049(16-2)	0.001(14-3) 0.031(16-24)	0.169(14-24) 0.065(16-3)
S_1^E	0.429(14-13) 0.344(14-1) 0.297(16-17) 0.337(16-15)	-0.055(14-2) -0.108(14-23) -0.0089(16-23) -0.050(16-2)	0.002(14-3) -0.040(16-24)	0.024(14-24) 0.065(16-3)

assumption, it is necessary to study the spectra of substances that model the half molecule of the indigoid dye.*

This idea is not new. The first qualitative data on the similarity between color changes in the thioindigoid dyes and the corresponding thioindogenides when the substituents R and R' [104,105] are varied were obtained a long time ago.

As a comparison of Figures 2 and 5 shows, the redistribution of electron density upon the light absorption by the half molecule in indigo and in benzylidene derivatives is closely similar. The sign of charge change is the same for the heteroatom and the oxygen and carbon of the carbonyl in positions 5, 6, and 7. The sign changes only for slight electron density redistributions at positions 2 and 4.

As Figure 6 shows, a comparison of ν_{max} for indigoid dyes and the corresponding nitrobenzylidene derivatives reveals a proportional response to a variation of the heteroatom. If the presence of a hydrogen bond in the indigo molecule and steric factors in the interaction of unbound atoms played an independent primary role in the formation of the coloration of the E-form of indigoid dyes, a considerable point scatter would be in evidence, and the point representing indigo would sharply drop out of the correlation,** which was not the case.

The presence of a proportional response in indogenides and indigoid dyes to heteroatom substitution indicates either the absence of a hydrogen bond in indigo or the absence of its influence on the effectiveness of the polarization of the electron cloud of the carbonyl of the heteroatom of the other half of the indigoid dye. The NH group behaves as a kind of "complex atom," being polarized by and polarizing according to the same laws that other heteroatoms (oxygen, sulfur, selenium) are governed by.

If a substituent R is present in the benzylidene part of the molecule as a substituent rather than a nitro group, similarity is maintained, as a rule, in the change of ν_{max} of indogenides and indigoid dyes (Figure 7). However, no explicit direct dependence is observed, while when R = N(CH$_3$)$_2$ similarity is disturbed: ν_{max} of thioindogenide is less than ν_{max} of the selenoindogenide. This violation is a consequence of the "saturation effect" [53,94,106]. The acceptor (carbonyl) which has a limited "capacity" cannot simultaneously interact vigorously with two donors (X and R) which are formally conjugated with it. The donor ability of the heteroatom X = O is small in aurones, and the position of ν_{max} is strongly influenced by the donor capacity of the substituent R: on transition from R = Cl to R = N(CH$_3$)$_2$ the shift is as much as 4000 cm^{-1} (see Figure 7). In thioindogenides and selenoindogenides the heteroatom's donor capacity increases, while as the influence of substituent R on ν_{max} diminishes upon tran-

---

*Isatin and benzo[b]thiophene-2,3-dione, contrary to the opinion of Klessinger and Lüttke [71], are unsuitable for modelling the spectral properties of the half molecule of the indigoid dye since in these substances a powerful acceptor (carbonyl) is directly added to the heteroatom, an acceptor that is absent from the structure of indigoid dyes. Not surprisingly, the model was rejected [71].
**A nitrobenzylidene derivative containing X = NH and comparable with indigo lacks a hydrogen bond.

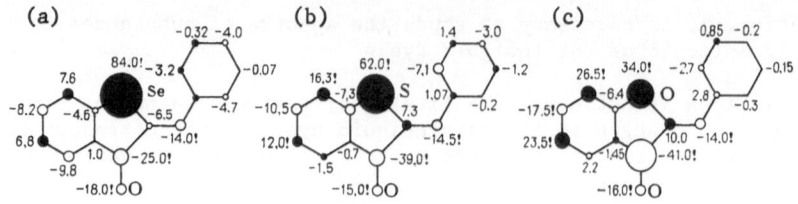

Fig. 5. Chart of the electron density redistribution for the $S_0 \rightarrow S_1$ transition in the indogenide series: a) selenoindogenide; b) thioindogenide; c) aurone. The charge of each individual sign is taken to equal 100%. The digits indicate the share (in %) of the redistributed charge (of the respective sign) on the atom. The exclamation points indicate central (key) chromophore atoms.

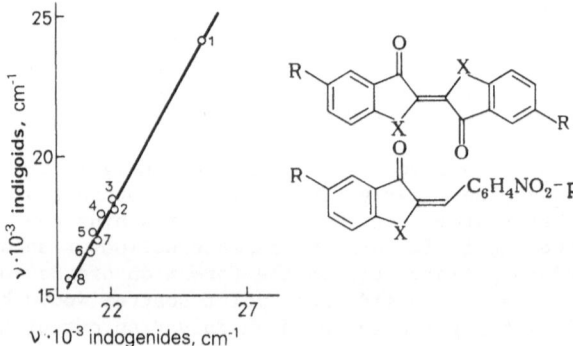

Fig. 6. Comparison of ν_{max} of the solutions of indigoid dyes and indogenides. 1) X = O, R = H; 2) X = S, R = H, in benzene; 3) X = S, R = H, in p-xylene; 4) X = S, R = CH$_3$; 5) X = Se, R = CH$_3$; 6) X = NH, R = H, in benzene; 7) X = NH, R = H, in p-xylene; 8) X = NCH$_3$, R = H.

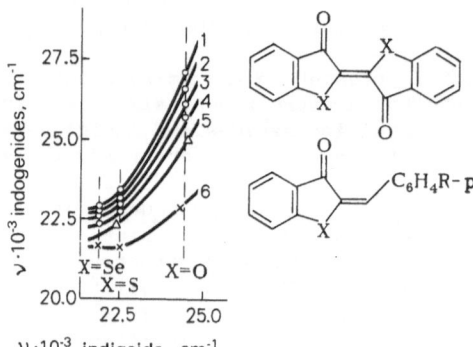

Fig. 7. Comparison between the positions of the long-wavelength absorption bands of indigoid dyes and indogenides upon variation of the heteroatom (X = O, S, Se) and substituent R: 1) H; 2) Cl; 3) OH; 4) OCH$_3$; 5) NH$_2$; 6) N(CH$_3$)$_2$.

sition from R = Cl to R = N(CH$_3$)$_2$ the shift is 1500 and 1200 cm^{-1}, respectively.

By introducing an amino group into position 5 of the benzo[b]thiophenone ring, the donor capacity of sulfur was enhanced to bring the be-

havior of thioindogenides closer to that of the selenoindogenides [107].
The introduction of a nitro group into position 5 of the benzo[b]thiophenone
ring made it possible to lower the sulfur's donor capacity to bring the
behavior of the thioindogenides closer to that of the aurones [107].   In the
molecule of the indigoid dyes a progressive "turning in upon itself" as the
heteroatom's donor capacity grew was experimentally established by electron
paramagnetic resonance (EPR).   The density of the unpaired electron in the
outer (annelated) benzene ring of the anion radicals of indigoids diminishes
in the following order: oxindigo ▸ thioindigo > selenoindigo [108].

The hypothesis mentioned above, to the effect that the electronic
structure of the half molecule fully determines the position of the absorption band maximum of a symmetrical dye, has found (so far in only a few
instances studied) remarkable confirmation.   Paramonov and Mostoslavskii
have observed (Figure 8) a proportional response of a long-wavelength absorption band to a change in structure of heterocyclic ketones and the corresponding symmetrical dyes.

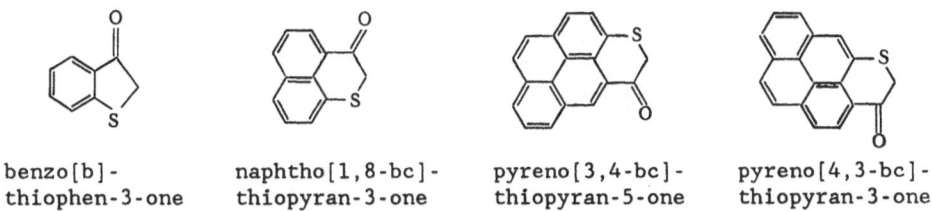

| benzo[b]- | naphtho[1,8-bc]- | pyreno[3,4-bc]- | pyreno[4,3-bc]- |
| thiophen-3-one | thiopyran-3-one | thiopyran-5-one | thiopyran-3-one |

The correlation equation is as follows: $\nu_{dye} = 1.115\nu_{ketone} - 12996.6$.
The correlation coefficient is very high at 0.999.   The slope is close to
unity; i.e., the distance between the energy levels of the ketones and the
corresponding symmetric dyes changes (on the energy scale) in practically
the same fashion.   It turns out that the skeleton of the indigoid dye is no
more sensitive to a structural change than the ketone modelling the half
molecule of the dye.   Inevitable differences in the distances between the
carbonyl and the heteroatom in thioindigo, perinaphthothioindigo, and dyes
based on pyrenothiopyrans do not upset the correlation.   Apparently, in this
dye series steric hindrances do not make themselves felt, and the capacity
of the half molecule to polarize and be polarized plays a principal role.

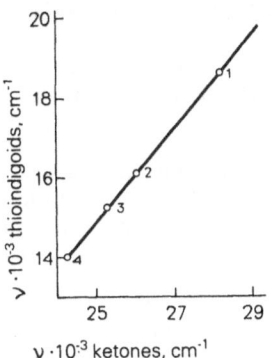

Fig. 8. Proportional response of ν_{max} of solutions in chlorobenzene:
benzo[b]thiophen-3-one (1); naphtho[1,8-bc]thiopyran-3-one (2);
pyreno[3,4-bc]thiopyran-5-one (3); pyreno[4,3-bc]thiopyran-3-one
(4) and the corresponding thioindigoid dyes: thioindigo, perinaphthothioindigo, 2,2'-bis(pyreno[3,4-bc]thiopyran)indigo,
2,2'-bis(pyreno[4,3-bc]thiopyran)indigo.

A comparison of the absorption maxima of the Z- and E-isomers of thio-indigoid dyes by the proportional response method makes it possible to establish a valid relation between them (Figures 9 and 10) [109].* Its linear character indicates a similar change in the interaction of unbound atoms of dyes substituted in different ways upon the transition from the E- to Z-isomer for every compound of this series. The slope of the straight line in Figure 10 clearly shows that in the case of the substituted forms of thioindigo the absorption band of the E-form gives a stronger response to the introduction of substituents (as was noted by Gendrikov in qualitative form [4]). On the other hand, the response of the E-form of the unsymmetrical dyes V to the introduction of substituents is weaker than that of the Z-form [109]. This confirms the somewhat different character of the chromophores of the E-isomer of dyes V and thioindigo.

The possibility of visual reading of information recorded on a photochromic material as well as the probability of a complete photochemical E → Z- and Z → E-conversion (i.e., purely optical recording and image erasure) depend on the distance between the absorption maxima of the Z- and E-forms (color contrast).

Tables 1 and 5-7 summarize data on the color contrast of certain indigoid dyes. Although for thioindigo derivatives the color contrast does not exceed 77 nm ($\Delta \nu \leq 3200$ cm^{-1}), evidence is available indicating the possibility of their use for data recording [7,8].

A sharp color contrast is observed [33,34,84] in N,N'-diacylindigoids (see Table 1). The replacement of the N,N'-methyl groups by N,N'-acyl groups causes a hypsochromic shift of λ_{max} of the E-isomer by 85 nm, and by 150 nm in the Z-isomer. The sharp increase in the "favorable polarization" of the unbound atoms in the E-form (due to the heteroatom's declining donor capacity) plays a lesser role than the change in the structure of the Z-form evidently associated with a deformation of the molecule. N-Benzoyl causes a somewhat smaller hypsochromic shift than does N-acetyl. 3,5-Dinitrobenzoyl, which is more electron-accepting and larger, shifts λ_{max} of the Z-form as much as the acetyl does, and λ_{max} of the E-form even more so. At the same time, ϵ_E and ϵ_Z and the ratio ϵ_E/ϵ_Z decrease. The color contrast in N,N'-dibenzoyl- and N,N'-bis(3,5-dinitrobenzoyl)indigo is practically the same (114 and 116 nm) and less than in the acetyl derivative (124 nm).

A sharp color contrast is exhibited by symmetrical (see Table 5) and asymmetrical (see Table 6) dyes based on naphtho[1,8-bc]thiopyran-3-(2H)-one. Their bands overlap little, which makes a practically complete E → Z-photoisomerization possible. Interestingly, despite the seemingly obvious assumption regarding an increased color contrast in perinaphthothio-indogenides [110] compared with the thioindogenides [96], a diminishing $\Delta \lambda_{E/Z}$ is observed.

However, in reality there is no contradiction. As in the case of the transition from thioindigo to perinaphthothioindigo, so upon transition from the benzylidene derivative benzo[b]thiophen-3(2H)-one to the benzylidene derivative naphtho[1,8-bc]thiopyran-3(2H)-one a bathochromic shift is observed in one form, in the other it is small. The difference is the fact that the maximum of the E-form is shifted in the case of dyes (the shift is originally long-wavelength), while in the case of thioindogenides − the maximum of the Z-form is shifted (which is originally short-wavelength). In the latter case the color contrast diminishes.

---

*An analogous relation has been established for aurones, thioindogenides, and selenoindogenides [106], for the ethyl esters of the substituted forms of cinnamic acid [109], and for a number of other substances.

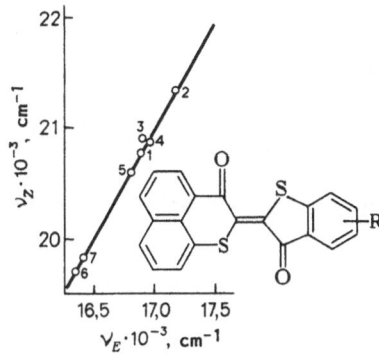

Fig. 9. Comparison of ν_{max} of the E- and Z-isomers of thioindigoid dyes.
The regression equation: $\nu_Z = 1.885\nu_E - 11097$. R = H (1);
6-OC$_2$H$_5$ (2); 4-CH$_3$, 6-Cl (3); 6-Cl (4); 5-NO$_2$ (5); 5,6-benzo,
7-Cl (6); 6,7-benzo, 1'-Cl (7).

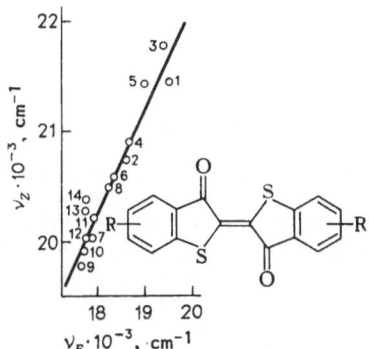

Fig. 10. Comparison of ν_{max} of the E- and Z-isomers of thioindigoid dyes.
The regression equation: $\nu_Z = 0.903\nu_E + 4071$. R = 5-NO$_2$ (1);
4-CH$_3$, 6-Cl (2); 6-OC$_2$H$_5$ (3); 6-Cl (4); 5-Br, 6-OC$_2$H$_5$ (5);
H (6); 4,5-benzo (7); 7-CH$_3$ (8); 4-CH$_3$, 5-Cl, 7-CH$_3$ (9); 6-NO$_2$
(10); 5-Cl (11); 5-Cl, 7-CH$_3$ (12); 4-Cl, 5-Cl, 7-Cl (13); 5-CH$_3$,
7-CH$_3$ (14).

On the basis of the above data the boundaries within which λ_{max} of the
E-isomers of indigoid dyes are located can be summed up as follows. The
most deeply colored thioindigo derivatives have a λ_{max} of about 590 mn (see
Table 3). Indigo derivatives with a sufficiently pronounced ability for E →
Z-photoisomerization fall short of this boundary: λ_{max} of N,N'-dibenzoyl-
indigo is 576 nm; for diacetylindigo — 562 nm. Thus in the class of
2,2'-bis(benzo[b]thiophene)indigo and in the class of 2,2'-bis(indole)indigo
there are no potentially useful photochromic substances with $\lambda_{max} > 590$ nm.*

Shifting absorption to the region of 630 nm allows the transition to
thioindigoid dyes having the naphtho[1,8-bc]thiopyran-3(2H)-one ring (see
Tables 5 and 6), and to the ~700 nm region — the transition to thioindigoid

---

*Although N,N'-dimethylindigo absorbs at 640 nm [35,39,111], its inability
to undergo a sufficiently high E → Z-photoconversion and the low thermal
stability of the photoinduced form disqualify it from being taken into
account.

dyes — derivatives of pyreno[3,4-bc]thiopyran-5-one; some of them have a λ_{max} of about 730 nm (see Table 7).

The rate of the thermal Z → E-isomerization in practical terms is the rate of the spontaneous disappearance of the image recorded on a photochromic material. The recording intensity is halved in an hour at a rate constant of thermal isomerization of about 10^{-4} sec^{-1}. For month-long data storage the rate constant of the Z → E-isomerization must be about 10^{-7} s^{-1}.

The fact of spontaneous dark thermal conversion of the Z-isomer of thioindigo and its substituted forms (in solution) into the E-isomer was first observed in 1951 [32]. The first quantitative data were obtained in 1962 and pertained to perinaphthothioindigo [5]. It was noted, in particular, that the dark reaction was more rapid in protic solvents than in aprotic ones. It was later demonstrated that the reaction proceeded in accordance with a first-order kinetic equation with the logarithm of the rate constant being linearly dependent on the "effectiveness" of the solvent S (on the Braunstein scale [112]). The use of a special technique made it possible to show that the thermal stability of dyes VI was higher than that of perinaphthothioindigo, that it depends on the structure of the dye and random microadmixtures, and that it is independent of the presence of oxygen [10]. The latter is a cogent argument in favor of a singlet mechanism of thermal isomerization. The activation energy is 85-100 kJ/mole (in benzene, chlorobenzene, and p-xylene at 60-110°C) [11]. Other indigoid dyes have much the same levels of activation energy [113].

Steric factors [115] and the field effect can have a strong influence on the activation energy value if a substituent containing unshared electron pairs [116] is in close proximity to the bond. Thus the dye Vc, R = H (Table 6), has a lower activation energy value than the other dyes studied by Mostoslavskii and Shapkina [11], and a low entropy factor in all solvents. The lower values of the frequency factor (down to 10^9 s^{-1}) compared with the normal (10^{12}-10^{14} s^{-1}) indicates a slower-than-usual entropy increase in the course of isomerization. It is possible that the half molecules in the Z-form are somewhat deformed because of considerable steric hindrances. Upon rotation about the central double bond these halves may assume a planar structure, thereby strengthening the overall structure.

If we are to judge the thermal stability of the Z-form of various indigoid dyes on the basis of the data cited in the papers [10,11,17,113], the following series with a rising stability of the Z-form emerges (in benzene): thioindigo < N,N'-diacetylindigo < 2-benzo[b]thiophene-2'-phenanthrene-indigo < 2-benzo[b]thiophene-2'-N-acetylindole-indigo < perinaphtho-thioindigo < 2-benzo[b]thiophene-2'-acenaphthene-indigo < 2-benzo[b]thiophene-2-naphtho[1,8-bc]thiopyran-indigo < 2'-naphtho[1,8-bc]thiophene-1'-acenaphthene-indigo.

In view of the great influence exerted on the isomerization rate by the solvent some of the members of the series may exchange places in another solvent. However, the last two classes of dyes remain the best. Judging by the rate constants of their thermal Z → E-isomerization, photochromic materials which are capable of image retention for weeks and even months can be obtained with their help.

The spectral and photochromic properties of the dyes of series IV-VI of naphtho[1,8-bc]thiopyran-3(2H)-one have been inadequately described, although their considerable color contrast, high molar extinction coefficient, and the high thermal stability of the Z-form cannot fail to attract attention. Indigo (see Figure 2) and perinaphthothioindigo (Figure 11) exhibit close similarity in the set of central chromophore atoms.

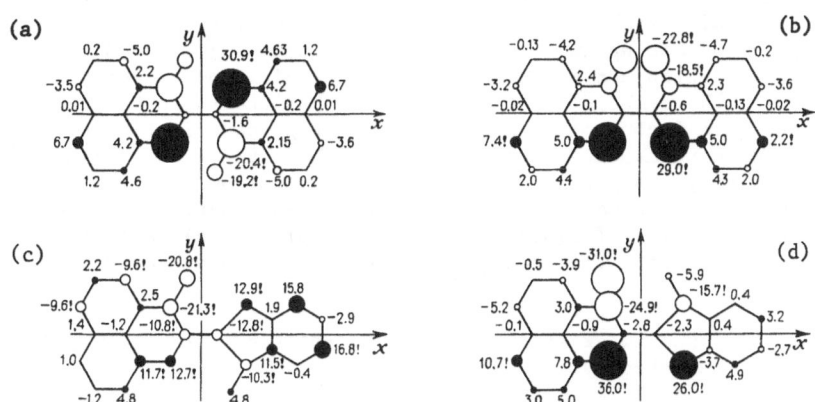

Fig. 11. Diagram of the electron density redistribution for the $S_0 \rightarrow S_1$ transition (see the caption to Figure 5): (a) E-perinaphthothio-indigo; (b) Z-perinaphthothioindigo; (c) E-V; (d) Z-V; X = S.

E-perinaphthothioindigo   E-V

In both the perinaphthothioindigo isomers and in Z-V (see Figure 11) the electron density shifts from the heteroatoms into the region of both carbonyl groups. In E-V the picture of charge redistribution is more complex. The borrowing of electron density occurs not so much from atoms of the thiopyran ring (including the heteroatom) but primarily from the carbon atoms and the sulfur atom of the benzo[b]thiophenone ring. An increase in electron density is observed in the region that encompasses the carbon of the thiophenone ring's carbonyl, the central double bond, and the carbonyl and individual carbon atoms of the molecule's thiopyran portion. Thus, the chromophore of E-V differs substantially from indigo's chromophore.

Compared with the arithmetical mean from the λ_{max} of E-thioindigo and E-perinaphthothioindigo, the λ_{max} of E-V is bathochromically shifted. Conversely, compared with the arithmetic mean from the λ_{max} of Z-thioindigo and Z-perinaphthothioindigo, the λ_{max} of Z-V is hypsochromically shifted. This ensures the abnormally great distance between the λ_{max} of the Z- and E-forms. The ϵ_{max} of E-V is also abnormally high. Chemical synthesis difficulties were in the way of a more detailed study (see Tables 5 and 6) of interesting photochromic dyes [4-6,10-12,18,19,48] that included the thiopyran ring. These difficulties were overcome by the authors of papers [120, 129]. A reliable procedure has been worked out for obtaining naphtho-[1,8-bc]thiopyran-3-one and its 7-derivatives − the half products for synthesizing symmetric and asymmetric thioindigoid dyes E-IV, E-V, and E-VI [95,129-133].

ON THE MECHANISM OF THE PHOTOISOMERIZATION OF INDIGOID DYES

Under the influence of light a great number of states can arise in a system consisting of a photosensitizer, the isomerized substance, and a quencher. These states differ in chemical structure, multiplicity, steric structure, and the reserve of vibrational energy. The number of possible monomolecular and bimolecular conversions is even greater.

In practically each paper devoted to photoisomerization the introduction of symbols takes up much space. This reflects objective difficulties in the identification and description of electronic monomolecular states. In particular, the symbols sometimes used in the literature to denote symmetry classes are inconvenient for describing E/Z-isomerization since a rotated molecule turns out to be asymmetric* while the Z-form belongs to a different symmetry group than the E-isomer. The following symbols form the basis of the system of notation used in the present work: S (singlets of uncharged particles); T (triplets of uncharged particles); K (cations); KR (cation-radicals); R (radicals); A (anions); AR (anion-radicals) (Table 10).

The double-bond twist angle θ is the coordinate in E/Z-isomerization. This angle is indicated by a superscript either in degrees or as discrete symbols: E ($\theta \sim 0°$), Z ($\theta \sim 180°$), P ($\theta \sim 90°$), B ($0° < \theta < 90°$). The superscript on the left indicates the role of the given substance in the system under study. The donor is marked d; the quencher, by the letter q. If the superscript is enclosed in parentheses, it means that the particle is not yet capable or is no longer capable of performing the indicated function. The subscript stands for the number of the level on the energy scale with a given geometry and in the field of its multiplicity. After the particle symbol, additional data can be given in parentheses (symmetry group, the energy reserve of the particle, etc.).

Franck–Condon states are asterisked, vibrationally excited states are marked with a tilde, and vibrational-equilibrium states are underlined. If the vibrational state of an excited particle is not marked in an original work, a symbol is used without any vibrational state index and without being underlined.

Photophysical and photochemical conversions are marked with an arrow between the starting and final particles. If the intermediate states that arise are not known or are omitted, a solid arrow is replaced with a dashed one. For instance, instead of three diagrams representing the formation of the Franck–Condon state of the first singlet, its transition to a vibrationally excited state, and, finally, a relaxed state:

$$\underline{S}_0^E \xrightarrow{h\nu} S_1^{E*}; \quad S_1^{E*} \to S_1^{E\sim}; \quad S_1^{E\sim} \to \underline{S}_1^E,$$

one can write

$$\underline{S}_0^E \overset{h\nu}{\dashrightarrow} \underline{S}_1^E$$

The mechanism of the E/Z-photoisomerization of indigoid dyes began to be studied much later than the mechanism of stilbene photoisomerization. The latter was subjected to a detailed study from a variety of angles [134-142] with the result that important data were obtained on the mechanisms of E/Z-isomerization about the C=C bond [143-150].

The available quantitative data on the photoisomerization of indigoid dyes are summarized in Table 11. Some of the characteristics for a number of compounds omitted from the table are given on the next page.

Until quite recently diametrically opposed views were held on the mechanism of photoisomerization in thioindigoid dyes. Fifteen years after the discovery of the E/Z-photoisomerization of indigoid dyes [32] the first quantitative data were obtained on the course of photoconversions [43,151, 152]. The value of $\phi_{Z \to E}$, as a rule, is much higher than $\phi_{E \to Z}$. The reason might be the different mechanisms of the two processes. An attempt was made [43] to attribute the greater ability of S_1^Z to isomerize to the inadequate "distinctness" of this energy level.

---

*If perpendicular states are realized, they should have optical isomerism.

Data is cited in the following order: Name of compound, $\phi_{E \to Z}$ ($\phi_{Z \to E}$); $\tau_{S_1}^E$, ns (underlined).

5,5'-Dineopentylthioindigo, $\underline{12}$; $\underline{0.4}^a$ [140]. 4-Methyl-6-chloro-6'-meth-oxythioindigo, 0.30^b [17]. 5,5'-Dichloro-7,7'-di(acetylamino)thioindigo, 0.05^b [17]. N,N'-Diacetylindigo, 0.096 [17]; 0.1^c [84,151,155] (0.13 [17]; 0.1^c [84,151]); ϕ_{f1} 0.21 [17]; $\underline{6.1}$ [17]; 5.8^c [155]. N,N'-Dibenzoylindigo, 0.08^c [155]; ϕ_{f1} 0.04 [17]; $\underline{8.0}^c$ [155]; $\underline{8.1}$ [17]. 5,5'-Dineopentylindigo, $\underline{0.017}$ [140]; $\underline{0.017}^a$ [140]; $\underline{0.037}^d$ [140]. 2,2'-Bis(thiophene)indigo, 0.63 (0.35) [90], ϕ_{f1} 0 [90]. N,N'-Dimethyl-5,5',7,7'-tetrabromoindigo, 0.009 [35,38]. 2-Benzo[b]thiophene-3'-(N-acetylindole)indigo, 0.1; ϕ_{f1} 0.07 [17]; $\underline{2.5}^e$ [155]. 2-Benzo[b]thiophene-2'-naphtho[1,8-bc]thiopyran-indigo, $0.024^{c,e}$; $\underline{6.1}^e$ [155]. 2-(5-Methylbenzo[b]thiophene)-2'-naphtho[1,8-bc]-thiopyran-indigo, 0.036 (0.28) [43]. 2-(6-Ethoxybenzo[b]thiophene)-2-naphtho[1,8-bc]thiopyran-indigo, 0.050 (0.41) [43]. 2-(6-Chlorobenzo[b]-thiophene)-2'-naphtho[1,8-bc]thiopyran-indigo, 0.043 (0.33) [43]. 2-(4-Methyl-6-chlorobenzo[b]thiophene)-2'-naphtho[1,8-bc]thiopyran-indigo, 0.030 (0.31) [43]. 2-Naphtho[2,1-b]thiophene-2-naphtho[1,8-bc]thio-pyran-indigo, 0.13 (0.29) [43]. 2-Naphtho[2,3-b]thiophene-2-naphtho-[1,8-bc]thiopyran-indigo, 0.023 [43]. 2-(9-Chloronaphtho[2,3-bc]thio-phene)-2'-naphtho[1,8-bc]thiopyran-indigo, 0.024 (0.32). 2,2'-Bis(ace-naphthene)indigo, 0, $\underline{2.6}$ [155]. 2-Benzo[b]thiophene-2'-acenaphthene-indigo, 0.12^e; $\underline{8.5}^e$ [155]. 2-Naphtho[1,8-bc]thiopyran-2'-acenaphthene-indigo, 0.47 (0.12) [17]; 0.035^e [15] (0.12 [17]); $\underline{1.7}^e$ [155]. N,N'-Dimethylindigo, 0.004-0.008 [35,38,79]; $\tau_{S_0}^Z$ $\underline{30}$ s [35]. N,N'-Dimethyl-5,5',7,7'-tetrabromo-indigo, $\tau_{S_0}^Z$ $\underline{125}$ s [35]. 2-(9-Chloronaphtho[1,2-b]thiophene)-2-naphtho-[1,8-bc]thiopyran-indigo, 0.069 [43].

---

^{a}In alcohol. ^{b}Deoxygenated solution. ^{c}In toluene. ^{d}In glycerine. ^{e}Oxygen-saturated solution.

While the S_1^E, S_1^{E*}, and $S_1^{E^\sim}$ were assumed to be planar and no spontaneous impulses towards rotation around the C=C bond should develop in the process $S_0^E \xrightarrow{h\nu} S_1^{E*} \to S_1^{E\sim} \to S_1^E$, a different situation was assumed for S_1^Z. The "tail" of the long-wavelength absorption band of the Z-isomer of thioindigo (which is absent from the E-isomer band) was attributed [152] to the low-intensity, nonadiabatic pure electron transfer $S_0^Z \xrightarrow{h\nu} S_1^Z$, as a result of which a strong-ly rotated relaxed particle S_1^Z emerges from a nearly flat molecule S_0^Z. The light absorption by the vast majority of the molecules S_0^Z (responsible for the formation of the bulk of the absorption band of the Z-isomer) occurs in accordance with the Franck—Condon rule without a configurational change and leads to almost planar particles S_1^{Z*}. In the course of the inevitable levelling off and relaxation $S_1^{Z*} \to S_1^{Z\sim} \to S_1^Z$ the half of the excited particle S_1^{Z*} will rotate through 90°; i.e., the first stage of the Z → E-photoisom-erization will be completed, which ought to ensure a smooth completion of this process.

The independence of $\phi_{E/Z}$ from the wavelength of the exciting light [152] and the valid relation (albeit within a narrow structural series) be-tween λ_{max} of the long-wavelength absorption band and $\phi_{E \to Z}$ [43] were noted.

Wyman and Zarnegar proposed a singlet mechanism for the direct E → Z-photoisomerization of thioindigo [90]:

$$S_0^E \xrightarrow{h\nu} S_1^E \to S_1^{P\sim} \to S_0^Z.$$

Table 10. Letter symbols of sensitizers, isomers, and quenchers whose appearance is possible on E → Z- or Z → E-photoisomerization

Type of particle	Sensitizer	E-isomer (θ ≈ 0°)	Twisted isomer configurations		Z-isomer (θ of about 180°)	Quencher
			perpendicular θ about 90°	acute angle 0° < θ < 90°		
S_0	$^{(d)}S_0,\ ^{(d)}\tilde{S}_0,\ ^{(d)}S_0^*$	$^{E}S_0,\ ^{E}\tilde{S}_0,\ ^{E}S_0^*$	$^{P}S_0,\ ^{P}\tilde{S}_0,\ ^{P}S_0^*$	$^{B}S_0,\ ^{B}\tilde{S}_0,\ ^{B}S_0^*$	$^{Z}S_0,\ ^{Z}\tilde{S}_0,\ ^{Z}S_0^*$	$^{q}S_0,\ ^{q}\tilde{S}_0,\ ^{q}S_0^*$
S_1	$^{d}S_1,\ ^{d}\tilde{S}_1,\ ^{d}S_1^*$	$^{E}S_1,\ ^{E}\tilde{S}_1,\ ^{E}S_1^*$	$^{P}S_1,\ ^{P}\tilde{S}_1,\ ^{P}S_1^*$	$^{B}S_1,\ ^{B}\tilde{S}_1,\ ^{B}S_1^*$	$^{Z}S_1,\ ^{Z}\tilde{S}_1,\ ^{Z}S_1^*$	$^{q}S_1,\ ^{q}\tilde{S}_1,\ ^{q}S_1^*$
S_2	$^{d}S_n,\ ^{d}\tilde{S}_n,\ ^{d}S_n^*$	$^{E}S_n,\ ^{E}\tilde{S}_n,\ ^{E}S_n^*$	$^{P}S_n,\ ^{P}\tilde{S}_n,\ ^{P}S_n^*$	$^{B}S_n,\ ^{B}\tilde{S}_n,\ ^{B}S_n^*$	$^{Z}S_n,\ ^{Z}\tilde{S}_n,\ ^{Z}S_n^*$	$^{q}S_n,\ ^{q}\tilde{S}_n,\ ^{q}S_n^*$
KR	$^{d}KR_0,\ ^{d}\tilde{KR}_0,\ ^{d}KR_0^*$	$^{E}KR_0,\ ^{E}\tilde{KR}_0,\ ^{E}KR_0^*$	$^{P}KR_0,\ ^{P}\tilde{KR}_0,\ ^{P}KR_0^*$	$^{B}KR_0,\ ^{B}\tilde{KR}_0,\ ^{B}KR_0^*$	$^{Z}KR_0,\ ^{Z}\tilde{KR}_0,\ ^{Z}KR_0^*$	$^{q}KR_0,\ ^{q}\tilde{KR}_0,\ ^{q}KR_0^*$
KR*	$^{d}KR_1,\ ^{d}\tilde{KR}_1,\ ^{d}KR_1^*$	$^{E}KR_1,\ ^{E}\tilde{KR}_1,\ ^{E}KR_1^*$	$^{P}KR_1,\ ^{P}\tilde{KR}_1,\ ^{P}KR_1^*$	$^{B}KR_1,\ ^{B}\tilde{KR}_1,\ ^{B}KR_1^*$	$^{Z}KR_1,\ ^{Z}\tilde{KR}_1,\ ^{Z}KR_1^*$	$^{q}KR_1,\ ^{q}\tilde{KR}_1,\ ^{q}KR_1^*$
T_1	$^{d}T_1,\ ^{d}\tilde{T}_1,\ ^{d}T_1^*$	$^{E}T_1,\ ^{E}\tilde{T}_1,\ ^{E}T_1^*$	$^{P}T_1,\ ^{P}\tilde{T}_1,\ ^{P}T_1^*$	$^{B}T_1,\ ^{B}\tilde{T}_1,\ ^{B}T_1^*$	$^{Z}T_1,\ ^{Z}\tilde{T}_1,\ ^{Z}T_1^*$	$^{q}T_1,\ ^{q}\tilde{T}_1,\ ^{q}T_1^*$
T_2	$^{d}T_n,\ ^{d}\tilde{T}_n,\ ^{d}T_n^*$	$^{E}T_n,\ ^{E}\tilde{T}_n,\ ^{E}T_n^*$	$^{P}T_n,\ ^{P}\tilde{T}_n,\ ^{P}T_n^*$	$^{B}T_n,\ ^{B}\tilde{T}_n,\ ^{B}T_n^*$	$^{Z}T_n,\ ^{Z}\tilde{T}_n,\ ^{Z}T_n^*$	$^{q}T_n,\ ^{q}\tilde{T}_n,\ ^{q}T_n^*$
AR	$^{d}AR_0,\ ^{d}\tilde{AR}_0,\ ^{d}AR_0^*$	$^{E}AR_0,\ ^{E}\tilde{AR}_0,\ ^{E}AR_0^*$	$^{P}AR_0,\ ^{P}\tilde{AR}_0,\ ^{P}AR_0^*$	$^{B}AR_0,\ ^{B}\tilde{AR}_0,\ ^{B}AR_0^*$	$^{Z}AR_0,\ ^{Z}\tilde{AR}_0,\ ^{Z}AR_0^*$	$^{q}AR_0,\ ^{q}\tilde{AR}_0,\ ^{q}AR_0^*$
AR*	$^{d}AR_1,\ ^{d}\tilde{AR}_1,\ ^{d}AR_1^*$	$^{E}AR_1,\ ^{E}\tilde{AR}_1,\ ^{E}AR_1^*$	$^{P}AR_1,\ ^{P}\tilde{AR}_1,\ ^{P}AR_1^*$	$^{B}AR_1,\ ^{B}\tilde{AR}_1,\ ^{B}AR_1^*$	$^{Z}AR_1,\ ^{Z}\tilde{AR}_1,\ ^{Z}AR_1^*$	$^{q}AR_1,\ ^{q}\tilde{AR}_1,\ ^{q}AR_1^*$

*If perpendicular states materialize, they must exhibit optical isomerism.

The planar S_1^E rotates first through 90° turning into S_1^P. In the second stage of the isomerization S_1^P completes its rotation through 180° with the degradation of electron energy and the formation of S_0^Z. The absence* of triplet intermediate particles in experiments involving pulsed photolysis was given as a straightforward argument in favor of the singlet mechanism of the photoisomerization of thioindigo.

In another paper the same authors [83] substantiated the singlet mechanism of thioindigoid photoisomerization by drawing on the data on triplet-sensitized isomerization which they assumed to proceed through a phantom-triplet (T_1^P). The reasoning was as follows. If T_1^P were an intermediate particle in direct and sensitized isomerization and if it were capable of only two conversions $T_1^P \rightarrow S_0^E$ and $T_1^P \rightarrow S_0^Z$, the ratio

$$k_{T_1^P \rightarrow S_0^Z} / k_{T_1^P \rightarrow S_0^E}$$

would coincide for direct and sensitized isomerization. However, a ratio equal to 1.22 was found for the direct isomerization, while for triplet-sensitized isomerization the ratio was found to be 0.37; i.e., the photoisomerization proceeded by a singlet mechanism.

Two years later Wyman changed his view radically. Kirsch and Wyman discovered [154] oxygen quenching of the E → Z-photoisomerization (but not of fluorescence and Z → E-photoisomerization) of thioindigo and 6,6'-diethoxythioindigo and proposed the following triplet mechanism of the E → Z-photoisomerization:

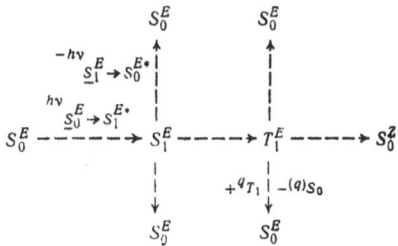

The emphasis was placed on the E-configuration of the intermediate triplet particle. The isomerization with simultaneous degradation of electronic energy was assigned to the $T_1^E \dashrightarrow S_0^Z$ stage. Five arguments were put forward in favor of the triplet mechanism. The inability of oxygen to quench the fluorescence of thioindigo (i.e., to quench the longest-lived particle S_1^E) shows that the quenching is performed not by a singlet but by a triplet intermediate particle T_1^E or T_1^P. Since Z → E-photoisomerization (both direct and triplet-sensitized) is not quenched by oxygen, the transoid particle T_1^E is subject to quenching rather than the particle T_1^P (common to both E → Z- and Z → E-isomerizations). The exclusion of air enhances the quantum yield of the triplet-sensitized isomerization by a factor of 2.66 and direct photoisomerization by a factor of 2.63, i.e., just about equally. Since the sensitized reaction proceeds by the triplet mechanism, this represents an indirect argument in favor of the triplet mechanism of direct photoisomerization. Singlet oxygen has been recorded during direct photo-

---

*Wyman and Zarnegar [90] made a reservation that the existence of particles with a lifetime of under 2×10^{-5} sec was not ruled out. We should note that the resolving power of the equipment available to Wyman and Zarnegar [90] was lower than what is required for a reliable recording of the absence of triplets: according to the ideas current at the time, it was 1/20 of the level required; by today's standards it was 1/200.

Table 11. Quantitative data on the photoisomerization[1] of indigoid dyes in benzene

Indigoid dye	$\phi_{E \to Z}$	$\phi_{Z \to E}$	ϕ_{fl}	$\phi_{S_1 \to T_1}^E$	$\phi_{S_1 \to T_1}^Z$	$\phi_{S_1 \to S_0}^E$	$\phi_{S_1 \to S_0}^Z$	$k_{S_1 \to S_U}^{E \to E}$	$k_{S_1 \to S_1}^P$	k_{fl}	k_{isc}	τ_S^E	$\tau_{T_1}^T$
								$k \cdot 10^{-8}$, s^{-1}				τ, ns	
Oxindigo[29]	0.63	0.35											
"Chromophore" thioindigo													
[90]	0.34	0.41	0.02					5.7	8,3	29		0.7	
[17]	0.042	0.45	0.65					9.5	17,29	49,61		13.3	
[90]	0.041	0.045	0.56					0.27	0.056	41		13.4	
[13]	0.055	0.63											
Thioindigo													
[153][2,3] or [4;4]	0.042	0.45	0.58										
[17][2]	0.11	0.45	0.71	0.23	(0.85)	0.06	(0.15)	4.5		53	17	13.3	
	0.11											13.4	300
[154][2]	0.11	0.45	0.58									13.3	
[13][5]	0.11	0.45	0.71	0.23								13.4	135
[151][6]	0.02	0.9											
[155][6]	0.03	0.25											
	0.041											13.0	
4,4'-Dichlorothioindigo													
[153][2]	0.07	0.34	0.68	0.22	0.50	0.10	0.50	7.8		53	17	12.9	580
5,5'-Di-tert-butylthio-indigo[154][2,3] or [4;4]	0.10	0.45	0.45	0.25	(0.76)	0.30	(0.24)	30		43	25	10.1	300
5,5'-Dibromothioindigo[17]	0.032		0.43									10.2	
6,6'-Diethoxythioindigo													
[17]	0.25	0.45	0.03					38.5	900,800	30,62		1.0	
[90]	0.19	0.37	0.03					7.4	3.3	33		0.9	
[17][2]	0.45												
[154][2]	0.45	0.45	0.04	1.02		0.57		0		44	1130	0.9	60
[153][2,3] or [4;4]	0.45	0.45	0.04	1.02	(0.80)	0	(0.20)			14	1130	0.9	140
[156, pp. 472-555][6]	0.20	0.38											
7,7'-Dichlorothioindigo													
[153][2,7,7]	0.07	0.34	0.64	0.23	0.49	0.13	0.57	10		49	18	13.0	140

Compound											
4-Methyl-6-chloro-6'-methoxythioindigo[17]	0.18	0.45	0.15				62,5		150,180	38,45		4.0
4,4',7,7'-Tetramethyl-thioindigo[153]²'³'⁷ $\rightarrow$ 4,4',7,7'-Tetramethyl-thioindigo[153]²'³'⁷	0.10	0.45	0.67	0.22	(0.40)	0.11	(0.10)	8.1	50	16	13.5	300
5,5'-Dichloro-7,7'-dimethylthioindigo[153]²'⁷'⁷	0.05	0.30	0.38	0.17	0.43		0.57	33	29	13	13.2	385
6,6'-Dichloro-7,7'-dimethylthioindigo²'⁷'⁷	0.13	0.29	0.62	0.40	0.43	0	0	0	67	43	9.3	200
Selenoindigo[90]	0.032	0.81	0.03	0.43		0.57	4.7	1.1	18			1.7
[36,152]⁶	0.025	0.275										
Indigo[17]	0.013	0.058	$2.8\cdot10^{-3}$				10^{4}		28,58			0.1
Perinaphthothioindigo[17]	0.0097	0.0038	0.07									
[156]⁶												
[155]⁶	0.009											4.0
2-Benzo[b]thiophene-2-naphtho[1,8-bc]thio-pyranindigo[43] [17]	0.377	0.30	0.35				100	9	58			6.0
	0.027	0.18										
2-Benzo[b]thiophene-2'-acenaphthene-indigo[17]	0.20	0.40	0.30				27,5	59,100	38,43			8.1
2-(5-Bromobenzo[b]thio-phen)-2'-acenaphthene-indigo[17]	0.19	0.40	0.26				42,5	5,13	33			8.0

1 Radiationless transitions are marked by straight arrows in the table and text. Two methods of calculating rate constants are used in the monograph[17]; both results are given; for details of the calculation the reader is referred to the original. Enclosed in parentheses are the values which Kirsch and Wyman [153] regarded as a rough estimate.

2 A de-oxygenated solution.

3 Mechanism A (see p. 82).

4 Mechanism B' (see p. 83). The mechanism of direct isomerization is indicated first, followed by the mechanism of sensitized isomerization.

5 Solution is oxygen-saturated.

6 In toluene.

7 Mechanism B (see p. 82). The mechanism of direct isomerization is indicated first; the mechanism of sensitized isomerization, second.

isomerization which should arise as a result of the triplet–triplet annihilation of T_1^E with a triplet oxygen molecule. The lifetime of the oxygen-quenched particle is estimated at 135 ns, which exceeds by 10-100 times the usual lifetime of S_1-states. This testifies to the triplet nature of the intermediate particle and is proof of the triplet mechanism of direct E → Z-photoisomerization. Oxygen is sometimes capable of accelerating E → Z-isomerization, but, Kirsch and Wyman claim [153,154], during thioindigoid quenching it does not change the ratio in which S_0^E and S_0^Z are formed from the triplet but only cuts the lifetime of the triplets.

Kirsch and Wyman [154] put forward alternative mechanisms for the Z → E-photoisomerization of photoindigo — a singlet mechanism:

$$S_0^Z \xrightarrow{h\nu} S_1^Z \dashrightarrow S_1^P \dashrightarrow S_0^E$$

and a singlet–triplet mechanism:

$$S_0^Z \xrightarrow{h\nu} S_1^Z \dashrightarrow T_1^P \dashrightarrow S_0^E$$

the latter with the qualification that the phantom-triplet T_1^P must be short-lived and, therefore, not quenched by oxygen.

Schulte-Frohlinde et al. [157] believe that it is impossible to study thioindigo by means of flash photolysis on EPA (ether–pentane–alcohol) glass because of its poor solubility. However, for its substituted forms and for those of selenoindigo, excited particles were discovered with λ_{max} bathochromically shifted compared with the long-wavelength absorption band of unexcited molecules and having a lifetime of 10^5 ns [157] at 77 K. This was seen [157] as a confirmation of the triplet mechanism of E → Z-isomerization of thioindigoids.

To interpret the experimental data on oxygen quenching of the photoisomerization of nine thioindigoids, Kirsch and Wyman [153] used three variants of the triplet–singlet mechanism (A, B, B′). The common part of the three variants includes the formation of T_1^P (from T_1^E) and its degradation:

$$S_0^E \xrightarrow{h\nu} S_1^E \dashrightarrow T_1^E \dashrightarrow T_1^P \dashrightarrow \alpha S_0^E$$
$$\downarrow$$
$$(1-\alpha)\,S_0^Z$$

Mechanism A postulates the ability of the phantom-triplet to be quenched by oxygen and assumes the existence of a phantom-singlet:

$$T_1^P \xrightarrow[-^{(q)}S_0]{+^q T_1} S_0^E \qquad S_0^Z \xrightarrow{h\nu} S_1^Z \dashrightarrow S_1^P \dashrightarrow \beta S_0^E$$
$$\downarrow$$
$$(1-\beta)\,S_0^Z$$

i.e., in this case Z → E-isomerization proceeds by the singlet mechanism. In mechanism B the phantom-triplet is not only oxygen-quenched but is also capable of arising from S_1^Z:

$$\downarrow$$
$$S_0^Z \xrightarrow{h\nu} S_1^Z \dashrightarrow T_1^P \xrightarrow[-^{(q)}S_0]{+^q T_1} S_0^E$$

In this case the Z → E-isomerization proceeds by the singlet–triplet mechanism.

In mechanism B' the phantom-triplet is incapable of being oxygen-quenched (the triplet of the E-isomer is quenched), but is capable of arising from S_1^Z:

$$S_0^Z \xrightarrow{h\nu} S_1^Z \dashrightarrow T_1^P \longleftarrow T_1^E \xrightarrow[-(q)S_0]{+^qT_1} S_0^E$$

An analysis of the kinetic equations involved, including oxygen-quenching, led to the simple relationship [153]

$$\alpha = K_{S-V}(s/i)^{-1}$$

Here α is the share of the hypothetical T_1^P which develops into S_0^E as a result of "natural dissociation," $S_0^Z \leftarrow T_1^P \to S_0^E$, in the absence of a quencher. The values of α and, before that, K_{S-V} and s/i were found separately for direct and triplet-sensitized isomerization. The Stern–Volmer quenching constant K_{S-V} was determined from the dependences of the relative quantum yield on the concentration of O_2; the ratio s/i (where s is the slope, i is the y-intercept), from linear dependences of the oxygen quenching ratio $[S_0^E]/[S_0^Z]$, achieved both directly and through the use of a sensitizer.

Since it was assumed that the oxygen does not change the triplet's configuration, the equality of α to unity would signify that what was quenched was not the phantom-triplet but T_1^E. If α was considerably less than unity, it was assumed that the phantom-triplet was quenched and that mechanism B materialized.* The approximate equality of α to unity in the triplet-sensitized isomerization, where the singlet mechanism was ruled out for obvious reasons, was believed to provide unequivocal evidence of mechanism B'. The approximate equality of α to unity in the direct isomerization was treated as proof of mechanism A if in the triplet-sensitized isomerization the ability of T_1^P to be quenched by oxygen was found. If, on the other hand, both in the direct and in the triplet-sensitized isomerization $\alpha \approx 1$, the choice between mechanisms A and B' in direct isomerization was impossible.

As a result, the "more phantom-triplet" mechanism B was assigned to both the direct and triplet-sensitized isomerization of five thioindigoids containing chlorine atoms (the possibility was mentioned of an increased intersystem crossing under the influence of the heavy atom). The same mechanism was attributed to the sensitized isomerization of 4,4',7,7'-tetra-methylthioindigo, which made it possible (see Table 11) to adopt for the direct isomerization of this dye mechanism A for the Z → E-part of the isomerization. Mechanism B' was adopted for the sensitized isomerization of thioindigo, 5,5'-di-tert-butylthioindigo, and 6,6'-diethoxythioindigo, while mechanism A or B' was assigned to direct isomerization, with the authors failing to make a definite choice between the two [153].

The fact that $\phi_{E \to Z}$ is considerably lower than $\phi_{Z \to E}$ for thioindigoids, as opposed to stilbene [43,151,153], was attributed by Kirsch and Wyman to the greater measure of stability of the E-structure in the thioindigoids on account of the electrostatic interaction between unbound carbonyl and sulfur with their oppositely polarized partial charges. A weakening of this interaction following the introduction of ethoxy groups into positions 6,6' increases $\phi_{E \to Z}$ by a factor of 10 (and k_{isc} by a factor of 50). The possibility of the $S_1^E \dashrightarrow S_0^E$ degradation is recognized without reservations only in the case of 5,5'-di-tert-butylthioindigo.

---

*Another qualitative feature of mechanism B was also an increase in $\phi_{Z \to E}$ as $\phi_{E \to Z}$ decreased.

An assumption was made that 5,5'-substitution exerts a greater influence on degradation than does 4,7-substitution and that, further, the
tert-butyl group contributes to energy dissipation.* The introduction of
chlorine atoms enhances the probability of the $S_1^Z \rightarrow T_1$ process developing.
For thioindigo and 5,5'-di-tert-butylthioindigo the equilibrium $T_1^E \rightleftharpoons T_1^P$ is
rejected while recognized for the rest (although it is not proved in the A–
B' scheme). The influence of structure on triplet energy is not interpreted. It is noted that the effect of substituents on isomerization is
analogous to that described in the article [43].

Their bias towards legitimacy of the singlet mechanism, which included
the equilibrium $S_1^E \rightleftharpoons S_1^P \rightleftharpoons S_1^Z$, caused Karstens et al. [159] to overlook the
quenching of E → Z-photoisomerization of thioindigo by oxygen and eventually
led them to neglect their own data on the monoexponentiality of S_1^E fluorescence decay. According to these data, the efficiency of E → Z-photoisomerization of thioindigo increases with rising temperature, while Z → E-
photoisomerization reaches its peak efficiency at 12°C. The hypothesis of
Karstens et al.[159] regarding the mechanism which includes the transition
of the substantially rotated S_1^E (52°; 248 kJ) to S_2^E (52°; 248 kJ):

$$S_0^E \ (0°; \ 0 \ kJ) \xrightarrow{h\nu} S_1^E \ (0°; \ 254 \ kJ) \ \text{--} \rightarrow S_1^E \ (52°; \ 248 \ kJ) \ \text{--} \rightarrow$$

$$\text{--} \rightarrow S_2^E \ (52°; \ 248 \ kJ) \ \text{--} \rightarrow S_2^P \ (90°; \ 198 \ kJ) \ \text{--} \rightarrow S_0^P \ (90°; \ 103 \ kJ) \ \text{--} \rightarrow$$

$$\text{--} \rightarrow S_0^Z \ (180°; \ 17.4 \ kJ)$$

did not survive for long (see [14]).

In 1978 Grelman and Hentschel [13] detected two short-lived particles
in the course of nanosecond laser photolysis of the E-isomer of thioindigo.
The lifetime of the first of these particles (15 ns at 25°C) coincided with
the lifetime of fluorescence. They identified this particle as S_1^E, and its
absorption (λ_{max} 460 nm) was attributed to the transition $S_1^E \text{--} \rightarrow S_n^E$. The
absorption maximum of the second particle (in the 570 nm region) was batho-
chromically shifted compared with the band $S_0^E \xrightarrow{h\nu} S_1^E$. The lifetime of the
second particle, designated ^3X, is 158 ns in methylene chloride and 279 ns
in benzene.

Particle ^3X appears in the course of the photolysis of E- and Z-isomer
solutions. The disappearance of the analogous particle (^3X) for 6,6'-dieth-
oxythioindigo is accompanied by a simultaneous gain in the absorption intensity of the Z-isomer (if the E-isomer solution is the starting material) or
the E-isomer (the starting solution — a mixture of isomers enriched in the
Z-isomer). Measuring the disappearance rate of ^3X in a degassed solution at
20 to −60°C made it possible to determine the activation energy and frequency factor of the Arrhenius equation $1/\tau = 5 \times 10^9 \exp(-16300/RT)$. The
lifetime of the ^3X particle decreased in the presence of oxygen. The rate
constant of its oxygen quenching in benzene was found from the Stern–Volmer
plot, 3.2×10^9.

The lifetime of the ^3X particle (which is longer than that of fluorescence) and its ability to be readily oxygen-quenched were evidence of its
being a triplet with the value of the quenching constant being typical of

---

*In the paper [158] it is assumed on the basis of quantum-chemical calcu-
lations that the main channel for energy dissipation, $S_1^{E^-} \rightarrow \underline{S}_1^E$, is the
oscillation of the C–H bond of aromatic nuclei both in the case of thio-
indigoid dyes and of other aromatic substances.

triplets. Grellmann and Hentschel [13] believe that ^3X is a perpendicular general triplet via which both $E \to Z$- and $Z \to E$-isomerizations proceed.* According to their data [13], oxygen-quenching not only causes $\phi_{E\to Z}$ to diminish but, contrary to evidence of the data [17,153,154], it increases $\phi_{Z\to E}$, which suggests the identity of ^3X and T_1^P.

A knowledge of ϕ_{f1}, $\phi_{E\to Z}$, and $\phi_{Z\to E}$ in the assumed absence of degradation of S_1^E and the sole isomerization path — through T_1^P — enables us to arrive at the following formula:

$$\phi_{S_1^E \xrightarrow{h\nu} T_1^P} \Big/ \phi_{S_1^Z \dashrightarrow T_1^P} = \phi_{T_1^P \dashrightarrow S_0^E} \left(1 - \phi_{S_1^E- \xrightarrow{h\nu} S_0^E}\right) \Big/ \Phi_{S_0^Z \dashrightarrow S_0^E}$$

and to compare the results obtained by the method of stationary concentrations and direct observation of T_1^P during laser photolysis. Grellmann and Hentschel found the change in $[T_1^P]$ by probing in the absorption maximum region of T_1^P, and the change in $[S_0^E]$ or $[S_0^Z]$ by probing in the region of their absorption bands suitable for the purpose and then extrapolating the results obtained towards zero time. The values found in the course of laser photolysis and experiments conducted by the photostationary concentration technique were close. Actually, Grellmann and Hentschel [13] replaced mechanisms A and B', proposed by Kirsch and Wyman [153], with mechanism B (see above) for thioindigo and 6,6'-diethoxythioindigo.

The monograph by Haucke and Paetzold [17] contains a wealth of experimental data on the absorption spectra, radiation and excitation, fluorescence, and $E \dashrightarrow Z$- and $Z \dashrightarrow E$-photoisomerization of indigoid dyes, both symmetric and asymmetric, including those that were researched for the first time [6,10,11,43]. Defending the singlet mechanism of photoisomerization (cf. [90,160]), Haucke and Paetzold [17] sought to repudiate the arguments made in the papers [13,153,154,157] favoring the triplet mechanism. However, the presence of a nonlinear section on the Stern–Volmer plot (at elevated oxygen pressures following the linear stepdown $\Phi_{S_0^E \dashrightarrow S_0^Z} \Big/ \Phi_{S_0^Z \xrightarrow{h\nu} S_0^E}$

by 6-8 times), notwithstanding the evidence cited in the paper [17], is not an obstacle in the way of calculating the lifetime of the quenched particle using the linear section of the plot. Nor does the second argument [17] eliminate the results [153] based on the unsubstantiated adoption of a concrete mechanism of oxygen-quenching.

The assumption of Haucke and Paetzold [17] that the short-lived absorption observed by Grellmann and Hentschel [13] and Schulte-Frohlinde et al. [157] is not triplet–triplet** but is due to extraneous radical particles and is not directly related to E/Z-photoisomerization is refuted by the appearance, upon quenching of the spectrally observable short-lived particle ^3X, of additional quantities of S_0^E (or S_0^Z) [13]. A more serious argument [17] in favor of the solute mechanism is the coincidence of the Stern–Volmer constants upon the quenching of fluorescence and of $E \dashrightarrow Z$-photoisomerization by dimethylaniline (or by other electron donors). Some doubt remains,

---

*Evidence against the identity of ^3X and T_1^P is the formation of S_0^E exclusively upon the oxygen-quenching of the ^3X particle; however, it could also be a consequence of the varying quencher behavior in different systems, and it did not occur to Grellmann and Hentschel [13] that ^3X could be identical with T_1^E.
**In principle, the attribution of absorption to a particular particle may well be erroneous. For instance, the absorption of 1-phenyl-2-(2-naphthyl)-ethylene, which was considered to be triplet–triplet, was recently identified as $S_1^E \xrightarrow{h\nu} S_n^E$, since the lifetime of the particle coincides with that of fluorescence, while the lifetime of the triplet determined by the method of quenching sensitized isomerization is roughly six times as long [162].

however, as to the effectiveness of quenching of thioindigoid triplet states by dimethylaniline because Haucke and Paetzold [17] do not cite any data on the quenching of sensitized isomerization by them, while the differences in quenching observed in the case of benzene and iodobenzene solutions may be due to the action of solvent on the quencher's molecule rather than on the indigoid. Inasmuch as thioindigoids are capable of yielding excimers with aromatic solvents [161], there is no reason to doubt that the quenching by diethylaniline proceeds with the intermediate formation of an excimer.

Haucke and Paetzold [17] share the view of other authors [53,153,163] that the interaction of nonbonding atoms is the main reason behind the different spectra of E- and Z-isomers. They explain the thermal stage on the way to E $\cdots\rightarrow$ Z-photoisomerization by the need to overcome the attraction of adjacent, oppositely polarized heteroatoms and carbonyls. The latter can only be true up to a point, in view of the presence of a thermally activated stage not only in the case of indigoid dyes but also in substances that lack the above-mentioned adjacent groups (e.g., in stilbene).

After the appearance of the article [13] Karstens et al. [14] began to favor the triplet mechanism of thioindigo photoisomerization. They have demonstrated that in rigid matrices (e.g., in polystyrene) that totally rule out E $\cdots\rightarrow$ Z-photoisomerization a spectrally recorded particle ^3X is formed from S_1. They concluded that ^3X was identical to T_1^E (and not to T_1^P, as was assumed in the article [13]). The affinity of the ^3X spectra in different solvents at different temperatures was reported. The possibility of ^3X being an anion-radical was rejected, and the high intensity of triplet–triplet absorption, which exceeded that of the transition $S_0^E \xrightarrow{h\nu} S_1^E$, was emphasized.

The lifetime $\tau_{T_1^E}$ of thioindigo diminishes with rising temperature and lengthens with increasing viscosity of the medium. In solution at 160°K $\tau_{T_1^E}$ retains the same order it has in polystyrene at 293°K.

The optical density $D_{T_1^E \xrightarrow{h\nu} T_n^E}$, observed during laser photolysis, is a value proportional to ϕ_{isc} (here the subscript "isc" stands for "intersystem crossing"):

$$\phi_{S_1^E \dashrightarrow T_1^E} = dD_{T_1^E \xrightarrow{h\nu} T_n^E} \left/ \left(\epsilon_{T_1^E \dashrightarrow T_n^E}^{l} \frac{dQ}{v} \frac{D_{S_0^E \xrightarrow{h\nu} S_1^E}}{\sum_i D_i} \right) \right.$$

$$\phi_{S_1^E \dashrightarrow T_1^E} = \frac{const}{dQ} dD_{T_1^E \xrightarrow{h\nu} T_n^E}$$

Because after a laser pulse of the same intensity $D_{T_1^E \xrightarrow{h\nu} T_n^E}$ is lower at a higher temperature in the experiment, it is clear [14] that ϕ_{isc} increases with declining temperature; concurrently ϕ_{fl} goes up; the effectiveness of the third path of energy loss in the S_1 state should fall (some think this represents a rotation about the central bond having a lower order in an excited singlet). It is assumed that the planarity of S_1^E favors intersystem crossing, while rotations about the central bond are competing processes to the loss of energy. The fact that the increase in ϕ_{isc} with declining temperature is accompanied by a fall in $\phi_{E\rightarrow Z}$ is attributed to a slowdown in the rotation stage of $T_1^E \rightarrow T_1^P$ (cf. [13]).

Karstens et al. [14] further assumed that the rate constant of the disappearance of T_1^E includes three terms (monoexponential decay):

$$k_{T_1^E} = k_{T_1^E \dashrightarrow S_0^E} + k_{T_1^E \dashrightarrow T_1^P} \exp\left[-15/(RT)\right] + k_{T_1^E \genfrac{}{}{0pt}{}{+^qT_1}{-(q)S_0} S_0^E}^{[^qT_1]}$$

and that the phantom-triplet is not spectrally detectable and is incapable of crossing to T_1^E. The activation energy of the transition $T_1^E \to T_1^P$ equals 15.1 kJ/mole; T_1^E is oxygen-quenched.

The following scheme is proposed for the isomerization, triplet–singlet both in the E → Z direction and in the reverse direction [14]:

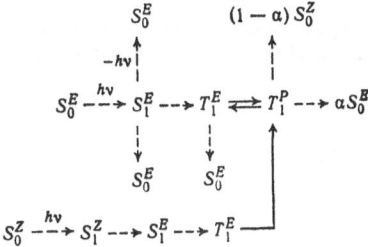

Although a phantom-triplet figures in the scheme, no arguments are given in favor of its existence.

Görner and Schulte-Frohlinde [15] confirmed that the observed triplet thioindigo has a planar E-structure since its spectrum in solutions and in polymethyl methacrylate are the same. Unfortunately, the simplest situation does not materialize whereby T_1^E would be converted into S_0^Z only. If it did, the increase of $\phi_{S_0^E \to S_0^Z}$ would be proportional to the increase of $\phi_{S_0^E \to T_1^E}$, which would be evidence that T_1^E is the intermediate product in the process $S_0^E \to S_0^Z$. The numerical value of $\phi_{S_0^E \to T_1^E}$ is not determined, and so it is impossible to compare it directly with the quantum yield from the T_1^E Z-isomer in the course of the E → Z-isomerization.

A comparison of the dependences of $\phi_{S_0^E \xrightarrow{h\nu} T_1^E}$ and $\phi_{S_0^E \xrightarrow{h\nu} S_0^Z}$ on temperature is possible. These dependences vary. While in a number of solvents $\phi_{S_0^E \xrightarrow{h\nu} T_1^E}$ is independent of temperature, $\phi_{S_0^E \xrightarrow{h\nu} S_0^Z}$ drops with declining temperature, obeying the Arrhenius equation [either on a single straight line, or on a broken line consisting of two straight lines with a break point (discontinuity) at temperature t_u^0]. An analogous temperature dependence of the rate constant in the disappearance of T_1^E (log $k_{T_1^E}$) is observed, with the difference that here the break point (t_v^0) is set at a much higher temperature.* Apparently, T_1^E has several directions of energy dissipation. Görner and Schulte-Frohlinde [15] consider these directions to be the reversible rotation of a half molecule $T_1^E \rightleftharpoons T_1^P$ and (if it is hindered by increased viscosity) degradation $T_1^E \to S_0^E$. On the whole, the photoisomerization scheme can be written as follows (see also the footnote** on the following page):

$$S_0^E$$
$$\uparrow$$
$$S_0^E \xrightarrow{h\nu} S_1^E \dashrightarrow T_1^E \rightleftharpoons T_1^P \dashrightarrow S_0^P \dashrightarrow S_0^E$$
$$\uparrow \quad \downarrow$$
$$S_0^Z \xrightarrow{h\nu} S_1^Z \dashrightarrow S_1^P \quad S_0^Z$$

It is triplet–singlet in the E → Z direction and singlet in the Z → E direction.

---

*This discrepancy is examined in detail in the case of the better-studied nitrostilbenes [164]. The explanation given in the paper does not seem to be satisfactory.

It is pointed out that the limiting stage of isomerization at room temperature is the conversion $T_1^P \dashrightarrow S_0^P$ and at the lower temperature t_u — the stage $T_1^E \dashrightarrow T_1^P$. At a viscosity of 5×10^8 Pa · sec, the $E \xrightarrow{h\nu} Z$ process does not go at all while $Z \xrightarrow{h\nu} E$ proceeds fairly noticeably. The photostationary ratio $[S_0^E]/[S_0^Z]$ increases with falling temperature.

Haucke and Paetzold [165] cast doubt (but do not rule out completely) on the participation of triplet states in direct E → Z-photoisomerization. The summary given in the formulation "it is accepted that direct E → Z-photoisomerization, at any rate initially, proceeds through a singlet state" is a good deal more guarded than the conclusion contained in a monograph that appeared a year earlier [17] which stated, among other things, "the arguments found in the literature in favor of the triplet mechanism proceed from the results of O_2 quenching and from flash photolysis data. These data are yet to be proved since $T \dashrightarrow T_n$-absorption as such is not identified."

A cogent argument against the participation of triplet states in $E \xrightarrow{h\nu} Z$-isomerization is the coincidence of the Stern–Volmer dependences of the quenching of this isomerization and fluorescence [165]. In iodobenzene, where because of the heavy atom effect an increase in the rate of conversion $S_1^E \dashrightarrow T_1^E$ is to be expected, a discrepancy of the Stern–Volmer quenching dependences of isomerization $E \xrightarrow{h\nu} Z$ and fluorescence was registered. This is attributed to the simultaneous quenching of the S_1- and T-states in the last-named process. The analogue of the indigoid dye with its rigid planar structure has a ϕ_{fl} of 0.93 and at room temperature exhibits a fine structure of the absorption spectrum and fluorescence; in thioindigo ϕ_{fl} is 0.65 and the curve of the fluorescence spectrum acquires a fine structure only when frozen. From these data and apparently from the fact that the planar structure of the particle [14] favors intersystem crossing it was concluded that the radiationless loss of energy as a result of rotation of the half molecule is the competing process to fluorescence in thioindigo and not intersystem crossing.* The absence of fluorescence in thioindigoids at low temperature is treated by Birckner et al. [166] as evidence of the absence of the conversion $S_1^E \to T_1^E$.

The authors of the aforementioned papers, while recognizing in principle the possibility of an alternative mechanism, were neatly divided into two camps: those who supported the triplet mechanism [13-15] and those who advocated the singlet mechanism [17,165] of thioindigo isomerization. Each of these mechanisms poses difficulties for an exhaustive interpretation of the experimental data. For instance, within the framework of the triplet mechanism it is difficult to explain the discrepancy of the temperatures t_u^0 and t_v^0 [15], i.e., the fact that under the effect of external influences the expenditure rate of T_1^E changes while $\phi_{E \to Z}$ remains unaffected in any way. Similarly, it is difficult to explain the coincidence of K_{S-V} of fluorescence quenching and E → Z-photoisomerization of thioindigo [17] [if diethylaniline is capable of quenching (in benzene) the triplet of thioindigo, which was not unequivocally proved].

––––––––––––––––
*The triplet–triplet absorption [13-15] and quenching [153,154] of $E \xrightarrow{h\nu} Z$ isomerization by oxygen registered on repeated occasions is not discussed in any serious way in the paper [165].
**Görner and Schulte-Frohlinde [15] do not rule out the participation of a singlet route in Z → E-isomerization:

$$S_0^E$$
$$\uparrow$$
$$\vdots$$
$$S_0^Z \overset{h\nu}{\dashrightarrow} S_1^Z \dashrightarrow S_1^P \dashrightarrow S_0^P \dashrightarrow S_0^Z$$

However, in the explanation of experimental data they do not use it, appealing exclusively to the scheme given above.

For the singlet mechanism the quenching of E $\dashrightarrow$ Z-photoisomerization by oxygen that does not affect either fluorescence or Z $\dashrightarrow$ E-photoisomerization accompanied by the formation of singlet oxygen [153,154] is difficult to explain. It is impossible to explain, from the standpoint of the singlet mechanism, not so much the very formation of triplet particles during photoisomerization, but the fact that for 6,6'-diethoxythioindigo the disappearance of the observable T_1^E is accompanied by the synchronous appearance of the ground states S_0^E and S_0^Z in an altered ratio as compared with the ratio observed before the laser pulse which populated T_1^E [13].

It is natural to suppose that both mechanisms may materialize in parallel. It is generally recognized [15,17,153,154] that oxygen does not quench fluorescence (and, apparently, isomerization proceeding via the singlet mechanism). Its ability to quench $E \xrightarrow{h\nu} Z$-isomerization (proceeding in all likelihood by the triplet mechanism) has been pointed out repeatedly. Under conditions when actinic light is absorbed only by the E-form or almost exclusively by the E-form, the extent of quenching of E $\dashrightarrow$ Z-photoisomerization by the maximum concentrations of oxygen may, in our supposition, characterize the contribution of the triplet mechanism to E $\dashrightarrow$ Z-photoisomerization: the fraction of the quantum yield which remains unchanged upon oxygen quenching provides the contribution of the singlet mechanism to E $\dashrightarrow$ Z-photoisomerization. Table 12 summarizes data which, if one accepts the above assumptions, show that the isomerization of dyes containing two six-membered rings proceeds via the singlet mechanism. The isomerization of dyes containing a single six-membered ring proceeds largely through the triplet states (up to about 90%). It is important to stress that the laser photolysis of dyes sensitive to oxygen revealed the absorption $T_1^B \xrightarrow{h\nu} T_n^E$. Dyes that isomerize by the singlet mechanism lack triplet—triplet absorption.

The tendency towards the determination, under various conditions, of the role of different photoisomerization mechanisms is clearly shown in the latest work done by Görner and Schulte-Frohlinde on substituted forms of stilbene [142].

THE EFFECT OF THE HETEROATOM OF AN INDIGOID DYE ON $\phi_{E \to Z}$. THE REASONS FOR THE ABSENCE OF PHOTOCHROMISM IN INDIGO

Among indigoid dyes possessing different skeletons and lacking substituents oxindigo has the maximum $\phi_{E \to Z}$ and indigo has the minimum; thioindigo and selenoindigo have higher $\phi_{E \to Z}$ than perinaphthothioindigo (see Table 11). Apparently, the stabilizing electrostatic interaction between the carbonyl and the heteroatom of the different halves of the molecule, which is the least in oxindigo on account of the high electronegativity of the oxygen atom [28], plays a substantial role in ensuring the configurational stability of the E-isomer. $\phi_{E \to Z}$ is comparatively high (20%) in 2-thionaphthene-2-acenaphthene-indigo possessing only one interaction $>S \cdots O = C<$, unlike thioindigo, which has two such interactions. The low $\phi_{E \to Z}$ in perinaphthothioindigo in this interpretation should be regarded as an indication of a stronger interaction of unbound atoms than in the case of thioindigo.

Indigo and other dyes (I-III, V) are not photochromic if the heteroatom X = NH. The reasons for this phenomenon are the subject of active debate at the moment. One of the more popular hypotheses in this connection explains the absence of photochromism in indigo (I, X = NH) by the stabilization of the only known isomer of indigo by hydrogen bonds (HBs) that hinder the transition of the E-form into the Z-form.

Table 12. Estimation of the relative contribution to the mechanism of E → Z-photoisomerization of the path, including and excluding triplet states

Indigoid dye	λ_{max}, nm $T_1^E \xrightarrow{h\nu} T_n^E$ transition	ϕ_{E-Z} without O_2	ϕ_{E-Z} P_{O_2} 1.1 MPa	Approximate contribution to E → Z-isomerization triplet path	Approximate contribution to E → Z-isomerization singlet path
Thioindigo	+	0.11	0.01	90	10
Perinaphthothioindigo	−	0.01	0.01	0	100
2-Benzo[b]thiophene-2-naphtho[1,8-bc]-thiopyranindigo	665	0.044	0.023	50	50
2-Naphtho[2,1-b]thiophene-2-[1,8-b]-thiopyranindigo	705	0.23	0.030	90	10
2-Benzo[b]thiophene-2-pyreno[3,4-bc]-thiopyranindigo	710	0.46	0.07	85	15

Unfortunately, the question of the presence of hydrogen bonds in indigo and of their intra- or intermolecular character has not been conclusively resolved to this day. It is shown in what follows that the facts adduced in support of the presence of hydrogen bonds in indigo are either not entirely reliable or may admit to a different interpretation. It is common practice to judge the presence of hydrogen bonds in indigo by the frequencies $\nu_{C=O}$ and ν_{N-H} [167,168]. However, in 4,4'- and 6,6'-diazaindigo [99] the low values of $\nu_{C=O}$ indicate the presence of hydrogen bonds, while the high values of ν_{N-H} are a sign of their absence, which raises the question of the suitability of $\nu_{C=O}$ and ν_{N-H} as criteria for the presence of hydrogen bonds not only in this particular instance but in others as well.

On the basis of their observation that the acid sulfate ester of the leuco form of indigo is not colored more deeply than the thioindigo ester, Dokunikhin and Levin assumed the existence of an intramolecular bond in the indigo molecule [169]. However, the transition of indigoid dyes to the leuco form is accompanied by the reduction of the carbonyl groups. A radical change in the chromophore materializes: the bidonor—biacceptor system makes the transition to a tetradonor system. Naturally, the character of the redistribution of the molecule's electron density during the long-wavelength transition changes, and this may result in a lessening of the influence of the heteroatom's nature on the position of the long-wavelength absorption band, irrespective of the presence or absence of hydrogen bonds in indigo.

According to X-ray crystallographic data [76,167,170,171], molecular hydrogen bonds are present in indigo, as a result of which all the molecules whose centers lie in the same plane become interlinked. According to the data cited in the paper by Gribova et al. [76] this accounts for the high melting point, poor solubility, reduced chemical activity, and crystal color differences of indigo vapors and solutions. However, these properties are more or less characteristic of thioindigo and some other indigoid dyes incapable of forming hydrogen bonds. The X-ray crystallographic data quoted in the papers [172-175] indicate that indigo crystals may (or may not) contain intra- and intermolecular hydrogen bonds.

The small (5 nm) hypsochromic shift upon the transition from indigo to N,N'-dideuteroindigo [176] is evidence of the presence of hydrogen bonds in dissolved indigo and of their effect on the long-wavelength absorption band. The replacement of hydrogen by deuterium should not affect the donor capacity of the nitrogen atom to any appreciable extent [177], while deuterium's ability to form hydrogen bonds is lower than that of hydrogen's [177]. It is not ruled out that this causes the absorption band to shift. One argument against the determining influence of the hydrogen bond on indigo's coloration is the bathochromic (rather than hypsochromic) shift that occurs upon the introduction of alkyl groups on the nitrogen atoms, which prevents hydrogen bonds.

Upon being heated to 390-392°C indigo's blue prisms make the transition coupled with dissociation to a purple-red melt (fusion), while indigo vapors are red in color (a fact known to Pliny, the Roman naturalist, as early as the first century A.D. [1]). A paler shade of the blue vapors is considered evidence of the absence of hydrogen bonds in indigo vapors and of their presence in crystals. The considerable color change occurring upon the transition from the vapor form to the solid state is observed in all indigoid dyes. Their absorption maxima in the vapor form are close together (thioindigo 508 nm, selenoindigo 530 nm, indigo 540 nm). On the contrary, in the solid state (in KBr tablets) the absorption maxima are wider apart (thioindigo 552 nm, selenoindigo 590 nm, indigo 660 nm). The shift upon the vapor → crystal transition is 44 nm for thioindigo, 60 nm for selenoindigo, and 120 nm for indigo. The greater shift (by 60 nm) of the spectrum band maximum in indigo than in selenoindigo may be attributed not only to the appearance of hydrogen bonds in indigo crystals, which are absent from indigo vapors, but also to indigo's greater polarizability under the effect of the condensation phase. Since the bathochromic shift upon the transition from indigo vapors to indigo solution is as much as 80 nm (in tetrachloro-ethane), the idea arises to use the method of proportional sensitivity in order to compare the polarizability of various indigoid dyes under the influence of intermolecular interactions.

The usual procedure, which consists of a consideration of the slope of the direct proportional sensitivity, could not be followed because the coefficients of proportional sensitivity r of indigo and thioindigo (0.53) and indigo and selenoindigo (0.59) are too low (Table 13). Conversely, the proportional sensitivity coefficient of selenoindigo and thioindigo (0.96) indicates a greater measure of affinity in their interaction with solvents. These data show the specificity of indigo's interaction with solvents.

Data have been reported in the literature showing that in red aniline solutions indigo is monomolecular, while in blue n-toluidine solutions it is dimeric [1]. Scheibe et al. [178] found it impossible to determine the degree of indigo association in solution because of its poor solubility, and after studying the spectra of styrylindigo in mixed solvents, they attributed the color change in indigo in different solvents to the polarization of the molecule by the solvent rather than to the equilibrium monomer-dimer. The authors of later papers [84,167] again associated the color change in indigo with its ability to form associates through hydrogen bonds. Again the authors of the papers [31,168] rejected this contention in favor of a purely solvatochromic explanation of the color change.

Inasmuch as the vapor state is, in a sense, a solution of a substance in the least polar solvent — a vacuum — and the crystal phase, the source of the maximum forces of intermolecular interaction, closes the "high-polarity" end of the "universal solvent scale" [54,179], it is advisable to calculate r of indigoid dyes, supplementing the data on fluid solvents with data on vapors and the solid state (separately and together). As the data presented in Table 13 indicate, r increases dramatically as a result. This shows the importance of a homogeneous polarizing interaction common to all indigoid dyes and embracing vapor, liquid solutions, and the solid state. The specificity of indigo manifests itself as slight deviations upon the action of solvents and may not be related to the presence of hydrogen bonds.

Table 13. Coefficients of proportional sensitivity R for pairs of indigoid and thioindigoid dyes (see formula E-I on page 45) and certain indogenides

Dye	Indigo				Thioindigo			
	L	with allowance for			L	with allowance for		
		V	S	V & S		V	S	V & S
Selenoindigo	0.59	0.91	0.86	0.94	0.96	0.99	0.97	0.98
Thioindigo	0.53	0.90	0.79	0.92	1.00	1.00	1.00	1.00
Oxindigo	0.98		0.96		0.86		0.90	
2-Benzylidenoindol-3-one	0.89				0.52			
2-Benzylidenobenzo[b]-thiophen-3-one	0.018				0.89			

Designations: L) liquid (fluid) solutions; V) vapor state; S) solid state.

If we rely on infrared spectra and relate frequency shifts to hydrogen bonds, we will find that indigo and 5,5'-dichloroindigo have hydrogen bonds in the solid state while the 7,7'-dichloro-substituted form lacks them. One would expect indigo and the 5,5'-disubstituted form to show an equal deepening of coloration upon the solvent—crystal transition and the 7,7'-disubstituted form to exhibit far less deepening of color, but this is not the case (Table 14). Upon the transition from solution in tetrachloroethane to crystals the shift is 40 nm for indigo, 14 nm for 7,7'-dichloroindigo, and 15 nm in 5,5'-dichloroindigo. The introduction of another two chlorine atoms into the 7,7'-dichloro-substituted form in positions 5,5', apart from eliminating the bathochromic effect of the solvent → crystal transition, results in a slight hypsochromic shift. The shift of λ_{max} upon the transition to the solid state of N-methylindigo (hydrogen bonds are possible) and in N,N'-dimethylindigo (no hydrogen bonds are possible) is practically the same. This indicates an insignificant influence exerted by hydrogen bonds on the deep coloration of solid indigo.

Summing up, it can be said that the presence of hydrogen bonds in indigo solution and in indigo crystals has not yet been reliably proved, and this detracts from the plausibility of the hypothesis regarding the stabilization of the E-form of indigo by means of hydrogen bonds. However, the absence of hydrogen bonds cannot be reliably established either. This allows one to attribute the photochromic behavior of indigo to the effect of hydrogen bonds.

Without rejecting the more usual hypotheses — a) the enhancement of the energy of hydrogen bonds upon photochemical excitation; b) the contribution of hydrogen bonds to the particularly rapid deterioration of electronic energy into vibrational and, later, translational energy (which strips the molecule of its energy before the E → Z-isomerization can begin) [180,181] — we list other possible reasons: c) because of the briefness of the molecule's excited state (notably a state requiring thermal activation), the entropy factor in the Arrhenius equation may decrease upon the transition from a reaction without energy activation to an analogous reaction with activation so abruptly as to make the reaction undetectable; d) the intramolecular catalysis, induced by the NH group localized in the region of the double bond, of the dark Z → E-conversion, which proceeds faster than the resulting Z-isomer can be registered; e) the increase in the torsional vibration quantum, which makes impossible the gradual rotation of the half molecule through "the accumulation of a number of torsional vibration quanta" since the energy exchange upon mutual collisions is the more hindered, the greater the size of the energy quantum transmitted.

Schulte-Frohlinde et al. [157] paid attention to the possibility, in principle, of the S_1^E- and T_1^E-states of the tautomer arising in indigo (in the event of hydrogen phototransfer of the NH group onto the carbonyl oxygen). These states differ substantially in structure from the S_1^E- and T_1^E-states of thioindigo, respectively. This assumption has been neither confirmed nor rejected.

The triplet state arising from the intersystem crossing of the S_1^E-tautomer and the triplet developing from intermolecular energy transfer onto S_0^E of indigo may be structurally different. But attempts at either direct or triplet-sensitized E → Z-photoisomerization of indigo have invariably failed.

Data obtained with the help of picosecond laser photolysis of 6,6'-dimethoxyindigo showed that in acetone the appearance time of the excited particle was 10 ps, its lifetime was 52 ± 6 ps, and a change in solvent viscosity did not affect these values [91]. It is assumed that the particle is the first singlet of the E-form, its observable absorption — the transi-

Table 14. The influence of substituents on the position of band maxima in the spectra of indigoid dyes

| Dye* | ν_{C-O}, cm^{-1} | ν_{N-H}, cm^{-1} | λ, nm | | | $\lambda_3-\lambda_2$, nm | $\lambda_3-\lambda_1$, nm |
			in ethanol (λ_1)	in tetra-chloroethane (λ_2)	crystals (λ_3)		
Indigo**	1626	3246	610	620	660	40	50
5,5'-Dichloroindigo	1629	3285	615	620	635	15	20
7,7'-Dichloroindigo	1653	3389	590	606	620	14	30
5,5',7,7'-Tetrachloroindigo	1659	3385	615	620	615	-5	0

* According to infrared spectrometry data, the first two compounds have a hydrogen bond and the other two do not [31].
** In vapor λ 540 nm.

tion $S_1^E \xrightarrow{h\nu} S_4^E$. Its disappearance path is a rapid deterioration into the ground state $S_1^E \to S_0^E$ [91]. In water the lifetime of the particle arising upon indigo's excitation is 0.20 ns. This excitation is believed to be accompanied by a phototransfer from the excited particle to the water or in the reverse direction [91]. Even these scant data show that indigo's behavior is different in different media; however, in all cases the lifetime of the excited particle is too short for E → Z-isomerization to occur. Fluorescence in thioindigo and stilbene is known to compete successfully with E → Z-isomerization; i.e., the rate constant of the latter is approximately 10^9 cm^{-1}, which is 1-2 orders of magnitude lower than the degradation rate of the excitation energy of indigo.

The long-discovered [96] ability of indogenide IX and the inability of indogenide X to undergo photoisomerization [103] are noteworthy.

IX   X

Apparently, the presence of carbonyl and the NH group in the molecule is not enough to prevent photoisomerization. This shows that Suhnel and Gustav [158] are in error in attributing the drastic shortening of the lifetime of the first indigo singlet (compared with thioindigo) to the special activity of the NH group in energy degradation, relying on the relatively narrow differences in the rate constant of energy degradation through the N–H (1.95×10^{12}) and C–H (1.07×10^{12}) bonds, ignoring, moreover, the predominance, by an order of magnitude, of the C–H bonds in the indigo molecule.

It is not ruled out that hydrogen bonds must form part of either intramolecular (A) or intermolecular (B or C) rings (to prevent isomerization).

A   B   C

Arguments are put forward in favor of the existence of ring A in indogenides X incapable of E → Z-photoisomerization [103]. Ring C may exist in 2-thionaphthene-3-indolinoindigoids, which for steric reasons may not form rings A and B, which "close" the molecule in the E-form. The peripheral position of ring C does not prevent it from precluding E → Z-isomerization, apparently, via the mechanism of electronic energy degradation at a rate far higher than the rate of isomerization.

THE INFLUENCE OF SUBSTITUENTS AND ENVIRONMENTAL PARAMETERS ON $\phi_{E \to Z}$

Substituents in a dye molecule may substantially alter $\phi_{E \to Z}$. For instance, in 6,6'-diethoxythioindigo the quantum yield is 5-7 times that in thioindigo. There is good reason to suppose that substituents that strongly influence the redistribution of electron density in the molecule or enhance the spin—orbital interaction are capable of changing the mechanism of

E/Z-photoisomerization. In such cases the introduction of a substituent largely determines the power (and at times also the direction) of the influence of external factors on $\phi_{E \to Z}$.

Upon excitation in the region of ν_{max} of the E-form the presence of an electron-donating substituent coupled with the carbonyl group causes $\phi_{S_0^E \dashrightarrow S_1^Z}$ to increase slightly (see Table 11). Mostoslavskii et al. [43] found a direct linear relation between $\phi_{E \to Z}$ and the frequency of the long-wavelength absorption band maximum of the E-isomer (Figure 12).

As already mentioned [62] the hypsochromic shift of the long-wavelength absorption band in the spectrum of indigoid dyes is caused by substituents which decrease the negative charge on the carbonyl or the positive charge on the heteroatom. Apparently, they also weaken the electrostatic attraction of these groups in the E-isomer stabilizing its configuration, and the electrostatic repulsion in the Z-isomer destabilizing its configuration. This is the qualitative chemical explanation of the dependence of the quantum yield on the nature of the substituent discovered by Mostoslavskii et al. [43]. If the energy differences between the ground and first singlet states $\nu_{S_0^E \dashrightarrow S_1^E}$ were a true argument of the functional dependence under consideration, all the points on the straight lines 1 and 2 in Figure 12 would lie on the same straight line. Actually, however, the shift of ν_{max} as a result of annelation of the benzene ring causes a greater change in $\phi_{E \to Z}$ than the same shift of ν_{max} caused by the introduction of substituents in position 6 or 5 (see Figure 12).

Apparently, the "true argument" is not the energy difference between S_0^E and $S_1^{E^-}$ but some much smaller energy value linearly related to this difference within a narrow series of substances but nonlinearly related with wide structural variations. Such a value could be the energy difference between two particles one of which turns into the other upon E → Z-photo-isomerization. The following particle pairs are possible here: S_1^E and S_n^E; S_1^E and T_1^E; S_1^E and T_n^E; S_1^E and T_n^P; S_1^E and S_1^P; $S_1^{E^-}$ and $S_1^{Z^-}$; S_1^P and S_0^P; T_1^E and T_1^P; T_1^P and S_0^P.

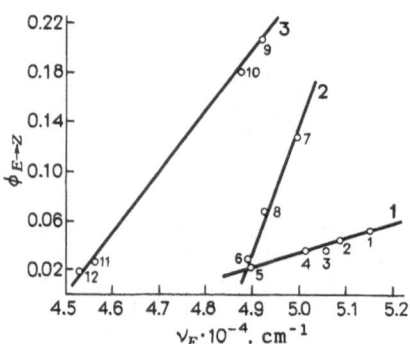

Fig. 12. The dependence of $\phi_{E \to Z}$ on ν_E for thioindigoid dyes having different cores. Substituted forms of 2-benzo[b]thiophene-2-naphtho[1,8-bc]-thiopyran-indigo: R = 6-OC$_2$H$_5$ (1), 6-Cl (2), H (3), 5-CH$_3$ (4), 5,6-benzo (5), 5,6-benzo, 7-Cl (6), 2-naphtho[1,8-bc]thiopyran-2'-naphtho[2,1-b]thiophene-indigo (7), 2-(9-Cl-naphtho[1,2-b]thio-phene)-2'-naphtho[1,8-bc]-thiopyran-indigo (8), 2-pyreno[4,3-bc]-thiopyran-2'-benzo[b]thiophene-indigo (9), 2-pyreno[4,3-bc]-thiopyran-2'-naphtho[2,1-b]thiophene-indigo (10), 2-pyreno[3,4-bc]-thiopyran-2'-benzo[b]thiophene-indigo (11), 2-pyreno[3,4-bc]thio-pyran-2-naphtho[2,1-b]thiophene-indigo (12).

This suggests a relation between this difference and the thermal stage of photochemical isomerization. If this assumption is correct, substances with the maximum quantum yield (and ν_{max}) will feature a minimal dependence of $\phi_{S_0^E \dashrightarrow S_1^{E \cdot}}$ on temperature. Therefore, the as yet unexplored temperature dependence of the quantum yield of substances belonging to the same series and exhibiting the widest ν_{max} differences is of great interest.

There is a strong possibility that the thermal stage represents the activation energy of rotation about the central bond, which in the excited state is closer in some dyes and further from the ordinary in others. The activation energy of rotation of the alkyl group about a simple bond is close to 12 kJ/mole, i.e., close to the activation energy of E → Z-photo-isomerization of thioindigoid dyes.

It is shown in the article [153] that the introduction of ethoxy groups into positions 6,6' of thioindigo has no effect on k_{fl} but increases k_{isc} by a factor of 50. This is attributed to the reconstruction of the molecule with an increased contribution of the resonance structure

and a weakening of resonance stabilization and the electrostatic interaction within the chromophore system of the E-isomer.

It is noted that the $S_1^E \to S_0^E$ process is facilitated by the presence of tert-butyl groups in positions 5,5' [153]. The assumption that these positions are especially effective for energy dissipation is considered unlikely by the authors. Chlorine atoms in position 6 enhance $\phi_{E \to Z}$ and decrease it in positions 4,7. Apparently, this is due to the donor ability of halogen [153] rather than the heavy-atom effect. It is well known that atoms with a large ordinal number, as a result of spin—orbital interaction, contribute to the "intermixing" of singlet and triplet states and should make intersystem crossing easier. Since in the triplet mechanism of E → Z-photo-isomerization the conversion $S_1^E \to T_1^E$ (or into T_n^E) is one of the stages, an increase of ϕ_{isc} to the detriment of the radiationless degradation of energy of S_1^E and fluorescence should increase $\phi_{E \to Z}$. Indeed, $\phi_{E \to Z}$ of 7,7'-diiodo-perinaphthothioindigo is eight times higher than $\phi_{E \to Z}$ of perinaphthothio-indigo. On the other hand, $\phi_{E \to Z}$ of 2-(7-iodonaphtho[1,8-bc]thiopyran)-2'-naphtho[2,1-b]thiophene-indigo is not higher than that of a dye lacking an iodine atom. Apparently, in such a dye interconversion proceeds with a sufficiently high yield, while quantum yield losses occur at other elementary stages of the process. Probably the introduction of a heavy atom into the molecule is a method more suitable for raising very low quantum yields rather than for improving quantum yields amounting to 10% and higher.

An increase in the viscosity of the medium hinders the rotation of the excited half particle about the central bond and causes $\phi_{E \to Z}$ to decline. In hard plastics (polystyrene [14] and polymethyl methacrylate [15]) practically no E → Z-photoisomerization occurs. Z → E-Isomerization is far less dependent on medium viscosity [17]. Apparently, this is due to the large size of the Z-isomer. It has been reported that the E → Z-photoisomerization of thioindigo adsorbed on aluminum oxide proceeds in the molecular layer and not in the polymolecular one [182]. A solution of 6,6'-dialkoxy-thioindigo in epoxy resin has been proposed as a photochromic material assuring nondestructive reading [8].

On the basis of a recent study of thioindigo by the picosecond photolysis method, Krysanov and Alfimov have assumed [183-185] that two different S_1^E particles arise upon the excitation of S_0^E: $S_0^E \overset{h\nu}{\underset{h\nu}{\rightleftarrows}} \overset{S_1^E}{} S_1^E$ (CT). It appears that vibrational relaxation, internal conversion, or a special kind of solvation of S_0^E at the moment of excitation may be responsible for the appearance of S_1^E (CT), and not dissociation or ionization from the higher levels. S_1^E can fluoresce and is responsible for the induced absorption band $S_1^E \dashrightarrow S_n^E$ in the 450-500 nm region. S_1^E (CT) cannot fluoresce and is noted for the high degree of charge transfer from sulfur atoms to the carbonyl oxygen, and it is responsible for the induced absorption band with a 603 nm maximum which appears within 7 ns and lasts for over 4 ns. Two hypotheses about the nature and mechanism of the appearance of the 603 nm band have been proposed: 1) S_1^E (CT) $\rightarrow S_n^E$ (CT); 2) S_1^E (CT) $\dashrightarrow T_1^E$ (CT) $\overset{h\nu}{\dashrightarrow} T^E$ (CT). The second variant, which allows a quick and reversible intersystem crossing, is possible if the charge separation results in the coincidence of the energies of S_1^E (CT) and T_1^E (CT) [183-185]. It is assumed [186] that the Z $\rightarrow$ E-photoisomerization of thioindigo proceeds through S_1^E. At 180°K. $\phi_{Z\rightarrow E}$ is close to unity.

Intermolecular charge transfer is an essential stage in the quenching of S_1 thioindigo by donors [187-188]. In nonpolar solvents a nonfluorescing charge-transfering exciplex arises, while an anion-radical of the thioindigo, recorded spectrally, is formed in polar solvents [187]. This is an argument in favor of our assumption mentioned above regarding the inability of donors to quench the T_1 thioindigo that arises spontaneously. It is believed [188-190] that ion-radical pairs or charge-transfering exciplexes may be the forerunners of the intersystem crossing $S_1 \rightarrow T_1$. Increasing attention is being focused on the derivatives of 2-naphtho[1,8-bc]thiophene-2'-naphtho[2,1-b]thiophene-indigo [185,188,191] investigated by us for the first time.

The rather surprising report [192] on the separation of both E- and Z-isomers during the synthesis of 2,2'-bis(5,6-benzindole)-indigo is argued for [193] with the aid of infrared, ultraviolet, and crystal spectra analyses; no mutual photochemical or thermal conversion of E- and Z-isomers was observed. X-ray crystallographic studies, which the stability of the Z-isomer should make easier, could help dispel doubts about the correctness of Z-isomer identification.

The study of indigo fluorescence with a picosecond time resolution confirms the ability of intra- and intermolecular hydrogen bonds to serve as a channel for a fast radiationless degradation of the electron energy of excitation as well as the dependence of the lifetime of S_1^E on the nature of the solvent [194].

The first reports have been published about the inclusion of thioindigoid dyes in the polymer chain and of their photoisomerization [195,196].

The photovoltaic effect of thin thioindigo and indigo films on Ni [197] should be mentioned in conclusion. The cause of the maximum manifestations of the effect observed at 400 nm and 590 nm is believed to be the $T_1 \rightarrow T_n$ transitions localized at approximately these wavelengths. No photochromic conversions in solid films occur. A change in the photoconductivity of zinc oxide through the formation of donor–acceptor complexes with indigo [198] was observed. Upon exposure to visible light a charge is injected from the indigo layer covering the polystyrene film. The charge injection increases the film's conductivity by a factor of 2-3, as compared with the level observed in the dark [199].

REFERENCES

1. D. I. Minaev, The Chemistry of Indigo and Indigoid Dyes, ONTI, Moscow—Leningrad (1934).
2. J. Martinet, Matieres Colorautes les Indigoides, Paris (1934).
3. G. M. Oksengendler and É. P. Gendrikov, Zh. Fiz. Khim., **33**(12):2791-2974 (1959).
4. É. P. Gendrikov, Author's abstract of Candidate's Thesis, Scientific-Research Institute of Organic Intermediates and Dyes, Moscow (1960).
5. M. A. Mostoslavskii, V. I. Izmail'skii, and M. M. Shapkina, Zh. Vses. Khim. Ova., **7**(1):108-109 (1962).
6. M. A. Mostoslavskii, S. I. Saenko, and M. M. Shapkina, Zh. Vses. Khim. Ova., **12**(6):702-703 (1967).
7. D. L. Ross, Appl. Opt., **10**:571-576 (1971).
8. U.S. Patent 3575872 (1971).
9. V. D. Paramonov, M. A. Mostoslavskii, and I. N. Shevchuk, Zh. Fiz. Khim., **52**(10):2676-2677 (1978).
10. M. A. Mostoslavskii and M. M. Shapkina, Zh. Fiz. Khim., **44**(12):2708-2711 (1970).
11. M. A. Mostoslavskii and M. M. Shapkina, Zh. Fiz. Khim., **44**(12):2975-2978.
12. M. A. Mostoslavskii, G. A. Yugai, and V. V. Yadrikhinskii, Zh. Org. Khim., **12**(8):1837-1838 (1976).
13. K. H. Grellmann and P. Hentschel, Chem. Phys. Lett., **53**(3):545-551 (1978).
14. T. Karstens, K. Kobs, and R. Memming, Ber. Bunsenges. Phys. Chem., **83**(5):504-510 (1979).
15. H. Görner and D. Schulte-Frohlinde, Chem. Phys. Lett., **66**(2):363-369 (1979).
16. H. Görner and D. Schulte-Frohlinde, Ber. Bunsenges. Phys. Chem., **82**:1102-1107 (1978).
17. G. Haucke and R. Paetzold, Photophysikalische Chemie Indigoider Farbstoffe, Nova Acta Leopoldina, **11**:123 (1978).
18. Z. V. Zhidkova, Zh. Prikl. Spektrosk., **16**(2):325-330 (1972).
19. Z. V. Zhidkova, Zh. Prikl. Spektrosk., **22**(5):861-864 (1975).
20. V. D. Paramonov, I. M. Byteva, V. V. Yadrikhinskii, and M. A. Mostoslavskii, Zh. Prikl. Spektrosk., **27**(1):181-183 (1977).
21. V. D. Paramonov, M. A. Mostoslavskii, V. F. Mandzhikov, et al., Zh. Prikl. Spektrosk., **27**(5):861-864 (1979).
22. G. N. Kagan, V. A. Kosobutskii, V. K. Belyakov, and O. G. Tarakanov, Khim. Geterotsikl. Soedin., **6**:794-798 (1972).
23. B. Pullman and A. Pullman, Quantum Biochemistry, Wiley (1963).
24. V. I. Minkin, B. Ya. Simkin, and R. M. Minyaev, The Theory of Molecular Structure (Electronic Shells), Vysshaya Shkola, Moscow (1967).
25. M.J.S. Dewar and R. C. Dougherty, The PMO Theory of Organic Chemistry, Plenum Press (1975).
26. V. I. Danilova, Ph.D. Thesis, Tomsk State Univ., Tomsk (1964).
27. J. N. Murrell, The Theory of the Electronic Spectra of Organic Molecules, London—New York (1963).
28. The Energy of Chemical Bond Breaking. Ionization Potential and Electron Affinity, Nauka, Moscow (1974).
29. H. Güsten, Chem. Commun., No. 3, 133-134 (1969).
30. H. Hermann and W. Lüttke, Chem. Ber., **101**(5):1715-1728 (1968).
31. W. Lüttke and M. Klessinger, Chem. Ber., **97**:2342-2357 (1964).
32. G. Wyman and W. R. Brode, J. Am. Chem. Soc., **73**(4):14871493 (1951).
33. J. Omoto, S. Imada, and H. Aoyama, Chem. Ind., No. 12, 415 (1979).
34. W. R. Brode, Recent Advances in the Chemistry of Colouring Matters, No. 4, 1-46 (1956).
35. G. R. Giuliano, L. D. Hess, and J. D. Margerum, J. Am. Chem. Soc., **90**(3):587-594 (1968).
36. R. Pummerer and Y. Marondel, Chem. Ber., **93**:2834-2839 (1960).

37. P. W. Sadler, Spectrochim. Acta, 16:1094-1099 (1960).
38. V. A. Izmail'skii, Problems of Aniline-Dye Chemistry. The Proceedings of the 8th Conference on Aniline-Dye Chemistry, Dec. 8-11, 1947, Izd. Akad. Nauk SSSR, Moscow—Leningrad (1950), p. 107.
39. R. Pummerer, F. Meininger, G. Schrott, et al., Lieb. Ann., 590:195 (1954); 602:228-232 (1957).
40. P. W. Sadler, J. Org. Chem., 21:316-318 (1956).
41. P. W. Sadler and D. G. O'Sullivan, Adv. Mol. Spectrosc., 2:542-547 (1962).
42. S. J. Holt and P. W. Sadler, Proc. Chem. Soc. London, 48:495-505 (1958).
43. M. A. Mostoslavskii, V. A. Yadrikhinskii, and O. K. Yadrikhinskaya, Zh. Fiz. Khim., 44(12):2975-2977 (1970).
44. P. W. Sadler and R. L. Warren, J. Am. Chem. Soc., 78(6):1251-1255 (1956).
45. N. S. Dokunikhin and Yu. E. Gerasimenko, Zh. Obshch. Khim., 30:635-638 (1960).
46. N. S. Dokunikhin and Yu. E. Gerasimenko, Zh. Obshch. Khim., 30:1987-1989 (1960).
47. S. K. Guha, I. N. Chatterjea, and A. K. Mirta, Chem. Ber., 94:2295-2305 (1961).
48. J. Formanek, Z. Angew. Chem., 41:1133-1141 (1928).
49. J. Weinstein and G. M. Wyman, J. Am. Chem. Soc., 78:2387-2390 (1956).
50. N. S. Dokunikhin and Yu. E. Gerasimenko, Zh. Obshch. Khim., 30:1231-1233 (1960).
51. N. S. Dokunikhin and Yu. E. Gerasimenko, Zh. Obshch. Khim., 31:219-223 (1961).
52. Yu. E. Gerasimenko, Author's abstract of Candidate's Dissertation, Moscow Institute of Chemical Technology, Moscow (1960).
53. G. A. Yugai, Author's abstract of Candidate's Dissertation, Scientific-Research Institute of Organic Intermediates and Dyes, Moscow (1973).
54. V. A. Palm, Fundamentals of the Quantitative Theory of Organic Reactions, Khimiya, Leningrad (1977).
55. A. I. Kiprianov, An Introduction to the Electronic Theory of Organic Compounds, Naukova Dumka, Kiev (1965).
56. M. Klessinger, Tetrahedron, 22:3355-3365 (1966).
57. R. Kuhn, Naturwissenschaften, 20:618-625 (1932).
58. V. A. Izmail'skii and E. E. Smirnov, Zh. Obshch. Khim., 7(2):523-536 (1937).
59. V. A. Izmail'skii, Proceedings of the IV Conference on Aniline-Dye Chemistry, 1939, Izd. Akad. Nauk SSSR, Moscow (1940).
60. G. D. Barabashova, V. A. Izmail'skii, and E. E. Milliaresi, Dokl. Akad. Nauk SSSR, Ser. Khim., 190:95 (1970).
61. G. M. Oksengendler and M. A. Mostoslavskii, Ukr. Khim. Zh., Issue 1, 69-72 (1960).
62. O. V. Sverdlova, Electronic Spectra in Organic Chemistry, Khimiya, Leningrad (1973).
63. P. P. Shorygin, M. F. Shostakovskii, E. N. Prilezhaeva, et al., Izv. Akad. Nauk SSSR, Otd. Khim. Nauk, No. 9, 1571-1577 (1961).
64. H. Hermann and W. Lüttke, Chem. Ber., 104(2):492-512 (1971).
65. F. Turecek and M. Procházka, Coll. Czech. Chem. Commun., 39(8):2073-2094 (1974).
66. W. Lüttke, H. Hermann, and M. Klessinger, Angew. Chem., Intern. Ed., 5:598-599 (1966).
67. U. Luhmann, F. G. Wentz, B. Kniereiem, and W. Lüttke, Chem. Ber., 111:3233-3245 (1978).
68. V. A. Izmail'skii, A. Ya. Kozhevnikova, and É. I. Fedorova, Dokl. Akad. Nauk SSSR, 183(2):341-344 (1968).
69. V. A. Izmail'skii and O. I. Lobova, Dokl. Akad. Nauk SSSR, 205(5):1100-1103 (1972).

70. K. D. Banerji, A. K. Mazumdar, and S. K. Guha, J. Indian Chem. Soc., 24:896-970 (1977).
71. M. Klessinger and W. Lüttke, Tetrahedron, 19(2):315-355 (1963).
72. H. L. Ammon and H. Hermann, J. Org. Chem., 43(24):4581-4586 (1978).
73. M. A. Mostoslavskii, Author's abstract of Candidate's Dissertation, Lensovet Leningrad Institute of Technology, Leningrad (1962).
74. W. R. Brode and G. Wyman, J. Res. Natl. Bur. Stand., 47:170-178 (1951).
75. M. A. Mostoslavskii and I. N. Shevchuk, Khim. Geterotsikl. Soedin., No. 1, (1970).
76. E. A. Gribova, G. S. Zhdanov, and G. A. Golder, Kristallografiya, 1:53-60 (1956).
77. G. S. Egerton, Nature, 103:389-390 (1959).
78. C. H. Bamford and C.F.H. Tipper (eds.), Comprehensive Chemical Kinetics. 1. The Practice and Theory of Kinetics, in: The Formation and Decay of Excited Species, New York (1969).
79. J. Weinstein and G. Wyman, J. Am. Chem. Soc., 78(16):4007-4010 (1956).
80. G. Wyman and A. F. Zenhaeusern, Ber. Bunsenges. Phys. Chem., 72(2): 326-328 (1968).
81. G. T. Bakulina, Author's abstract of Candidate's Dissertation, Scientific-Research Institute of Organic Intermediates and Dyes, Moscow (1964).
82. G. Wyman, Chem. Commun., 21:1332-1334 (1971).
83. G. Wyman, B. M. Zarnegar, and P. G. Whitten, Phys. Chem., 77(21): 2584-2586 (1973).
84. W. R. Brode, E. G. Pearson, and G. M. Wyman, J. Am. Chem. Soc., 76(4):1034-1036 (1954).
85. G. Wyman and A. F. Zenhaeusern, J. Org. Chem., 30(7):2348-2352 (1965).
86. S. I. Pekar, Adv. Phys. Sci., 50:197-252 (1953).
87. V. I. Permogorov, L. A. Serdyukova, and M. D. Frank-Kamenetskii, Izv. Akad. Nauk SSSR, Ser. Fiz., 32(9):1561-1563 (1968).
88. H. Suzuki, Electronic Absorption Spectra and Chemistry of Organic Molecules. An Application of Molecular Orbital Theory, New York (1967).
89. V. A. Izmail'skii and M. A. Mostoslavskii, Dokl. Akad. Nauk SSSR, 139(3):601-604 (1961).
90. G. M. Wyman and B. M. Zarnegar, J. Phys. Chem., 77(6):831-837 (1973).
91. T. Kobayashy and P. M. Rentzepis, J. Chem. Phys., 70(2):886-892 (1979).
92. V. D. Paramonov, M. A. Mostoslavskii, I. N. Shevchuk, et al., Ukr. Khim. Zh., 45(8):768-771 (1979).
93. G. M. Oksengendler and Yu. E. Gerasimenko, Spectrophotometric Analysis of Thioindigoid Dyes, in: Organic Half-Products and Dyes, Goskhim-izdat, Moscow (1961), Issue 2, p. 219.
94. M. A. Mostoslavskii and M. D. Kravchenko, Khim. Geterotsikl. Soedin., No. 1, 58-60 (1968).
95. G. A. Yugai, M. A. Mostoslavskii, and T. V. Denisova, Khim. Getero-tsikl. Soedin., No. 10, 1326-1329 (1970).
96. M. A. Mostoslavskii and V. A. Izmail'skii, Zh. Obshch. Khim., 31:17-28 (1961).
97. W. Orville-Thomas, Internal Rotation in Molecules, Halsted Press (1973).
98. H.E. Simons, D. S. Blomstrom, and R. D. Vest, J. Am. Chem. Soc., 84(24):4756-4771 (1962).
99. K. Nakanishi, Infrared Spectra and the Structure of Organic Compounds, Mir, Moscow (1965).
100. J. Suhnel and K. Gustav, J. Prakt. Chem., 320:917 (1978).
101. M.J.S. Dewar, Molecular Orbital Theory of Organic Chemistry, McGraw-Hill (1969).
102. E. B. Knott, J. Soc. Dyers Colourists, 67:302-306 (1951).

103. M. A. Mostoslavskii, M. D. Kravchenko, and I. N. Shevchuk, Zh. Fiz. Khim., 44(4):1008-1012 (1970).
104. S. K. Guha, J. Ind. Chem. Soc., 12:659-664 (1935).
105. S. K. Guha, J. Ind. Chem. Soc., 15:359-364 (1938).
106. G. A. Yugai and M. A. Mostoslavskii, Khim. Geterotsikl. Soedin., No. 8, 1032-1035 (1976).
107. G. A. Yugai, M. A. Mostoslavskii, Yu. L. Yagupol'skii, et al., Khim. Geterotsikl. Soedin., No. 8, 1148-1149 (1972).
108. F. Bruin, F. W. Heineken, and M. Bruin, J. Org. Chem., 28:562-564 (1963).
109. M. A. Mostoslavskii and S. I. Saenko, Reakts. Sposobn. Org. Soedin., 4(2):323-329 (1967).
110. M. A. Mostoslavskii, M. M. Shapkina, and S. I. Saenko, Khim. Geterotsikl. Soedin., No. 3, 465-467 (1967).
111. W. Lüttke, H. Hermann, and M. Klessinger, Angew. Chem., 78:638-639 (1966).
112. M. A. Mostoslavskii and M. M. Shapkina, Proceedings of the Conference on Problems of the Application of Correlational Equations in Organic Chemistry, Vol. 2, Izd. Tartusk. Gos. Univ., Tartu (1963), pp. 119-125.
113. M. Erler, G. Haucke, and R. Paetzold, Z. Phys. Chem. (Leipzig), 258(2):315-320 (1977).
114. Ken'Ichi Higasi et al., Quantum Organic Chemistry, Wiley (1965).
115. Advances in Stereochemistry, Goskhimizdat, Moscow (1961).
116. G. M. Oksengendler and É. P. Gendrikov, Zh. Obshch. Khim., 29:3898-3901 (1959).
117. E. A. Gastilovich, K. V. Tskhai, and D. N. Shigorin, Opt. Spektrosk., 41(4):566-572 (1976).
118. P. Friedlender and N. Woroshzow, Ann., 38:1-23 (1912).
119. J. Harley-Mason and F. G. Mann, J. Chem. Soc., 404-415 (1942).
120. G. M. Oksengendler and É. P. Gendrikov, Ukr. Khim. Zh., 25(2):206-209 (1959).
121. Inventor's Certificate 566841 (1977).
122. C. C. Cook and F. K. Sutcliffe, J. Chem. Soc., 8:957-960 (1968).
123. C. C. Cook and F. K. Sutcliffe, Chimia, 135-139 (1968).
124. German Patent 362551 (1922); Frdl., 14:932 (1926).
125. French Patent 628617 (1927); Zbl., 1:1463 (1928).
126. German Patent 445218 (1927); Frdl., 15:624 (1928).
127. U.S. Patent 1623410 (1927); Zbl., 11:339 (1927).
128. German Patent 414084 (1925); Zbl., 11:861 (1925).
129. M. A. Mostoslavskii, S. I. Saenko, and G. A. Yugai, Zh. Vses. Khim. Ova., 13(4):462-463 (1968).
130. M. A. Mostoslavskii, S. I. Saenko, et al., Khim. Geterotsikl. Soedin., No. 8, 1036-1038 (1970).
131. M. A. Mostoslavskii, S. I. Saenko, and V. L. Belyaev, Khim. Geterotsikl. Soedin., No. 8, 1034-1036 (1973).
132. M. A. Mostoslavskii, L. F. Gorbas, and E. B. Georgieva, Khim. Geterotsikl. Soedin., No. 5, 637-640 (1973).
133. Inventor's Certificate 309032 (1969).
134. G. M. Wyman, Chem. Rev., 55:625-657 (1955).
135. N. P. Kovalenko, M. V. Alfimov, and Yu. B. Shekk, The Spectral Luminescent Properties and Photoisomerization of Diarylethylenes, Preprint, Chernogolovka (1977).
136. J. Saltiel, J. D'Agostino, E. D. Magarity, et al., Org. Photochem., 3:1-113 (1973).
137. J. Jortner, S. A. Reice, and R. M. Hochstrasser, Adv. Photochem., 7:149-309 (1969).
138. R. Gregory and D. F. Williams, J. Phys. Chem., 83(20):2652-2662 (1979).
139. R. M. Hochstrasser, Pure Appl. Chem., 52:2683-2691 (1980).

140. P. Wirth, S. Schneider, and F. Dorr, Ber. Bunsenges. Phys. Chem., 81(11):1127-1998 (1977).
141. F. E. Doany, B. J. Greene, and R. M. Hochstrasser, Chem. Phys. Lett., 75(2):206-208 (1980).
142. H. Görner and D. Schulte-Frohlinde, J. Phys. Chem., 83(24):3107-3118 (1979).
143. A. R. Olson and F. L. Hudson, J. Am. Chem. Soc., 55(4):1410-1424 (1933).
144. G. N. Lewis, T. T. Magel, and D. Lipkin, J. Am. Chem. Soc., 62(11): 2973-2980 (1940).
145. M. A. Mostoslavskii, Zh. Fiz. Khim., 34(11):2405-2407 (1960).
146. J. Saltiel, D.W.L. Chang, E. D. Megarity, et al., Pure Appl. Chem., 41(4):559-579 (1975).
147. J. B. Birks, Chem. Phys. Lett., 38(3):437-440 (1976).
148. T. Forster, Z. Elektrochem., 56:719-720 (1952).
149. G. H. Ting and D. S. McClure, J. Chin. Chem. Soc. (Taipei), 18:95-105 (1971).
150. R. H. Dyck and D. S. McClure, J. Chem. Phys., 36:2326-2345 (1962).
151. J. Blanc and D. L. Ross, J. Phys. Chem., 72(8):2817-2824 (1968).
152. D. L. Ross, J. Blanc, and F. G. Matticoli, J. Am. Chem. Soc., 92(19): 5750-5752 (1970).
153. A. Kirsch and G. M. Wyman, J. Phys. Chem., 81(5):413-419 (1977).
154. A. Kirsch and G. M. Wyman, J. Phys. Chem., 79(5):543-544 (1975).
155. G. Hauke, M. Érler, R. Paetzold, et al., Vestn. Mosk. Gos. Univ., Ser. Khim., 17(2):190-194 (1976).
156. D. L. Ross and J. Blanc, in: Photochromism, ed. by G. H. Brown, Wiley, New York–London (1970), pp. 472-555.
157. D. Schulte-Frohlinde, H. Germann, and G. M. Wyman, Z. Phys. Chem. (Frankfurt am Main), 101:115-121 (1976).
158. J. Suhnel and K. Gustav, Mol. Photochem., 8(6):437-458 (1977).
159. T. Karstens, K. Kobs, R. Memming, and F. Schropel, Chem. Phys. Lett., 48(4):540-544 (1977).
160. N. B. Shekk, Khim. Vys. Energ., 7(3):221-226 (1973).
161. E. Birckner, G. Haucke, and R. Paetzold, Z. Chem., 19(7):258-259 (1979).
162. J. Saltiel and D. W. Eaker, Chem. Phys. Lett., 75(2):209-213 (1980).
163. D. A. Rogers, J. D. Margerum, and G. M. Wyman, J. Am. Chem. Soc., 79(10):2464-2468 (1957).
164. D. Schulte-Frohlinde and H. Görner, Pure Appl. Chem., 51(2):279-297 (1979).
165. G. Haucke and R. Paetzold, J. Prakt. Chem., 321(6):978-986 (1979).
166. E. Birckner, G. Haucke, and R. Paetzold, Z. Chem., 19(6):219 (1979).
167. N. S. Shigorin, N. S. Dokunikhin, and E. A. Gribova, Zh. Fiz. Khim., 29(5):867-876 (1955).
168. M. Klessinger and W. Lüttke, Chem. Ber., 99:2136-2145 (1966).
169. N. S. Dokunikhin and E. K. Levin, Dokl. Akad. Nauk SSSR, 35:110-112 (1942).
170. E. A. Gribova, Dokl. Akad. Nauk SSSR, 102:279-281 (1955).
171. E. A. Gribova, in: Organic Half-Products and Dyes, Vol. 2, Goskhimizdat, Moscow (1961), p. 222.
172. H. V. Eller, Bull. Soc. Chim., France, 106:1429-1433 (1955).
173. H. V. Eller, Bull. Soc. Chim., France, 106:1433-1438 (1955).
174. H. V. Eller, Bull. Soc. Chim., France, 106:1438-1444 (1955).
175. H. V. Eller, Bull. Soc. Chim., France, 106:1444-1449 (1955).
176. R. Trauzeddel and H. Hubner, Isotopen Praxis, 7(3):106-107 (1971).
177. V. A. Terent'ev, The Thermodynamics of the Hydrogen Bond, Izd. Saratovsk. Univ., Saratov (1973).
178. G. Scheibe, H. Dorfling, and J. Assmann, Lieb. Ann., 544:240-252 (1940).

179. V. A. Palm and J. A. Koppel, in: Advances in Linear Free Energy Relationships, ed. by N. B. Chamau and J. Chorter, Plenum Press, New York–London (1972), p. 203-233.
180. G. M. Wyman and B. M. Zarnegar, J. Phys. Chem., 77(10):1204-1207 (1973).
181. W. M. Gelbart, K. F. Freed, and S. A. Rice, J. Chem. Phys., 52(5): 2460-2473 (1970).
182. H. D. Breuer and H. Jacob, Chem. Phys. Lett., 73(1):172-174 (1980).
183. S. A. Krysanov and M. V. Alfimov, Chem. Phys. Lett., 76(2)221-224 (1980).
184. S. A. Krysanov and M. V. Alfimov, Dokl. Akad. Nauk SSSR, 258(3):665-668 (1981).
185. S. A. Krysanov and M. V. Alfimov, Chem. Phys. Lett., 82(1):51-54 (1981).
186. R. Memming and K. Kobs, Ber. Bunsenges. Phys. Chem., 85(3):238-242 (1981).
187. R. Memming and K. Kobs, J. Phys. Chem., 85(19):2771-1777 (1981).
188. R. Memming and K. Kobs, Chem. Phys. Lett., 80(3):475-478 (1981).
189. T. V. Leshina, K. M. Salikhov, R. Z. Sagdeev, S. G. Belyaeva, V. I. Maryasova, P. A. Putrou, and Yu. N. Molin, Chem. Phys. Lett., 70(2):228-232 (1980).
190. T. V. Leshina, S. G. Belyaeva, V. I. Maryasova, R. Z. Sagdeev, and Yu. N. Molin, Chem. Phys. Lett., 75(3):438-448 (1980).
191. M. A. Mostoslavskii, V. D. Paramonov, and V. F. Mandzhikov, Ukr. Khim. Zh., 47(4):440 (1981).
192. T. M. Ivanova, G. M. Ostapchuk, L. V. Shagalov, and V. N. Berdyugin, Khim. Geterotsikl. Soedin., No. 4, 501-504 (1981).
193. L. B. Shagalov, G. M. Ostapchuk, V. N. Eraksina, N. N. Suvorov, and T. M. Ivanova, Khim. Geterotsikl. Soedin., No. 7, 996 (1980).
194. S. Schneider, E. Litt, R. Hefferle, and F. Doerr, Nuovo Cimento Soc. Ital. Fis. B., 63B(1):411-419 (1981); Chem. Abstr., 95:186187f (1981).
195. C. P. Klages, FRG Patent 3007296 (Cl 6 03C1/733); Chem. Abstr., 95: 178691a (1981).
196. L. F. Gorbas and M. A. Mostoslavskii, Summaries of the Papers Presented at the All-Union Conference on "The Synthesis and Industrial Uses of Dyes and Intermediate Products," Sept. 9-11, 1981, Rubezhnoe (1981).
197. S. C. Dahlberg and C. B. Reinganum, J. Chem. Phys., 75(5):2429-2431 (1981).
198. T. Kavaguchi, S. Ichikawe, Yuko Juen, Y. Niina, and S. Nagegawa, Tokyo Gakugei Daigaku Kigo Dai-4-bumon, 32:85-89 (1980); Chem. Abstr., 94: 113192y.
199. J. K. Kulshrestha and A. P. Strivastava, Indian J. Pure Appl. Phys., 19(5):478-480 (1981).

Chapter 3

PHOTOCHROMIC COMPOUNDS WITH N=N AND C=N CHROMOPHORES

D. Fanghänel, G. Timpe, and V. Orthman

K. Schorlemmer School of Advanced Technical Studies
Mersebury, GDR

Like compounds with the C=C chromophore, numerous groups of compounds with the N=N or C=N chromophore are photochromic. Their photochromism is based on $E \underset{h\nu'}{\overset{h\nu}{\rightleftharpoons}} Z$-isomerization [1-3] and is light-induced in both directions. The special features of systems incorporating N=N and C=N bonds, as compared with iso-π-electronic systems with the C=C bond, are attributable to the presence of an unshared electron pair on the nitrogen atom.

This chapter examines the photochromism of azo compounds, aromatic diazo compounds, as well as azomethines, oximes, oxime ethers, azines, and hydrazones. The quantitative characteristics of the spectra and photochromism of those compounds for which data are available on the E- and Z-isomers are tabulated. Literature published before 1980 has been used, but no data are given on the spectra of the E- and Z-isomers with N=N and C=N chromophores already available in review literature and surveys [1].

THE PRINCIPAL PATHS OF PHOTOCHEMICAL AND THERMAL E/Z-ISOMERIZATION

Thermal and photochemical E/Z-isomerization is known for many unsaturated compounds containing C=C, C=N, or N=N double bonds. In most cases, the E-isomer is thermodynamically more stable, absorbs at longer wavelengths, and has a higher extinction coefficient than the Z-isomer.

Most E → Z-photoisomerizations are reversible. Since there is often no wavelength region where only one of the two isomers absorbs upon monochromatic irradiation, a photostationary state is established whose position is determined by the equation

$$[S_0^Z]/[S_0^E] = (\phi_{Z \to E}/\phi_{E \to Z})(\epsilon_E/\epsilon_Z) \qquad (1)$$

The possibility of a thermal isomerization depends on the energy levels of the ground states of the isomers as well as on the height of the energy barriers of the transitions between them. Three cases can be singled out (Figure 1, curves 1-3, respectively):

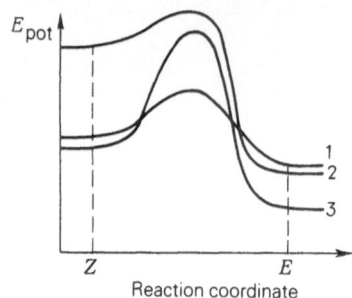

Fig. 1. Potential curves for thermal E/Z-isomerization in the ground state (see explanation in the text): 1) $Z \underset{\Delta}{\overset{\Delta}{\rightleftarrows}} E$; 2) $Z \overset{\Delta}{\longrightarrow} E$; 3) $Z \overset{}{-\times\rightarrow} E$.

1. The energy levels of the ground states of both isomers differ insignificantly, and the activation barriers between them are relatively small. Thermal isomerization is then possible in both directions, and for this reason equilibrium is established thermally between both isomers. This behavior is exhibited by 1-phenyl-3,3-bis(trimethylsilyl)triazenes [4].

2. The energy of the Z-isomer is considerably greater than that of the E-isomer, for which reason the thermal equilibrium is almost completely shifted towards the E-isomer. When the activation barrier between the two isomers is not too high, thermal Z → E-conversion can be observed, while no thermal E → Z-isomerization occurs. An example of this behavior is provided by azobenzenes [5].

3. The activation barrier between the two isomers is so high as to rule out thermal isomerization. This is observed in the case of various azoalkanes or stilbenes [6].

The mutual conversion of E/Z-isomers may proceed basically by the following three pathways:

1. By a rotational mechanism, as a result of rotation about the X=Y bond:

In this process the π-bond is broken homolytically or heterolytically, as a result of which parts of the molecule may freely rotate about the X–Y axis. The transition state (energy maximum) lies at an angle of rotation of about 90° and is therefore nonplanar.

2. By an inversion mechanism, as an inversion of substituents at the X=Y bond:

Isomerization by the inversion mechanism is only possible given the presence, apart from a substituent, of an unshared electron pair or an unpaired electron in one of the atoms forming the double bond. Through overhybridization the unshared electron pair transfers from the sp²- to the p-orbital and the substituent may vibrate in the plane of the molecule from

106

one position to the other. The elongated configuration along the X=Y–a
bonds corresponds to the transition state of the isomerization.

3. By a dissociation–recombination mechanism – the double or ligand
bonds are broken and the subsequent recombination gives rise to another
isomer:

Isomerizations by tautomeric transfer, as is sometimes the case in
hydrazones, or through the addition of a catalyst to the double bond may be
classified with the rotational ones since isomerization proceeds at the
moment of the double bond's transition into an ordinary bond.

The dissociation–recombination mechanism is discussed only in the context of dimers of nitroso compounds [7], O-alkyl ethers of oximes [8], as
well as for aliphatic-aromatic azo compounds [9]. In systems with the C=C
bond (e.g., stilbene) the only possible isomerization path is believed to be
rotation; for systems with C=N and N=N bonds it is necessary to choose between the theoretically possible mechanisms of rotation or inversion. For
Z → E-thermal isomerization the choice is made on the basis of the results
of research into various influences on isomerization kinetics. The accelerating steric influence of the groups near the inverting atom and the slight
influence of solvent polarity on the isomerization rate [5,10] serve as
criteria for the inversion mechanism. Protic solvents, unlike aprotic ones,
reduce the inversion rate because of the formation of hydrogen bonds by the
unshared electron pair of the inverting atom.

The criteria for the polar rotational mechanism include the rapid increase in the isomerization rate in polar solvents as well as the inhibition
of the reaction by sterically large groups next to the double bond around
which the rotation takes place. Besides, for a polar rotational mechanism
one would expect a strong influence by the electronic nature of substituents. A nonpolar rotational mechanism with a rupture of the π-bond into a
biradical requires very high levels of activation energy for thermal reactions, and more often than not is ruled out for energetic reasons.

Most of the systems having N=N and C=N double bonds isomerize thermally
by the inversion mechanism. Detailed quantum-chemical calculations of
diazenes and the most elementary azoalkanes and azomethines also confirm
that for energetic reasons inversion is more favorable than rotation [11,
12]. Only in the case of some of the diazo compounds and azomethines
strongly polarized by azobenzene substituents is the polar rotational mechanism discussed [5].

There is no unequivocal proof in the literature of the mechanisms of
photochemical isomerizations of compounds having N=N or C=N bonds. Dissociation–recombination, as a rule, is ignored, and it is not possible to
draw a line of distinction between the mechanisms of inversion and rotation
without a special investigation.

The mechanism of rotation or inversion during photoisomerization about
the N=N or C=N bond was discussed for azomethine dyes of the benzoylacet-

anilide or pyrazolone type [13]: for compounds with electron-donating substituents — the rotation mechanism, and with electron-accepting substituents
— the inversion path. The quantum-chemical calculations for diimide and
azomethane favor inversion in the state $^3(\pi, \pi*)$ and rotation in the states
$^3(n, \pi*)$ and $^1(n, \pi*)$.

Direct and Triplet-Sensitized Photoisomerization

Direct and triplet-sensitized E/Z-photoisomerization may proceed via
three different paths [14]: adiabatic isomerization occurs on the potential
surface of the excited state responsible for isomerization (Figure 2); nonadiabatic isomerization occurs upon the deactivation of the excited state
into the ground state. The path from one isomer to the other lies through
the potential surface of both the excited and the ground states (Figure 3).
E/Z-Isomerization may proceed through the "hot" ground state populated
following light-induced excitation and deactivation. In this process the
light only helps to supply the energy necessary for overcoming the thermal
threshold of activation in the ground state (Figure 4).

With the exception of certain stilbenes [14], where the adiabatic path
of the Z → E-conversion is possible, as a rule, E/Z-isomerizations proceed
nonadiabatically. The potential surface of the excited state responsible
for the isomerization has a minimum between the energy levels of both
isomers which corresponds to the relatively energy-poor excited intermediate
state X*.

For systems with N=N and C=N bonds it is impossible in most cases to
tell unequivocally whether this intermediate state belongs to the inversion
path (linear arrangement) or to the rotation path (twisted state). Deacti-

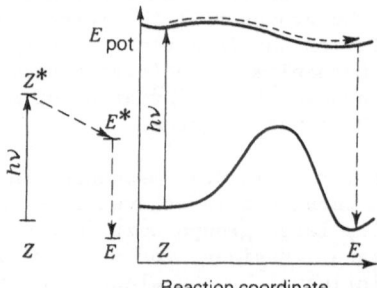

Fig. 2. Yablonskii's diagram and potential curves for adiabatic E/Z-photo-
isomerization.

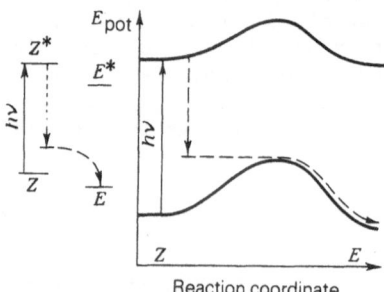

Fig. 3. Yablonskii's diagram and potential curves for nonadiabatic E/Z-
photoisomerization.

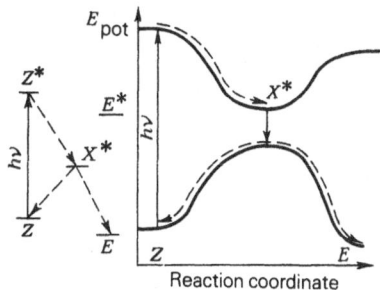

Fig. 4. Yablonskii's diagram and potential curves of E/Z-isomerization
through the hot ground state.

vation takes place from the intermediate state X* onto the potential surface
of the ground state. This surface has a maximum in the minimum of the
energy of the excited state, and further deactivation may proceed both
towards the E- and the Z-isomer.

Figure 5 shows the possible pathways of energy transfers between various states capable of participating in E/Z-isomerizations, while Figure 6
shows schematically the corresponding potential curves. The chain of investigations consists in ascertaining the determining stages of deactivation in
the course of photoisomerization.

If the direct radiationless deactivation of the S*- and T-states into
the ground state of a corresponding isomer is small compared with the transitions to the excited intermediate states X*, then $\phi_{Z \to E} + \phi_{E \to Z}$ can reach
unity. In sensitized photoisomerization the sum of the quantum yields upon
quantitative energy transfer from the sensitizer to the substrate corresponds maximally to ϕ_{isc} of the sensitizer. Often the transition of the excited state of the E-isomer into X* or the transition of X* to the S_0-state
of the Z-isomer has a low threshold of thermal activation that determines
the temperature dependence of $\phi_{E \to Z}$.

Although the general ideas set out above (see Figures 5 and 6) on the
course of E/Z-isomerization are now beyond doubt, the nature of the excited
state responsible for nonsensitized isomerization often remains unclear. It
has been shown that for stilbene isomerization may proceed through S- or
through T-states, depending on the nature of the substituents in the nuclei
and the conditions. The reaction can be channelled via the triplet path
using the internal or external heavy-atom effect. In the case of such
substituted forms of stilbene it has been possible, with the help of pulsed
spectroscopy, to detect triplet states directly [15]. Because of the absence of fluorescence and unsuccessful attempts to detect intermediate

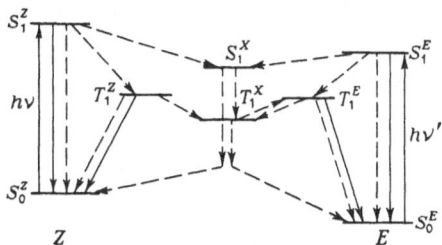

Fig. 5. Yablonskii's diagram for photochemical E/Z-isomerizations
(radiationless transitions are marked ---→).

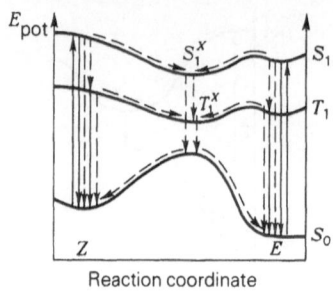

Fig. 6.   The course of potential curves in E/Z-photoisomerization.

states by the fast spectroscopy technique, it is impossible to resolve the
question of the multiplicity of these states in systems with N=N and C=N
bonds.   Often the triplet mechanism is accepted as the most likely one.

In triplet sensitization the reaction proceeds through the T_1^X state
(see Figures 5 and 6).   A comparison of the ratios of deactivation into the
E- and Z-forms by triplet-sensitized and direct photoisomerization was used
by many authors as a criterion of the unsensitized reaction path.   Varying
deactivation ratios suggest a singlet mechanism of direct photoisomerization; equal ratios — the deactivation in both cases of one and the same
intermediate state.   The latter could correspond to the triplet path of
direct photoisomerization.   However, the coincidence of the ratios of deactivation between direct and photosensitized isomerization per se is not a
proof of the triplet mechanism of the nonsensitized reaction.   Deactivation
ratios depend on the arrangement with respect to each other of the minima on
the potential surfaces of the excited states responsible for isomerization,
and the maximum of the ground state.   If their positions on the reaction
coordinate coincide, then upon deactivation the maximum of the potential
surface of the ground state is reached precisely and the subsequent deactivation into the E- or Z-form should have approximately the same probability.   With the positions of the minima of the potential surfaces of the S_1-
and T-states being the same, the same region on the surface of the ground
state becomes populated.

This leads to the equal probability of the deactivation paths in both
instances.   Only if the positions of the minima of the potential surfaces
differ substantially, something that is difficult to establish with any
certainty, can differences arise in deactivation ratios.

The deactivation ratio $k_{X*\to E}/k_{X*\to Z}$ for states of X* can be determined
experimentally.   In direct photoisomerization it corresponds to the ratio
$\phi_{Z\to E}/\phi_{E\to Z}$ and can be found from the photostationary state through equation
(1) if $\phi_{Z\to E} + \phi_{E\to Z}$ reaches unity.   Usually, along with deactivation through
X* other deactivation processes have a role to play which can lead to differences between the ratio of quantum yields and the ratio of deactivation
of the X* state.   In sensitized photoisomerization, sensitizers with a
sufficiently high triplet energy which transfer energy to both isomers with
equal efficiency should be used to determine the deactivation ratios of the
T_1^X state.   The ratio $k_{X*\to E}/k_{X*\to Z}$ then corresponds to the ratio of isomer
concentrations in the photostationary state.

The ratio ϕ_E/ϕ_Z for sensitized photoisomerization also coincides with
the deactivation ratio.   However, the deactivation ratios arrived at in this
way hold true only if the triplet states of both isomers almost quantita-
tively convert themselves into the state T_1^X ($\phi_{Z\to E} + \phi_{E\to Z} = \phi_{isc\ sens.}$) upon

the quantitative transfer of triplet energy, and other deactivation processes have no role to play.

To investigate the position of the photostationary state depending on $E_{T\,sens.}$, the E_T of the E- and Z-isomers are determined. If $E_{T\,sens.}$ lies above the E_T of both isomers, the energy transfer to both isomers proceeds with equal probability and the position of the photostationary state will be determined by the deactivation ratio only. Therefore, the position of the photostationary state in this region is independent of the E_T of the sensitizer. If the $E_{T\,sens.}$ is less than the E_T of the Z-isomer but greater than the E_T of the E-isomer, the transfer will predominantly proceed to the E-form and the photostationary state will shift towards the Z-form. The point in Figure 7 at which the curve turns upwards corresponds approximately to the E_T of the Z-isomer. Finally, if the $E_{T\,sens.}$ is lower than the E_T of the E-isomer, energy transfer to both isomers is only possible to a severely limited extent. The shift of the photostationary state to the direction of the Z-form in this region stops. The maximum of the curve in Figure 7 corresponds approximately to the E_T of the E-isomer.

The shift towards the E-form occurring during the subsequent diminution of $E_{T\,sens.}$ (see Figure 7) indicates that the rate constant of the energy transfer to the E-isomer accompanied by a decrease in sensitizer energy drops faster than the rate constant of the energy transfer to the Z-isomer. It is, indeed, surprising that sensitizers with such a low E_T should be effective at all. For an explanation of this let us examine the cross section of the potential surface (see Figure 6). As the twist of the ground state increases, the differences between the energies of the transitions $S_0 \rightarrow S_1$ diminish. The slow energy transfer may be attributed to the population of the high vibrational levels of the ground state in which rotation about the double bond [17] can be minimally possible.

Photocatalytic Isomerizations

Photocatalytic isomerization often occurs in the presence of bromine, iodine, metal-carbonyl complexes, and univalent copper salts. Isomerization initiated by iodine or bromine has been detected in the case of stilbene [18], azo compounds [19], and diazo compounds. In these processes there is only the conversion of the Z-isomer into the E-isomer, and no back reaction occurs. Iodine irradiation with light of 546 nm wavelength causes iodine dissociation into atoms, with the latter initiating isomerization apparently by addition to the double bond [18,20]. In the case of azo compounds isomerization also occurs upon the excitation of the Z-isomer and iodine complex, with the iodine absorbing in a shorter-wavelength region [19]. In a radical — the product of the addition of a halogen atom — the double bond disappears so that the molecule should, in theory, revolve freely about a new bond. Consequently, thermal equilibrium is established between the E- and Z-isomers, which in certain cases is almost entirely shifted towards the

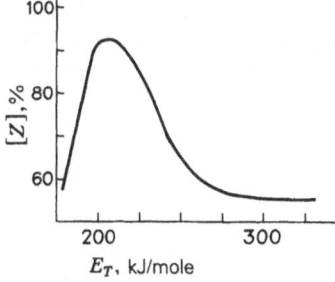

Fig. 7. The dependence of Z-isomer content in the photostationary mixture on the energy of triplet sensitizer in the case of stilbene [38].

E-isomer. The cleavage of the adduct by a halogen gives the E-isomer and a catalyst atom which can initiate a new addition so that a process of the radical-chain type materializes.

Photoisomerization catalyzed by metal-carbonyl compounds also proceeds in the same fashion. Irradiation is conducted in the region of the absorption of the metal-carbonyl complex and unsaturated compound. The intermediate state which is responsible for the isomerization contains a σ-bond with the metal. Consequently, the double bond here is also broken, and the molecule can revolve freely around the new bond [21].

Recently, it has become possible to detect E/Z-isomerization photocatalyzed by electron transfer. Upon the excitation of the charge-transfer complex from the E-stilbene and the fumaronitrile, both unsaturated compounds isomerize. It is accepted that the isomerization proceeds through the triplet state of both the components which form upon the deactivation of the geminal radical-ion pair arising upon light excitation. Experiments conducted by the chemically induced dynamic nuclear polarization method confirm this reaction mechanism [22]. Z/E-Isomerization has been detected upon the electrochemical reduction of fumaric or maleic acid derivatives into the corresponding anion-radicals [23]. As the last two examples show, Z/E-isomerizations may be initiated by oxidation—reduction or electron transfer reactions. Although triplet sensitizers can, in principle, also participate in the electron transfer reaction, this aspect of sensitized Z/E-isomerizations has not yet been studied.

PHOTOCHROMIC AZO COMPOUNDS

Electronic Absorption Spectra

Although the spectra of azo compounds have been studied comprehensively (see surveys [11,24,26-28]), most of the data in the literature deal with the E-isomer spectra, while often little or nothing is available on the spectra of unstable Z-isomers.

In the spectra of XRN=NR'Y azo compounds $n \rightarrow \pi*$- and $\pi \rightarrow \pi*$-transitions are observed. Their position and intensity depend on the nature of the residues R and R' and the substituents X and Y, on the steric configuration (E or Z) due to the azo group, and on the nature of the solvent.

As the linear combinations of the wave functions of the MOs (molecular orbitals) of the unshared nitrogen atom pairs of the azo group show, n_+ and n_- orbitals arise upon their overlapping which corresponds to the transitions $n_+ \rightarrow \pi*$ and $n_- \rightarrow \pi*$ (Figure 8) [24,27]. The theoretical study of the cleavage of the n_+ and n_- levels [29-36] as well as experimental investigation by photoelectronic spectroscopy [37] gave a scheme of energy levels for azomethane (Figure 9) and a relative series of transition energies: $\Delta E(n_+ \rightarrow \pi*) < \Delta E(n_- \rightarrow \pi*) < \Delta E(\pi \rightarrow \pi*)$. The cleavage of the n_+ and n_- levels is between 1.6 and 3.5 eV in the case of other compounds. The constancy of the n- and π-level cleavage, which amounts to about 2.5 eV, is noteworthy.

In simple E-azo compounds the $n \rightarrow \pi*$ ($n_+ \rightarrow \pi*$) transition is symmetry-forbidden, and for this reason is of low intensity; for the Z-isomer the $n \rightarrow \pi*$ ($n_- \rightarrow \pi*$) transition is permitted, and it is of a high intensity [27, 38-41] (Table 1). Because of the mixing of the $n \rightarrow \pi*$- and $\pi \rightarrow \pi*$-transitions, the intensity of the $n \rightarrow \pi*$-transitions of the E-isomer of azoaromatic compounds is greater than expected on account of the forbidden overlap [27,38].

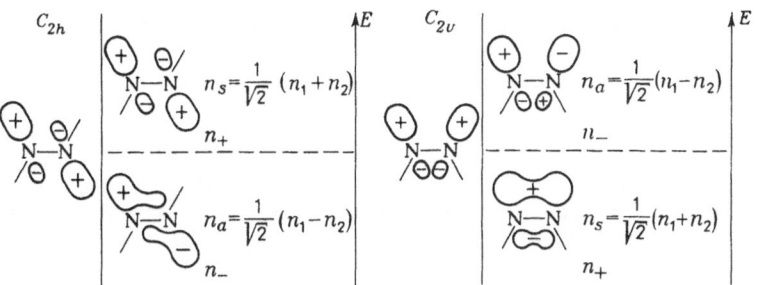

Fig. 8. The cleavage of n-orbitals of the E- and Z-isomers of azo compounds.

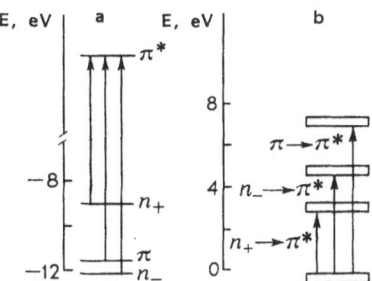

Fig. 9. Levels of MO energy (a) and a plot of the levels of terms (b) of
E-azomethane [47].

A characteristic yellow coloring of aliphatic and cycloaliphatic azo
compounds is caused by the long-wavelength n → π*-transition in the 320-460
nm region. The n → π*-transition band of aromatic monoazo compounds lies in
the 430-460 nm region, and its position is not dependent on the nature of
the substituents in the aromatic residue nor on the nature of the solvent
[38,44]. With an increase in steric hindrances in the aryl residue and its
rotation out of the molecule's plane, the intensity of the n → π*-transition
in E-azobenzenes increases [5,45]. In nonpolar solvents in the spectra of
aliphatic and cycloaliphatic azo compounds and monosubstituted and severely
sterically hindered polysubstituted azobenzenes, the n → π*-transition is
distinctly separated from the π → π*-transition [38,45]. In azobenzenes
with polarizing substituents and in most azo dyes, the band of the n →
π*-transition is overlapped by the bathochromically shifted π → π*-band
[44].

In azo compounds with a direct bond between the nitrogen and the phos-
phorus or silicon of the organophosphorus or organosilicon residue,

Table 1. The position and intensity of band maxima of n → π*-transi-
tions of solutions of E-azo compounds $R-N=N-R^1$ in dichloromethane

R,R'	λ, nm (log ε)	R,R'	λ, nm (log ε)
$(CH_3)_3C$, $P(CH_3)_2$	425(1.83) [64]		
C_6H_5CO, COC_6H_5	469(1.62)*[73]	$(CH_3)_3C$, $Si(OCH_3)_3$	507(0.85) [64]
$(CH_3)_3C$, $Ge(CH_3)_3$	480(1.08) [64]	$(C_6H_5)_2PO$, $PO(C_6H_5)_2$	568(1.36)* [73]
		C_6H_5, $Si(CH_3)_3$	575(1.41) [64]
$(CH_3)_3C$, $Si(CH_3)_3$	500(0.95) [64]	$(CH_3)_3Si$, $Si(CH_3)_3$	784(0.70)**[64]

*In dioxane; **in hexane.

the n → π*-transition band shifts nearly to 800 nm [29,43] (see Table 1).
There is a linear dependence between the energy of the n → π*-transition and
the potential of azo compound ionization [11,43], with the exception of compounds with para-substituents having atoms with unshared electron pairs
[43].

The Z-isomers of aliphatic azo compounds generally absorb in a longer-
wavelength region than the corresponding E-isomers. Differences in wavelengths Δλ of absorption bands were measured in the case of pairs of stable
E- and Z-isomers. They ranged from −22 nm (azo-2-propane) to −87 nm (azo-
1-adamantane) (Table 2). Quantum-chemical calculations of cyclic and noncyclic Z-azoalkanes revealed a decrease in the energy of the n → π*-tran-
sition with increasing N=N−C angle [56], which is in good agreement with the
experimental data [27]. For azobenzenes, azopyridines, and phenylazopyr-
idines, as distinct from aliphatic azo compounds, the n → π*-transitions of
the E-isomers lie in a longer-wavelength region than those of the Z-isomers;
Δλ is between 10 and 20 nm (see Table 2).

The band of the n → π*-transitions of solutions and gaseous aliphatic
and aromatic E- and Z-isomers of azo compounds lacks a fine structure [51,
74-76]. Only in the spectra of stressed cyclic Z-azo compounds is a vibrational structure of the bands of n → π*-transitions often observed even at
room temperature [51,65,77,78] (Figure 10). This azo compound band, as
expected, displays a slight blue shift on going from nonpolar to polar
solvents [65,78]. Upon the protonation of aromatic azo compounds this band
disappears completely [51,70,80] but remains upon the protonation and alkylation of cyclic azoalkanes [81].

The bands of π → π*-transitions in aliphatic E- and Z-azo compounds lie
in a wavelength region shorter by 400 nm and for this reason make no substantial contribution to their visible coloration.

In aromatic monoazo compounds π → π*-transitions are bathochromically
shifted as compared with aliphatic ones; they lie near E-isomers of over
320 nm (Tables 2 and 3). The Z-isomers display a hypsochromic shift in the
long-wavelength n → π*-transition with a simultaneous fall in intensity; Δλ
ranges from 30 to 75 nm (Table 2 and Figure 11). The largest values of Δλ
(75 nm) were found for mesoionic phenylazotriazolopyridazines; for the E-
and Z-isomers of azobenzenes and azopyridines the values of Δλ are about
45 nm [83-85]. A substantial spectral shift is observed in fluorinated
4-aminoazobenzene and its N-methyl derivative upon their irradiation with
polychromatic light [86].

E-Azobenzene and its para-substituted forms in crystals [87,88] and
in solution are nearly planar [89]; in crystals of sterically hindered

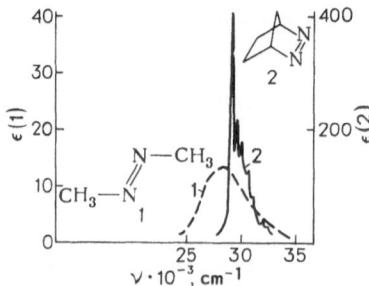

Fig. 10.   The bands of n → π*-transitions in the spectra of E-azomethane in
           isooctane (1) and 2,3-diazabicyclo[2.2.1]hept-2-ene in hexane (2)
           [51].

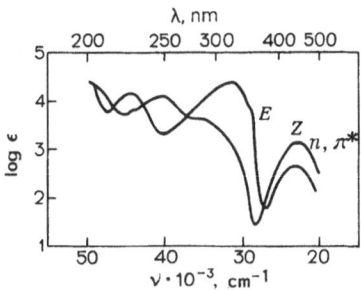

Fig. 11. Electronic spectra of E- and Z-azobenzene in hexane [312].

E-2,2'-dichloroazobenzene, aryls are rotated at an angle of 13° [90]. In gaseous E-azobenzene the phenyl group is also rotated through about 30° relative to the N—C bonds [91], which is in good agreement with the calculated data [92,93]. Because of the steric hindrance caused by ortho-hydrogen atoms, the phenyl groups in Z-isomer crystals are parallel to one another and rotated at an angle of 53° out of the plane of the azo group [94, 95]. This results in weaker conjugation between the azo chromophore and the aryl residues with the consequent hypo- and hypsochromism observed in the $\pi \rightarrow \pi*$-band in the Z-isomer. The spectra of azobenzenes and azoaryls were analyzed thoroughly by means of quantum-chemical calculations [47,71,96,99].

E/Z-Isomerization in azobenzenes has but a slight influence on the length of the C—N (E 1.45 Å; Z 1.41 Å) and N=N bonds (1.24 Å) [87,88,91,94, 95]. Z-Azobenzene has 50.4 kJ/mole more energy than the E-isomer, as established by quantum-chemical calculations that took into account all the valence electrons [51,100]; Z-azobenzene has also a higher alkalinity than the E-isomer [101]; for their R_f values see [102].

The long-wavelength $\pi \rightarrow \pi*$-transition of E-azobenzene has a noticeable fine structure in nonpolar solvents which is more pronounced in solid matrices at low temperatures [51,74]; the long-wavelength $\pi \rightarrow \pi*$-transition of the Z-isomer also remains structureless at 77°K as well [77]. Rigid cyclic Z-azo compounds — o,o-azodiphenylmethane, benzo[c]cinnoline, and other o-diazines — on the other hand, have structured bands of the $\pi \rightarrow \pi*$- transitions [51,74]. The fine structure of the $\pi \rightarrow \pi*$-band is peculiar to the Z- and E-isomers of 1,4-dihydronaphtho[1,8-de][1,2]diazepine [68].

The position of the long-wavelength $\pi \rightarrow \pi*$-transition of azoaromatic compounds is heavily dependent on the nature of the substituents and the type of aromatic residue (see Tables 2 and 3). In monosubstituted forms of E- and Z-azobenzenes both donor and acceptor substituents induce a bathochromic shift [67,103-106] which is especially well pronounced in the E-series in compounds with substituents having a low ionization potential: NR_2, SR, O⁻, NR⁻ [47,106,107]. An especially strong bathchromic shift of the $\pi \rightarrow \pi*$-transition is caused by polarizing substituent pairs in positions 4,4' of azobenzene [108-113]; this transition makes a considerable contribution [114]. Substituent effects are well pronounced in phenylazonaphthalenes [115-117] and in 5-phenylazotropolones [118]. The strong hypsochromic shift is noteworthy in 4,4'-substituted forms of azobenzenes with polarizing substituents when the dialkylamino group forms an integral part of the crown ether complexed by metal ions [119]. In azobenzenes with an electron donor at position 4 and in the presence of other polarizing substituents, the bathochromic effect grows in the series 3' < 2' < 4' or 3',5' < 2',6' < 2',4' [116] (see Table 3). In compounds with a single acceptor at position 4 and in the presence of several donors, the bathochromic effect increases in the series 2',4' < 3',4' < 2',5' [109,110,120]. This impact of substitu-

Table 2. Band maxima in the electronic spectra of the E- and Z-isomers of azo compounds and quantum yields of their photoisomerization

Compound	λ, nm (log ε)				Solvent; t,°C	$\phi_{Z\rightarrow E}$ (λ, nm)	$\phi_{E\rightarrow Z}$ (λ, nm)
	π → π*		n → π*				
	Z	E	Z	E			
XC$_6$H$_4$N=NC$_6$H$_4$Y (reduced X:Y): H; H [5,83]	270(3.71)	315(4.35)	437(3.11)	448(2.65)	heptane methylcyclohexane + isohexane; 23	0.40(313), 0.40(436)	0.09(313), 0.25(436)
H; H [267]					hexane; 4	0.44(313), 0.68(436)	0.10(313), 0.20(436)
H; 4-NO$_2$ [5,213]		328		440	methylcyclohexane + isohexane; -30	0.38(313), 0.36(436)	0.07(313), 0.14(436)
4-MeO; 4-NO$_2$ [5]	~335	~365	~455		"	0.50(365), 0.55(436)	0.10(365), 0.17(436)
4-(CH$_3$)$_2$N,3-CH$_3$;4-NO$_2$ [5]		420	not established		propanol + iso-propyl alcohol	0.62(313), 0.71(436), 0.41(313), 0.48(436)	0.02(313), 0.027(436)*, 0.008(313), 0.0035(436)**
4-(CH$_3$)$_2$N;4-NO$_2$ [5]		470	"		"	0.70(313), 0.84(436), 0.55(313), 0.84(436)	0.16(313), 0.20(436)*, 0.073(313), 0.06(436)**
2,4,6-Tri-CH$_3$;H [5]	~245	~325	~455	~455	methylcyclohexane + isohexane; 23	0.35(313), 0.38(510)	0.10(313), 0.16(510)

116

Compound	~260	~315	~450	~490	Solvent		
2,2',4,4',6,6'-Hexa-CH$_3$ [5]	~260	~315	~450	~490	methylcyclohexane + isohexane; 23	0.33(313), 0.44(510)	0.09(313), 0.20(510)
2,2',4,4'-Tetra-CH(CH$_3$)$_2$ [29]		343(4.29)		462(2.86)	methylcyclohexane + isopentane (3:1)		0.112(365)
H;4-Me [105,268]	296(3.74)	333(4.37)	466(3.22)	451(2.83)	ethanol	0.51(313), 0.57(436)	0.36(313)
4,4'Di-CH$_3$ [105,268]	296(3.80)	335(4.35)	439(3.31)	439(2.87)	ethanol[3]*	0.84(313), 0.38(436)	0.23(313), 0.26(436)
H;4-F [105,268]	287(3.68)	323(4.31)	433(3.11)	441(2.71)	"	0.39(313), 0.55(436)	0.34(313), 0.38(436)
H;4-Cl [105,268]	289(3.76)	328(4.37)	434(3.20)	443(2.78)	"	0.37(313), 0.62(436)	0.33(313), 0.28(436)
H;4-Br [105,268]	289(3.79)	329(4.43)	437(3.24)	445(2.80)	"	0.51(313), 0.66(436)	0.24(313), 0.26(436)
H;4-I [105,268]	292(3.81)	331(4.45)	435(3.23)	445(2.85)	"	0.41(313), 0.65(436)	0.32(313), 0.29(436)
H;4-C$_2$H$_5$O [105,268]	317(3.88)	348(4.42)	440(3.39)	434	"	0.51(313), 0.56(436)	0.52(313), 0.31(436)
H;4-COOMe [105,268]		330(4.47)	447(2.97)	466(2.81)	"	0.27(313), 0.58(436)	0.54(313)
2-MeO;H [166]	wide band	310	430	360	methylcyclohexane; -100	0.7(313), 0.36(436)	0.09(313), 0.06(436)
2-OH;H [166]	360,300	320	450	390	"	0.45(313), 0.3(436)	0.05(313), 0.09(436)

Table 2. (Continued)

Compound	λ, nm (log ε)				Solvent; t,°C	$\phi_{Z\rightarrow E}$ (λ, nm)	$\phi_{E\rightarrow Z}$ (λ, nm)
	$\pi \rightarrow \pi^*$		$n \rightarrow \pi^*$				
	Z	E	Z	E			
2,2'-OCH₃ [139]	280	310,370	380	460	methylcyclohexane + methylcyclopentane	0.17(313), 0.35(365)	0.08(313), 0.115(365)⁴*
4-Methoxynaphthalene-1-azobenzene[247]	329	386	453	402	methylcyclohexane; -18		
3-Oxynaphthalene-2-azobenzene[202]⁵*	325	355,282 370,295	455	433	methylcyclohexane + isohexane	0.07(313), 0.08(436)	$\phi_{E\rightarrow Z}/\phi_{Z\rightarrow E}$
3-Methoxynaphthalene-2-azobenzene[202]⁶*		350,270, 280	430	450	"	0.6(313), 0.26(436)	$\phi_{E\rightarrow Z}/\phi_{Z\rightarrow E}$ at -50°C
(reduced R):							
CH₃[84]⁷*	330	393(4.33)	467	488(3.30)	methanol		0.13(405)
Cl[84]⁷*	330	402(4.33)	472	483(3.34)	"		0.09(405)
COOEt[84]⁷*	328	403(4.37)	481	510(3.36)	"		0.07(405)
α-Pyridine-N=N-C₆H₄X-p (reduced X):							
H [83]⁸*	269(3.77)	310(4.27)	437(2.96)	456(2.48)	heptane		
F [269]		318(4.19)		446(2.53)	"		0.15(333)
Cl [269]		324(4.30)		457(2.87)	"		0.11(333)

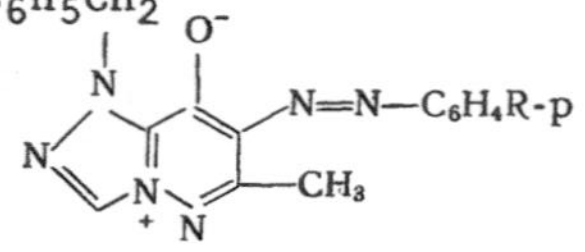

(reduced R):

$C_6H_5CH_2$ — N, O⁻, N=N—C₆H₄R-p, CH₃, N, N⁺, N (pyrazolo ring structure)

118

Compound	λ (log ε)	λ (log ε)	λ (log ε)	λ (log ε)	Solvent	Quantum yield
Br [269]		325(4.30)		454(2.80)	"	0.08(333)
I [269]		330(4.36)		456(2.72)	"	0.06(333)
NO_2 [269]		323(4.32)		478(2.60)	"	0.02(333)
CH_3O [269]		343(4.31)		448(2.83)	"	0.09(333), 0.36(365)[9]*
$(CH_3)_2N$ [269]		275(3.95), 401(4.43)			cyclohexane	0.28(297), 0.26(436)[10]*
β-Pyridyl-N≡N-C_6H_4X-p (reduced X):						
H [83][8]*	271(3.75)	317(4.34)	440(3.04)	449(2.58)	heptane	0.16(333)
H [269]		318(4.30)		450(2.59)	"	0.15(333)
Cl [269]		326(4.33)		452(2.75)	"	0.15(333), 0.29(365)[9]*
CH_3O [269]		349(4.40)		442(2.91)	"	0.40(405)[12]*, 0.30(436)[13]*
$(CH_3)_2N$ [270]		410(4.44), 410(4.50)		410	cyclohexane	0.02[11]*
Pyridine-4-azobenzene [83]*	inflection (band)	309(4.30)	433(2.99)	453(2.90)	heptane	
2-Azopyridine [83][8]*	266(3.88)	306(4.18)	448(3.03)	466(2.36)	"	
3-Azopyridine [83][8]*	280(3.79)	317(4.30)	441(2.97)	450(2.53)	"	
4-Azopyridine [83][8]*		286(4.25)	431(2.79)	460(2.36)	"	
Benzeneazomethane	395				isooctane	0.51(365), 0.60(436)
4-Hydroxybenzeneazomethane [56][15]*	395					0.42(365)[14]*, 0.53(436)

Table 2.   (Continued)

Compound	λ, nm (log ε) n → π* Z	λ, nm (log ε) n → π* E	Solvent	Data on the mutual conversion of Z- and E-isomers
Azomethane [27]	368(2.38)	352(1.40)	hexane benzene	$\phi_{Z\to E}$ 0.45 (n→π*), $\phi_{E\to Z}$ 0.42 (n→π*), $\phi^E_{N_2}$ 0.089, $\phi^Z_{N_2}$ 0.008
Azoethane [27] [16*]				$\phi_{Z\to E}$ 0.3 (n→π*), $\phi_{E\to Z}$ 0.3 (n→π*)
Azopropane [27,35a,82]	380(1.85)	355(0.97)	gas phase	ϕ heavily dependent on pressure; T_τ 162°C [17*]
	380(2.15)	358(1.17)	isooctane	$\phi_{Z\to E}$ 0.50 (n→π*), $\phi_{E\to Z}$ (0.52) (n→π*), $\phi_{Z\to E}$ 0.60 (365 nm), 0.59 (405 nm), $\phi_{E\to Z}$ 0.48 (365 nm) 0.44 (404 nm), $\phi^E_{N_2}$ 0.025 (365 nm), 0.016 (405 nm), $\phi^Z_{N_2}$ 0.025 (365 nm), 0.012 (405 nm),
			water	$\phi_{E\to Z}$ 0.51 (n→π*), $\phi_{E\to Z}$ 0.53 (n→π*)
Azo-tert-butane [271,272]	447	368	hydrocarbon	T_τ −11°C, ϕ^E_{dec} 0.46
Azonorbornene [271,272]	423(1.94)	364	"	T_τ 86-102°C, ϕ^E_{dec} 0.0008
Azoadamantane [271,272]	455(1.97)	368	"	T_τ 30.6°C, ϕ^E_{dec} <0.004
Naphtho[1,8-cd]-1,2-bis-azacyclo-1-heptene [68] [18*]	403(2.79) 382(3.93) (π→π*)	403(2.33) 366(3.78) (π→π*)	ethyl ether	$Z \overset{\Delta}{\to} E$: $\Delta H^{\ddagger}$ 92.1 kJ/mole, $\Delta S^{\ddagger}$ 2.5 J/(mole K)

Compound			Solvent	Remarks
(2-Phenylpropane-2)-azo-2-norbornene [273]	416	366(1.46)	isooctane	Z obtained by photolysis of E at −125°C
2-Azo-2-methyl-3-butyne [274]	442	366	polymethyl methacrylate	T_τ −110°C
2-Azo-2-methyl-3-butyne [274]	420	350	"	T_τ −110°C
2-Methyl-3-butene-2-azo-tert-butane [274]	444	366	"	T_τ −110°C
Azobisisobutyronitrile [274]	390	341		T_τ −60°C
2-Azo-2-phenylpropane [274]	434	360		T_τ −100°C
2-Methyl-3-butene-2-azomethane [274]	380	363		T_τ −20°C
2-Methyl-3-butene-2-azo-2-propane	385	364		T_τ −20°C
1,2-Bisazacyclo-1-octene [275,276]	389(2.04)	372(1.54)	isooctane	
	373(2.03)	363(1.58)	methanol	
3,7-Dimethyl-1,2-bisazacyclo-1-octene [275]	387(2.11)	371(1.60)	isooctane	
Z-3,7-Diphenyl-1,2-bisaza-cyclo-1-octene [276]	381(2.04)	371(1.83)	chloroform	
E-3,7-Diphenyl-1,2-bisaza-cyclo-1-octene [276]	390(1.79)	363(1.62)	"	
Azoxybenzene [207,277][19*]	327,239		ethanol	$\phi_{Z\rightarrow E}$ 0.6 (330–370 nm), $\phi_{E\rightarrow Z}$ 0.1 (330–370 nm) ethanol: E_a (Z $\overset{\Delta}{\rightarrow}$ E) 103.6, log A 13.32 heptane: E_a (Z $\overset{\Delta}{\rightarrow}$ E) 80.1, log A 10.38
2-Azoxypropane [278][19*]	232(3.90)	220±3		

Table 2. (Continued)

Compound	λ, nm (log ε) $n \to \pi^*$		Solvent	Data on the mutual conversion of Z- and E-isomers
	Z	E		
Cyclohexanazoxymethane [278]^{19*}	232(2.78)	220±3		

* -110°C.

** -150°C.

3* Values of φ obtained during work with liquid filters.

4* -80°C.

5* The spectra determined at -75°C; E_a 55-58 kJ/mole.

6* The spectra determined at -125°C; E_a 96 kJ/mole.

7* Z→E; $\Delta H^{\ne}$ (kJ/mole): 75 (R = CH_3); 79.4 (R = Cl); 67 (R = COOEt); $\Delta S^{\ne}$ (J/mole·K): respectively -68; -62.2; -100.

8* Z→E; E_a (kJ/mole): -90.5 (pyridine-2-azobenzene); 93.0 (pyridine-3-azobenzene); 91.8 (pyridine-4-azobenzene); 88.0 (2-azopyridine); 94.7 (3-azopyridine); 89.7 (4-azopyridine); log A (A in sec^{-1}): respectively 10.1; 10.6; 11.1; 9.0; 10.9; 18.1; $\Delta G^{\ne}$ (kJ/mole): respectively 107; 106.3; 102.3; 109.8; 106.1; 106; $\Delta S^{\ne}$ (J/mole·K): respectively -60.8; -50.4; -41.5; -83.3; -44.8; -60.8.

9* In alcohol.

10* φ in tert-butyl alcohol. ϕ_{red} 0.003 in methanol, 0.065 in isopropyl alcohol.

11* Found for the photosteady-state.

12* In dioxane.

13* In triethylamine.

14* ϕ_{N_2} 0.

15* λ_{max} of the hydrazo form also at 395 nm; the extinction coefficients of the three forms are inseparable.

16* Determined in frozen matrices.

17* T_τ - the temperature at which the semiconversion time of the Z-isomer is 10 min.

18* The spectral data have been borrowed from the figure. The π→π* band in E- and Z-isomers has a fine structure.

19* The spectral data of the π→π*-transition are cited.

ents is in agreement with the results of quantum-chemical calculations performed on the basis of perturbation theory [47].

The spherical substituents CH_3 and Cl in positions 2' and 6' of azo-benzenes with 4,4'-polarizing substituents cause a distinct hypsochromic shift and a broadening of the $\pi \to \pi^*$-band on account of severe steric hin-drances caused by the ortho-hydrogen atoms of the neighboring phenyl nucleus [121,122]. Analogous relations are observed also for sterically hindered diazo compounds. The position of the long-wavelength $\pi \to \pi^*$-transition band in the spectra of polysubstituted forms of azobenzenes can be calculated with the help of substituent increments as well [123].

Most of the azoheterocyclic compounds can be included in the group of aromatic azo compounds with polarizing substituents. The lower the ion-ization potential of a particular heterocycle, the longer is the wavelength region in which such an azo compound absorbs. Data have been reported for azo compounds: nitrogen- [83,103,124-128], sulfur- [129,130], selenium- [129], sulfur- and oxygen- [125], sulfur- and nitrogen- [125, 130, 131], and selenium- and nitrogen-containing [95] heterocycles. The absorption band for such relatively simple systems extends from 286 nm for the $\pi \to \pi^*$-tran-sitions of E-4,4'-azopyridine [83] to 625 nm in the case of the 3-(p-di-methylaminophenylazo)benzo-1,2-dithiolium salt [130]. In bisazoaromatic compounds the bands of the $\pi \to \pi^*$-transitions lie in the same region as in the corresponding monoazo derivatives [132-135].

In azo compounds capable of azo–hydrazone tautomerism, especially in o-hydroxyazoaromatic compounds and potentially tautomeric azoheterocycles, tautomeric equilibria are superimposed on E/Z-photoisomerization [47,136-138].

With falling temperature, a bathochromic shift and a gain in the inten-sity of the $\pi \to \pi^*$-band in the spectrum are observed [23,139]. Temperature changes in the spectra of azoaryls with o,o'-substituents were attributed to equilibria between the various rotamers [139]. The long-wavelength band of the $\pi \to \pi^*$-transitions of azoaryls exhibits a positive solvatochromism [79, 140]. Because of the increasing dipole moment upon excitation, the batho-chromic shift is especially well pronounced in azoaryls polarized by sub-stituents and in other polar azoaryls [79,141-143].

Protonation in the azo group in a strongly acidic medium results in a large bathochromic shift of the $\pi \to \pi^*$-band [70,108,111-113,120,133,136, 144-147] (Table 4), especially in azobenzenes that contain donor substitu-ents. In Z-azobenzenes the acichromic effect is more strongly pronounced than in the E-isomers [101,148]. The asymmetry of the protonated azo group was established following numerous experimental and theoretical studies [80,148-152]. In amino-substituted forms of azobenzenes, protonation occurs primarily in the azo and, later, in the amino groups [70,136,148,153,154]. The protonation-induced bathochromic shift of the $\pi \to \pi^*$-transition is stronger, the closer the structure of the azobenzene to a polyene [e.g., in 4,4-bis-(dialkylamino)azobenzene]; azo compounds with a pronounced poly-methine structure (heavily polarized) display a negative acichromism [112, 113,147,155,156] (see Table 4) (cf. the negative acichromism of diazo com-pounds). The acichromism of azo compounds was analyzed within the framework of perturbation theory by means of a simple molecular orbital (MO) model [157]; see also [158].

Luminescence

Most of the E- and Z-forms of azo compounds display no luminescence [6,51]. The exceptions are sterically stressed E-azobenzenes as well as

Table 3. The influence of substituents on the position of maxima of long-wavelength bands of the $\pi-\pi^*$-transitions of aromatic and heteroaromatic E-azo compounds $XC_6H_4N=NH_6H_5$, $YC_6H_4N=NC_6H_4Z$ in different solvents.

X	λ, nm (log ϵ)	Y; Z	λ, nm (log ϵ)
H [103,104]	317(4.34), cyclohexane	4-NO₂; 4-N(CH₃)₂ [69]	476, ethanol
4-CH₃ [67]	323(4.31), hexane	4-NO₂; 2-OH, 5-CH₃ [109]	545, ethanol, potassium hydroxide solution
4-F [103,104]	319(4.36), cyclohexane	4-NO₂; 2-OH, 5-Cl [109]	418, acetic acid
4-Cl [67]	323(4.30), hexane	4-NO₂ [119];	477, methanol, acetonitrile
4-Br [103,104]	326(4.40), cyclohexane	4-NO₂ [119];	357, acetonitrile
4-I [103,104]	331(4.32), cyclohexane	2,4-di-CN; 4-N(C₂H₅)₂ [116]	514(4.60), ethanol

Substituent [ref]	λ (log ε), solvent
4-OCH₃ [103,104]	342(4.41), cyclohexane
4-NH₂ [67]	362(4.50), hexane
4-SCH₃ [103,104]	362(4.44), cyclohexane
3-NH₂ [67]	370;314(3.48, 4.16), hexane
2-NH₂ [67]	406(3.95), hexane
2,6-di-CN; 4-N(C₂H₅)₂ [116]	503(4.52), ethanol
4-CN; 4-N(C₂H₅)₂ [116]	466(4.51), ethanol
3,5-di-CN; 4-N(C₂H₅)₂ [116]	478(4.53), ethanol
3-CN; 4-N(C₂H₅)₂ [116]	466(4.45), ethanol
2-CN; 4-N(C₂H₅)₂ [116]	462(4.48), ethanol
4-N(CH₃)₂ [67]	398(4.38), hexane
4-NO₂, 2-Cl; 4-N(C₂H₄CN)(C₂H₅) [121]	475(4.60), methanol
4-NO₂ [38]	328(4.38), hexane
4-NO₂, 2,6-di-Cl, 4-N(C₂H₄CN)(C₂H₅) [121]	417(4.49), methanol
3-NO₂ [38]	317(4.22), hexane
4-N(C₂H₄CN)(C₂H₅) [121]	

p'-O₂NC₆H₄N=N– [structure] [122]   518(4.55), methanol

125

Table 4. Acichromism of azo compounds p-$[CH_3]_2NC_6H_4N=NC_6H_4X$-p' in ethanol + HCl[112,113], $(\Delta\lambda = \lambda_{max}(H^+) - \lambda_{max})$

X	$\lambda_{max}(H^+)$ (λ_{max}), nm	$\Delta\lambda$	X	$\lambda_{max}(H^+)$ (λ_{max}), nm	$\Delta\lambda$
$(CH_3)_2N$	664(450)	214	Cl	522(417)	105
CH_3S	555(420)	135	NO_2	508(475)	33
CH_3O	555(407)	148	$C(CN)=C(CN)_2$	586(580)*	-6
CH_3	533(407)	126			
H	518(407)	111		464(564)**	-100

* In benzene + hydrochloric acid. (In formula of the azo compound $N(C_2H_5)_2$ instead of $N(CH_3)_2$[155].
** For 4-nitrophenyl azo-5-dimethylamino-2-selenophene in methanol + hydrochloric acid[155].

alkylamino- and dialkylaminoazobenzenes, which exhibit faint luminescence in solid matrices at low temperatures [67,69]. A lessening of the energy differences of the $\pi \to \pi^*$- and $n \to \pi^*$-transitions in aminoazobenzenes and a rotation of aryl residues about sterically stressed azobenzenes are believed to make possible a mixing of the $^1(n, \pi^*)$- and $^1(\pi, \pi^*)$-states. As a result, symmetry-forbidden $n \leftarrow \pi^*$-fluorescence arises in the above-mentioned E-azobenzenes. Fluorescence appears in sterically hindered azobenzenes at 77°K and no E → Z-photoisomerization is observed.

Cyclic aliphatic azo compounds often display fluorescence, as well as o-diazines with a fixed Z-structure [27,65,77,159-163]. For these Z-compounds the mixing of states of the same nature is forbidden, for instance, $^1(n, \pi^*)$ and $^3(n, \pi^*)$, so that the deactivation of the n,π^*-singlet through the n,π^*-triplet becomes impossible, which leads to an increased probability of $n \leftarrow \pi^*$-fluorescence [51]. The value of ϕ_{fl} increases with increasing rigidity of the Z-azo compounds. In protic solvents, the lower values of ϕ_{fl} and τ_{fl} for aliphatic azo compounds than in, for instance, acetonitrile are often observed because of the formation of hydrogen bonds and because of the detachment of hydrogen in aliphatic hydrocarbons [65,164]. The formation of a diazenium salt [27,81] is believed to quench fluorescence in sulfuric acid [65].

Both o- and p-hydroxyazoaryls, capable of azo–hydrazone tautomerism, display good fluorescence which is attributed to the hydrazo form [165, 166]. Because of an increase in the rigidity of the molecule in boron complexes of o-hydroxyazoaryls, the fluorescence intensity goes up sharply; one suggestion has been to use such complexes as working media for lasers [167]. o-Hydroxyazobenzene is also used for the fluorometric identification of metal ions [168]. Protonated azo compounds fluoresce in a way similar to the complexes [169]. In both cases a decrease in the energy of the $\pi \to \pi^*$-transition as compared with the $n \to \pi^*$-transition makes $\pi \leftarrow \pi^*$-emission possible (Table 5).

Since azo compounds do not fluoresce, the energies of their triplet levels were determined by means of quenching and sensitization (Table 6). For aliphatic azo compounds the E_S (n, π^*) were equal to 290-380 kJ/mole, while the E_T (n, π^*) were 80-120 kJ/mole lower. Data on the energies of other levels of azo compounds were obtained by means of photoelectronic spectroscopy [29,36,37,43,125-127,171-176]. The lowest ionization potentials of azoalkanes lie in a region between 8 and 9.75 eV, while for trimethylsilyldiazene — even lower (> 7.37 eV).

Compound	ϕ_f	τ_f, ns	ϕ_{N_2}
(bicyclic azo structure)	0.56 gas phase [77] 0.41 acetonitrile [65] 0.20 [123]; 0.02 methanol	330 [65]	0.022
(bicyclic azo structure)	0.015 gas phase [77]		1
(bicyclic azo structure, CH3)	0.44 [177]	500 [177]	
(bicyclic azo structure, CH3, H3C)	0.53 acetonitrile [177] 0.7 acetonitrile	599 [177] 740	0.15
(bicyclic azo structure, Cl, Cl)	0.033 [177]	58 [177]	0.020
(bicyclic azo structure, Br, Br)	0.032 [177]	60 [177]	0.022
(H3C)(H3C)(CH3)C—C(CH3) N=N	0.01 hexane [65,180]		0.52 benzene
(tricyclic azo structure)	0.11 hexane [178] 0.44 acetonitrile	410	
(cubane-type azo structure)	0.2 hexane [178]	310	0.02
(polycyclic azo structure)	0.43 acetonitrile	410	
(polycyclic azo structure)	0.13 hexane [75,179] 0.66 acetonitrile	410	0.01
CHCl, H3C, CH3, H3C, CH3, N=N	0.01 [163]		
H3C—C—CH3, H3C, CH3, H3C, CH3, N=N	0.15 acetonitrile	225	0.7

Table 6. The energy (kJ/mole) of the lower n,π^*-triplet and singlet states of certain azo compounds

Compound	E_S	E_T
E-(CH$_3$)$_2$CH-N=N-CH(CH$_3$)$_2$ [6]	<285	226
2,3-diazabicyclo[2.1.2]hept-2-ene [6]	353	251±5
2,3-diazabicyclo[2.2.2]oct-2-ene [6]	318	228±5
E-C$_6$H$_5$-N=N-C$_6$H$_5$ [51,170,181]		<250,<190,>220
E-n-, iso-, sec-, tert-azobutane [182]		220±5
CH$_3$-N=N-CH$_3$ [51,170,181]		220÷230

The Mechanisms of E/Z-Photo- and Thermal Isomerization and Photo- and Thermal Dissociation

Although for the most elementary azo compounds — diazene and methyl-diazene — only E-isomers are known to date, it has been possible to obtain a Z-isomer from E-azomethane photochemically at low temperatures [208]. Most aliphatic Z-azo compounds are thermally unstable and decompose homolytically at room temperature to form nitrogen and alkyl radicals [27,87,209]. Thermal Z → E-isomerization in aliphatic azo compounds is not feasible, as a rule, since the activation energy of Z-isomer degradation (decay) is usually lower than the isomerization barrier in the ground state. The thermal and photostability of aliphatic azoalkanes increase in compounds with a bicyclic alkyl residue, e.g., in azo-1-norbornene [27]:

In this compound the activation energy of Z → E-isomerization is lower than the activation energy of decay, and exothermic thermal Z → E-isomerization can thus be observed (see Chapter 5). In other azoalkanes with polycyclic alkyl fragments, for instance, in

in which the activation energies of thermal dissociation and isomerization of the Z-isomer are about equal, both conversion paths are observed [27]. An inversion mechanism is discussed for the thermal isomerization of such azoalkanes [21,27,210].

Azoisopropane is especially convenient for the study of the photo-isomerization of aliphatic azo compounds since its Z-isomer is thermally relatively stable. The direction and nature of its photoconversions in the gas phase display a definite dependence on pressure [82]. At low pressure practically no photoisomerization of azoisopropane is observed, only a photodecay with $\phi \sim 1$. With an increase in pressure the ϕ of photodissociation drops to 0.5 as a result of photoisomerization. The latter predominates in solution, with photodegradation playing a subordinate role only; $\phi_{Z \to E}$ and $\phi_{E \to Z}$ amount to about 0.5 both in water and in isooctane; in water $\phi_{Z \to E}$ is 0.54 and $\phi_{E \to Z}$ is 0.50 [17].

The use of triplet sensitizers such as benzaldehyde and benzene in the gas phase induces the photodissociation of Z-azoisopropane with $\phi \sim 0.1$, and no isomerization takes place. Conversely, in solution photodissociation can

be neglected upon sensitization by acetophene and benzophene, and isomerization predominates with $\phi_{Z \to E} \sim 0.7$ and $\phi_{E \to Z} \sim 0.06$.

For azomethane in benzene at 25°C, $\phi_{Z \to E}$ is 0.45 and $\phi_{E \to Z}$ is 0.42 [35a];
in frozen matrices the values are 0.38 and 0.5, respectively [206]. Azo-
ethane also isomerizes in frozen matrices at 77°K with $\phi_{Z \to E} \sim \phi_{E \to Z} \sim 0.3$,
which indicates the absence of a high thermal barrier of photoisomerization
[27,211].

Quantum-chemical calculations show that for the $S_1(n, \pi*)$- and
$T_1(n, \pi*)$-states of diimine and azomethane, no energy barrier of rotation
exists, which allows photoisomerization to proceed through these states
[19,20,27,28,123,210,212-217]. The minimum of the potential curve of the
$S_1(n, \pi*)$-state at an angle of rotation of 90° touches the potential curve
of the ground state; the potential curve of the $T_1(n, \pi*)$-state lies even
lower and intersects the potential curve of the ground state at two points.
The differences between the results of the direct and triplet-sensitized
photoisomerization of azoisopropane may be due to the forward reaction
proceeding in a singlet state and the sensitized reaction — in a triplet
state [82].

The competition between photodissociation and photoisomerization occurring on the collision of excited states with other particles is considered in various schemes of azoalkane photolysis. Under the simplest
mechanism the vibrationally excited state $S_1(n, \pi*)$ disintegrates either
directly or undergoes deactivation into S_0 as a result of collisions [218-
225]. More complex schemes are as follows:

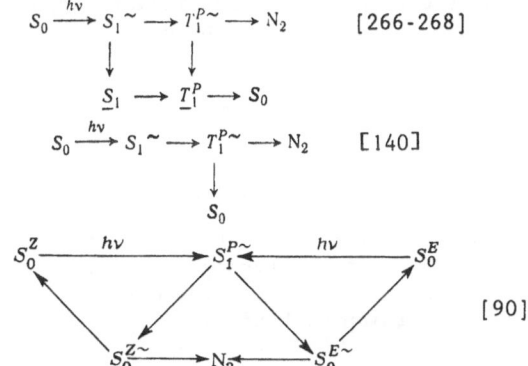

Unfortunately, none of these schemes describes the entire set of available experimental data, which makes it difficult to draw a final conclusion
in favor of any of them [11].

On the whole, a roughly analogous mechanism is proposed for the triplet-sensitized photolysis of azoalkanes [27,225,227,231,233]. In the gas
phase the vibrationally excited T-state reacts as follows:

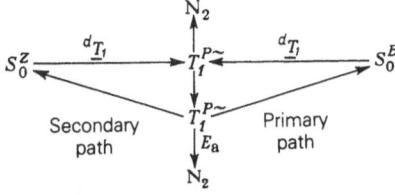

The efficiency of the photodissociation of azoalkanes in solutions is
heavily dependent on the nature of the substituents at the azo group
(Table 7). The quantum yields of triplet-sensitized photodissociation are
lower than in direct photolysis [35a,225,227,229,230,234,235]. These re-

sults can be explained with the help of the above-cited schemes. The direct photolysis of azoalkanes is poorly efficient (see Table 2), but the thermal degradation of photochemically generated Z-isomers occurring in solution increases the total degradation yield [11].

$$\underset{R}{\overset{R'}{\diagdown}}N{=}N \underset{h\nu'}{\overset{h\nu}{\rightleftharpoons}} \underset{R}{\overset{}{\diagdown}}N{=}N\underset{R'}{\overset{}{\diagup}} \overset{\Delta}{\longrightarrow} \begin{cases} \overset{a}{\longrightarrow} R{-}N{=}N\cdot + R' \\ \quad\quad\downarrow a \\ \overset{b}{\longrightarrow} R\cdot + R'\cdot + N_2 \end{cases}$$

This photolysis scheme explains the equal ratios of recombination and disproportionation of tertiary alkyl radicals upon the photolysis and thermolysis of the corresponding azoalkanes since both processes result in one and the same radical pair in the ground state [236].

Although path a is in better agreement with the experimental findings [237-239], according to the data cited in the paper [240], the thermolysis of symmetrical azoalkanes proceeds as a synchronous process with the simultaneous homolytic cleavage of both C–N bonds (path b).

The isomerization of the optically active N-phenyl-N'-(2-phenyl-2-

butyl)diazene [11] has been studied in detail. In diazene, $E \xrightarrow{h\nu} Z\cdot$ and

$Z \xrightarrow{\Delta} E$-isomerizations (−78°C) do not result in racemization. At 25°C the Z-form undergoes thermal degradation, and a partially racemized E-form appears along with the degradation products. A dissociation–recombination mechanism follows from the data on the thermal Z → E-isomerization, while for the photolysis of the E-compound, photoisomerization with subsequent thermal degradation of the Z-form is accepted.

$$\underset{Ar}{\overset{R}{\diagdown}}N{=}N \underset{h\nu'}{\overset{h\nu}{\rightleftharpoons}} \underset{Ar}{\overset{}{\diagup}}N{=}N\underset{R}{\overset{}{\diagdown}}$$

$$Ar\cdot + R\cdot + N_2 \longleftarrow Ar{-}N{=}N\cdot + R\cdot$$

The mechanism of the photochemical E/Z-isomerization of aromatic azo compounds has been studied most thoroughly in the case of azobenzene and its substituted forms. Their photoisomerization may proceed both by a rotational mechanism and by a inversion mechanism.

Unlike stilbenes, azobenzenes upon isomerization have so far failed to exhibit triplet states or any other short-lived intermediate particles. Since azobenzenes fluoresce poorly and only under conditions precluding

Table 7. Quantum yields of the photodecomposition of azoalkanes R–N=N–R in solution[6]

R	ϕ_{N_2}	ϕ_{N_2} (sens)
CH$_3$ [6,241,242]	0.15	0.009
C$_2$H$_5$ [243]	0.023	
C$_3$H$_7$	0.007	
iso-C$_3$H$_7$ [6,244]	0.025	0.001
tert-C$_4$H$_9$ [242]	0.46	0.016
C$_6$H$_5$C(CH$_3$)$_2$	0.36	0.31
NCC(CH$_3$)$_2$ [6,245]	0.44	0.11
CH$_2$=CHC(CH$_3$)$_2$ [6,246]	0.57	0.54
CH≡CC(CH$_3$)$_2$ [6,246]	0.47	0.27
C$_2$H$_5$OOCC(CH$_3$)$_2$ [236]	0.42	0.06

130

photoisomerization [51,67], this severely limits the possibility of studying photoisomerization paths. Besides, because of the presence of $\pi \rightarrow \pi^*$- and $n \rightarrow \pi^*$-transitions, azobenzenes have more electronically excited states than stilbenes which could participate in the isomerization. That is why the present knowledge of the mechanisms of azobenzene isomerization is less detailed and less reliable. The available data on them relies primarily on the study of the impact of various factors on the quantum yields of isomerization. The quantum yields of azobenzene photoisomerization are independent of concentration, of the intensity of the irradiated light, and, within a single absorption band, also of the wavelength of light [32]; nor does the nature of the solvent have any appreciable influence either [185]. See also the works of Fischer et al. [5,186,187].

For substituted azobenzenes $\phi_{Z \rightarrow E} + \phi_{E \rightarrow Z}$ (see Table 2) is very much less than unity; $\phi_{Z \rightarrow E}$ is almost independent of the nature of the substituents and, like stilbenes, is independent of temperature and, more often than not, also of whether the $\pi \rightarrow \pi^*$- or the $n \rightarrow \pi^*$-transition gets excited or not; it amounts to 0.45. The yield $\phi_{E \rightarrow Z}$ is likewise almost independent of the nature of the substituents; with $n \rightarrow \pi^*$-excitation the values of ϕ are higher ($\phi_{E \rightarrow Z}$ 0.20) than upon $\pi \rightarrow \pi^*$-excitation ($\phi_{E \rightarrow Z}$ 0.10). With falling temperature $\phi_{E \rightarrow Z}$ drops sharply, which suggests the existence of a thermal activation barrier in going from the S_1 E-isomer to the S_0 Z-isomer. In azobenzenes this activation threshold is 8.4 kJ/mole. The temperature dependence of $\phi_{E \rightarrow Z}$ obeys the Arrhenius equation far from ideally [5]. An increase in the viscosity of the reaction medium also causes $\phi_{E \rightarrow Z}$ to fall [188]. Heavily substituent-polarized azobenzenes (e.g., 4-dimethylamino-4'-nitroazobenzene) exhibit a number of differences: the $\pi \rightarrow \pi^*$-bands are substantially bathochromically shifted as compared with azobenzenes, so much so that the $n \rightarrow \pi^*$-bands cannot be identified; $\phi_{Z \rightarrow E}$ compared with azobenzene is somewhat larger, so that the sum of the quantum yields approaches unity [23] (see Table 2). In principle, analogous regularities were obtained during the study of the photoisomerization of azonaphthalenes [189] and azopyridines [23a].

The photoisomerization of azobenzenes was eventually sensitized with 3-acetylpyrene ($E_T \sim 188$ kJ/mole), β-acetonaphthone ($E_T \sim 248$ kJ/mole), and triphenylene ($E_T \sim 279$ kJ/mole) [181]. In all three cases from 1.5 to 1.8% of the Z-isomer corresponded to the photostationary state. This showed that the energy of the azobenzene triplet (the E- and Z-isomer) did not exceed 190 kJ/mole (see Table 6). The ratio [E]/[Z] in the photostationary state corresponds to the deactivation ratio T_1^X. The value obtained (60) differed considerably from the ratio of the quantum yields of direct photoisomerization (upon $\pi \rightarrow \pi^*$-excitation $\phi_{Z \rightarrow E}/\phi_{E \rightarrow Z} \sim 4$; upon $n \rightarrow \pi^*$-excitation $\phi_{Z \rightarrow E}/\phi_{E \rightarrow Z} \sim 2$), which points to different reaction paths in both cases and so to the singlet path of direct photoisomerization. However, Fischer [190] was unable to corroborate these results. He used naphthalene ($E_T \sim 251$ kJ/mole) as well as triphenylene as sensitizers and discovered that roughly 25% of the Z-isomer was in the stationary state. The resulting deactivation ratio (4) was in good agreement with the ratio of the quantum yields of direct photoisomerization, which suggested an isomerization through triplet states in both cases.

For the benzyl-sensitized photoisomerization of azobenzene, $\phi_{Z \rightarrow E}$ was 0.49 and $\phi_{E \rightarrow Z}$ was 0.027 [191]. They were independent of the azobenzene concentration within $(1-8) \times 10^{-5}$ mole/liter. The deactivation ratio of the triplet-excited intermediate state, equal to 18, differed considerably from the ratio of the quantum yields of direct photoisomerization. While $\phi_{Z \rightarrow E}$ was in good agreement with the value of $\phi_{Z \rightarrow E}$ of direct photoisomerization, the value of $\phi_{E \rightarrow Z}$ in sensitized photoisomerization was found to be lower.

In later works, the same quantum yields were found for direct azoben-

zene photoisomerization [185,192] as formerly [32]. The photochemical
isomerization of azobenzene can be sensitized by chrysene ($E_T \sim 239$ kJ/
mole), phenanthrene ($E_T \sim 260$ kJ/mole), and naphthalene ($E_T \sim 251$ kJ/mole)
and by benzyl ($E_T \sim 222$ kJ/mole) and diacetyl ($E_T \sim 233$ kJ/mole). Bengal
rose ($E_T \sim 165$ kJ/mole), eosin ($E_T \sim 184$ kJ/mole), fluorescein ($E_T \sim 195$
kJ/mole), and methylene blue ($E_T \sim 143$ kJ/mole), with their much lower triplet energy, are also sensitizers. With the use of sensitizers of a high
triplet energy $\phi_{E \to Z}$ was 0.03, with a low triplet energy $\phi_{E \to Z}$ was 0.5. No
such difference is observed upon the Z → E-transition or in the use of sensitizers with differing triplet energies.

On the basis of these results, the isomerization scheme presented in
Figure 12 was proposed. Both for the Z- and the E-isomers the existence of
two triplet states with different energy levels has been postulated without
drawing any conclusions regarding the n,π^*- or π,π^*-nature of these states.
It is assumed that isomerization proceeds only from the lower T_1-state of
both isomers. Since no difference between the two triplet states was experimentally established in the case of the Z-isomer, the qualitative deactivation of T_2^Z into the T_1^Z state was postulated.

In the case of the E-isomer with the use of sensitizers of a high
triplet energy, the T_2^E-state is predominantly populated. It becomes deactivated by a mere 5% into the T_1^E-state and by 95% into the S_0^E-state, which
accounts for the small quantum yields of isomerization. Conversely, with
the use of low-triplet-energy sensitizers the T_1^E-state gets populated, which
isomerizes with a ϕ of 0.5. Upon direct $n \to \pi^*$-excitation, intersystem
crossing is more likely than upon $\pi \to \pi^*$-excitation; this may account for
the different values of $\phi_{E \to Z}$ upon $n \to \pi^*$- and $\pi \to \pi^*$-excitations.

Yablonskii's diagram presented in Figure 12 is in agreement with the
values of the quantum yields measured in direct and sensitized photoisomerizations but contradicts the above-mentioned general principles regarding
the deactivation of T_1-states. Photoisomerization will definitely proceed
nonadiabatically. Therefore, the potential curve of the T_1-state should
feature a minimum corresponding to the triplet-excited intermediate state
T_1^X. The T_1-states of the Z- and E-isomers should quantitatively cross into
the T_1^X-state, which then, with equal probability, becomes deactivated into
the S_0-states of both isomers. This more exact refinement of Yablonskii's
diagram brings it into line with the general principles of describing Z/E-
photoisomerizations without changing the basic conclusions drawn.

The study of 1-phenylazo-2-hydroxynaphthalene by flash photolysis in
the picosecond region failed to prove the presence of triplet states [138],
but it did establish that the conversion E $\overset{h\nu}{\to}$ Z at 21.5°C in methylcyclo-
hexanone (η 0.729 mPa $\cdot$ s) is completed within 14 ± 3 ns. In a more viscous
(η 9.63 mPa $\cdot$ s) methylcyclohexane–cyclohexanol system (5:9), the conversion

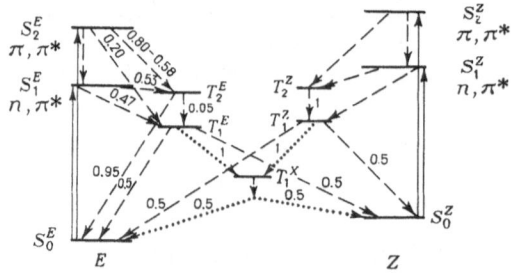

Fig. 12. Yablonskii's diagram for the E/Z-photoisomerization of azobenzene
[192] (see the caption to Figure 5). The numbers on the arrows
represent the values of the quantum yields. The dots indicate the
formation of T_1^X and its deactivation.

$E \xrightarrow{h\nu} Z$ takes 107 ± 29 ns to complete. Apparently, the triplet states through which isomerization can proceed are extremely short-lived.

Since none of the results obtained points to a fundamentally different behavior of azoaromatic compounds as compared with stilbene, the photoisomerization of azo compounds may be satisfactorily described also with the help of the mechanism of stilbene isomerization. Since the $\pi,\pi*$-states of azobenzene are similar to the $\pi,\pi*$-states of stilbene, we may accept the similar potential curves for the $\pi,\pi*$-states, depending on the angle of rotation about the N=N double bond. In accordance with this, photoisomerization should have proceeded through the nonplanar S_1 or T intermediate state which underwent radiationless deactivation into the S_0 Z- and E-forms (see Figure 12).

In keeping with the selection rules [193], radiationless transitions between the $n,\pi*$- and $\pi,\pi*$-states are more efficient than transitions between states of the same electronic configuration. Thus $n \rightarrow \pi*$-excitation would result preferably in a $\pi,\pi*$-triplet and vice versa. Assuming that photoisomerization proceeds through the $n,\pi*$-triplet, the almost equal quantum yields obtained for $Z \rightarrow E$-isomerization upon $n \rightarrow \pi*$- and $\pi \rightarrow \pi*$-excitation could be due to an efficient transition from the $\pi,\pi*$-singlet to the $\pi,\pi*$-triplet or to a similar course of the potential curves of the $n,\pi*$- and $\pi,\pi*$-triplet states in this region or, alternatively, to isomerization in the S_1-state.

The fall in $\phi_{E \rightarrow Z}$ at low temperature may be due to the existence of a thermal activation barrier upon the transition $S_1(E) \rightarrow T_1(E)$ or upon the transition of the triplet (singlet) state of the E-isomer into the triplet (singlet)-excited intermediate state T_1^X (S_1^X).

The $\pi \rightarrow \pi*$-excitation of azobenzenes in isopropyl alcohol results in, apart from isomerization, photoreduction into hydrazobenzenes and aromatic amines [102,194]. The photoreduction is accelerated by the addition of a small quantity of oxygen and is quenched by 1,3-pentadiene [194]. The quantum yields of reduction amount to 10^{-4}-10^{-6} (cf. [195]). Fluorene and aromatic carbonyl compounds sensitize the reduction. The sensitization mechanism may consist in energy transfer and chemical sensitization. In the latter case, the ketyl radical arising from the excited sensitizer should reduce azobenzenes in the ground state [102,194].

Photocyclization is another side reaction of photoisomerization. The photocyclodehydrogenation of stilbenes [196], diphenylamines [197], and Schiff's bases [198] proceeds in organic solvents. Under these conditions azobenzenes do not undergo photocyclization since their $n,\pi*$-level is below the $\pi,\pi*$-level. Photocyclodehydrogenation occurs only upon level inversion [199]. This takes place upon protonation or complex formation with Lewis acids. Thus, benzo[c]cinnoline is formed upon the irradiation of azobenzene in concentrated H_2SO_4 as a result of consecutive photocyclization and dehydrogenation [200,201]:

133

98% H_2SO_4 can act on its own as a dehydrogenating agent, reducing to SO_2. Already in a doubly diluted H_2SO_4 environment and in the presence of Lewis acids, the starting monoprotonated azobenzene acts as a dehydrogenating agent. As a result, a gram-mole of hydrazobenzene, which in an acidic medium may rearrange to benzidine, is formed per gram-mole of benzo[c]cinnoline. For azobenzene in an 18 N solution of H_2SO_4, ϕ_{cycl} is 0.014 at 25°C and λ_{exc} is 436 nm, and under these conditions $\phi_{Z\to E}$ is 0.26 and $\phi_{E\to Z}$ is 0.18. Since photoisomerization is more efficient than cyclization, the latter proceeds from the photostationary state of isomerization and, probably, from the S_1-state of the protonated azobenzene.

The iodine-induced photoisomerization of azobenzenes is known to exist [41,43]. Upon irradiation with 313-436 nm wavelength light at −78°C the absorption of the complex of iodine and Z-azobenzene is observed along with the formation of E-azobenzene; the E-isomer does not convert into the Z-isomer. Photolysis at 546 nm results in the dissociation of the iodine into atoms, which initiate Z → E-isomerization; the k of the reaction of I⁎ with Z-azobenzene amounts to 2×10^6 liters/(mole · s), which is considerably more than the analogous value for the reaction of I⁎ with Z-stilbene [6.6×10^4 liters/(mole · s)] [43].

Only the isomerization of the Z-isomer to the E-isomer proceeds thermally. The energy of activation for azobenzenes is between 96 and 100 kJ/mole and is not appreciably influenced by either electronic or steric substituent effects [5]. Only in the case of pronounced donor and acceptor substituent pairs in azobenzenes such as 4-nitro-4'-dialkylaminoazobenzene does the activation energy decline [23]. Azobenzenes and azonaphthalene also exhibit abnormal behavior with OH groups; the lower activation energy in this case is due to azo—hydrazone tautomerism [202].

For the thermal isomerization of azobenzene, azopyridine, and azonaphthalene the inversion mechanism is assumed as being the more likely [12,15,83,203]. According to the data cited by Hofmann [204], in the transition state of the isomerization of azoarenes the aryl residue in the inverting nitrogen atom is rotated through 90° to the molecule's plane.

The thermal and photoisomerizations of azoarenes are possible also if the azo compound is in a polymer matrix or is covalently linked with a polymer [205]. But the study of the reaction mechanisms is made more difficult in this case since the matrix's properties influence appreciably the course of isomerization [206].

For azoxybenzene [207] $\phi_{Z\to E}$ is 0.6 and $\phi_{E\to Z}$ is 0.1 (in ethanol and heptane), and they are independent of the wavelength of actinic light in the 330-370 nm region. The presence of oxygen has no influence, something that, according to the data cited by Rhee and Jaffe [207], suggests a singlet mechanism without ruling out a triplet mechanism. For parallel thermal Z → E-isomerization the activation energies amount to 104.7 kJ/mole in ethanol and 79.5 kJ/mole in heptane. These values are of the same order as those for azobenzenes. The azoxybenzenes lack luminescence, and no excited states can be registered by means of flash photolysis. The mechanism of azobenzene isomerization can be extended to cover the isomerization of azoxybenzenes. Photorearrangement into hydroxyazobenzenes or hydrazones, which are tautomeric to them, occurs simultaneously as a side process.

The Application of the Photochromism of Azo Compounds

Depending on the structure of azo compounds, their isomerization may result in either a deepening or a fading of the visible coloration on account of the shift of long-wavelength absorption. The E → Z-isomerization of azoalkanes and azoaryls with a slight substituent-induced polarization

results in a deepening of color: in Z-azoalkanes the long-wavelength
n → π*-transition is more intense and bathochromically shifted compared with
the E-isomer; in Z-azoaryls the intensity of the n → π*-transition is higher
than in the E-isomers, and despite the fact that it lies in a shorter-wavelength region than in the E-isomer, it overlaps the n → π*-transition of the
latter, which is objectively interpreted as a deepening or fading of color
intensity upon E → Z-isomerization. In substituent-polarized azoaryls the
π → π*-transition is long-wavelength; in the Z-isomer it lies in a shorter-
wavelength region and is less intense than in the E-isomer and thus upon
E → Z-isomerization results in a fading of color (see Table 2).

The rate of thermal Z → E-isomerization of azo compounds depends on
their structure and the nature of the medium and varies over a wide range;
$\Delta H^{\neq}$ amounts to 35-100 kJ/mole. The activation energies of the thermal
Z → E-isomerization of azoaryls having polarizing donor and acceptor substituent pairs ($\Delta H^{\neq}$ 35-40 kJ/mole) are particularly low. Therefore, light-
induced changes in optical density in such systems can be detected only by
means of fast spectroscopy or by delaying (inhibiting) the reaction at a low
temperature. The combination of photochemical E → Z-isomerization with fast
thermal Z → E-reaction along with azo—hydrazone tautomerism in isomers makes
possible the efficient conversion of light energy into heat; it is this that
accounts for the fastness to light (photostability) of azo dyes [247,248].

Thermal Z → E-isomerization can also be substantially accelerated with
the help of acidic catalysts [83,249]. Compounds with a high activation
energy of thermal Z → E-isomerization, e.g., azo-1-norbornene, may furnish
the basis for developing a solar energy storage system which would incor-
porate E → Z-photoisomerization and a catalytic thermal exothermal Z → E
back reaction [11,122]. Porphyrin-like macrocycles with an azoaromatic side
chain [250] can be used for the same purpose, as can the azobenzenes themselves [251].

The number of cycles in reversible E/Z-isomerization upon π → π*-exci-
tation impairs the photoreduction into anilines [102,194] and the cyclodehy-
drogenation into benzo[c]cinnolines [101,199,201] of protonated azoarenes
and azoarene complexes with Lewis acids, having low-energy π → π*-transi-
tions as well as the radical dissociation of alkylazoarenes, azoalkanes, and
azocycloalkanes. This thermal dissociation of Z-azoalkanes into nitrogen
and alkyl radicals can be used for the photoinitiation of radical reactions.

If solvents with a low donor capacity of hydrogen atoms are chosen for
azoarene photoisomerization to suppress a side photoreduction, systems with
a slow thermal Z → E back reaction (azobenzene, 2,2',4,4'-tetraisopropylazo-
benzene) could be used as actinometers; the favorable factors here are the
large number of reaction cycles and the easy spectrometric determination of
concentrations of the E- and Z-forms [252,253]. On the other hand, in the
quantitative spectrometric determination of alcohols in the form of 4-(4-ni-
trophenylazo)benzoates, E → Z-photoisomerization distorts the values of
optical density with an analytic wavelength which impairs the analysis
results [266].

Polymers with a covalently added azochromophore which are photochromic
both in solution and in a matrix are of special interest [205,254-261]. At
a temperature below the glass point of the polymer, thermal Z → E-isomerization proceeds, as a rule, more rapidly than in solution. Two independent
Z → E-isomerizations are observed whose apparent activation energies correspond to the forward (E_a ~ 27 kJ/mole) or rotatory (E_a ~ 69 kJ/mole) motion
of segments in the polymer chain. Thus, the activation energies of thermal
Z → E-isomerization are determined by the mechanisms of polymer relaxation
rather than by isomerization about the N=N bond. What is more, azo com-

pounds placed in polymer matrices exhibit a behavior analogous to polymers with a covalently added azochromophore.

Unfortunately, $\phi_{E \to Z}$, as the behavior of model compounds demonstrates, is much lower in solid matrices than in solutions [261]. It was established during the study of substituted azobenzenes and 1-methoxy-4-(4'-X-phenyl-azo)naphthalenes [247] in polyamide films that compounds with X donor sub-stituents as well as those that contain polarizing substituent pairs undergo thermal Z → E-isomerization more rapidly than unsubstituted compounds. This is in agreement with their behavior in solution. An analogous picture is observed for the copolymerizates of methacrylic acid and azobenzene deriva-tives [262].

The photochromism of azo compounds in polymers or of azo compounds with polymer residues makes it possible to obtain important information about the microstructures of polymers [205,259-261]. E → Z-Photoisomerization changes the properties of liquid-crystal azo compounds [263,264]. If during ir-radiation Z-isomers are accumulated, a transition of the liquid-crystal phase into the isotropic phase occurs; this transition is thermally rever-sible. Upon photolysis in a liquid-crystal environment the reflection spectra of azo compounds shift hypsochromically [265].

PHOTOCHROMIC ARYLDIAZO COMPOUNDS

It was demonstrated as early as the late 19th century that aryldiazo-sulfonates and aryldiazocyanides exist in the form of two isomers. The hypothesis formulated at the time, to the effect that a less stable form indicated the Z-isomer while a more stable form indicated the E-isomer, is now generally accepted [279,280]. The Z- and E-isomers have been discovered in aryldiazosulfonates [279,281-285], aryldiazocyanides [280,282,286-288], aryldiazosulfides [288-295], 1-aryl-3-heterocyclidenetriazenes [296], aryl-diazoesters [297-302], aryldiazotates [303-305], 1-phenyl-3,3-bis(trimethyl-silyl)triazenes [4], and symmetric triazatrimethinecyanines [306] (Table 8). From the first five groups of compounds, as well as from 1-phenyl-3,3-bis(trimethylsilyl)triazenes and symmetric triazatrimethinecyanines, both isomers were isolated. In the case of diazosulfonates some authors failed to isolate the Z-isomers, while not ruling out completely their existence, however [307,308]. The presence of Z-isomers of aryldiazoesters follows from the results of the kinetic investigation of the reactions between aryldiazonium salts and nucleophiles.

The following were not strictly identified: the Z- and E-isomers of 1-(arylsulfonyl)-3-(3-ethylbenzothiazolin-2-ylidene)triazenes [310], aryl-diazoamines(triazenes) [311,312], arylpentazenes [297,313], aryldiazosul-fones [314-317], dialkyl esters of arylazophosphoric acid [318], arylazo-carboxylic acids [318], and aryldiazoisocyanides [287]:

$$\overset{S}{\underset{N-C_2H_5}{\bigcirc}} =N-N=N-SO_2C_6H_4X\text{-}p, \qquad XC_6H_4-N=N-NRR',$$

$$XC_6H_4-N=N-N=N-NRR', \quad XC_6H_4-N=N-SO_2R, \quad XC_6H_4-N=N-PO(OR)_2,$$
$$XC_6H_4-N=N-COOH, \qquad XC_6H_4-N=N-NC$$

However, a reversible shift of the ultraviolet spectrum was discovered upon the irradiation of phenylazocarboxylic acid amide [319], which indi-cated the formation of a Z-isomer from the E-form. Upon the irradiation of aryldiazoamino compounds, changes occur in the electronic spectra which could also be induced by Z/E-isomerization [311,312]. The existence of very thermally unstable Z-aryldiazosulfones [317] is not completely ruled out.

Later, Majer et al. [320] upon the irradiation of diazoamino compounds at low temperature, as well as Fanghänel and Link [321,322] for aryldiazosul-fones, succeeded in finding by means of flash photolysis evidence for the existence of a thermally unstable Z-isomer. The assumption followed from this that the Z- and E-isomers exist for all aryldiazo compounds. However, the stability of some of the Z-forms is so small that they either rapidly change thermally into E-isomers or disintegrate. That is the reason why they are undetectable.

Upon the synthesis of aryldiazo compounds from the diazonium salt and the corresponding azo component (nucleophile), first a less stable Z-isomer is formed, which can then thermally change into the E-form [4,290,296,301, 303,304].

$$\text{Ar}-\overset{+}{N}\equiv N + :X^- \rightleftharpoons \left[\text{Ar}-N \underset{-\delta X}{\overset{+\delta}{\cdots}} N \right]^{\neq} \longrightarrow \text{Ar}-N=N\overset{}{\underset{X}{}} \overset{\Delta}{\longrightarrow} \overset{}{\underset{\text{Ar}}{}}N=N-X$$

The question that arises in this connection is why the less stable Z-isomer is the first to form. No exhaustive answer to this question is to be found in the literature. However, according to Zollinger [309], in the transition state arising upon the reaction of a diazonium cation with a nucleophile, the unshared electron pair of the β-N atom is in the E-position relative to the aromatic ring, which is why the nucleophile's attack should be directed towards the Z-position and produce a Z-isomer. If the transition state is shifted along the reaction coordinate towards the initial products, the repulsion forces between the nucleophile and the aromatic ring are weak and a Z-isomer will be formed. This variant occurs for small-volume nucleophiles. In the event of a late transition state (e.g., in a diazo reaction) a powerful repulsion occurs between the aromatic systems, with the immediate formation of an E-isomer.

In a recent work [309a] the Z-addition of a nucleophile is attributed to the conservation of orbital symmetry in this process (in contrast to the symmetry-forbidden E-addition). The one-stage formation of the E-isomer is possible in the event of the reaction proceeding through the Z-transition state. However, it is impossible at this time to choose between two-stage and one-stage formation of the E-isomer for a particular system.

Electronic Spectra

$\pi \rightarrow \pi^*$-Transitions at 300-400 nm and $n \rightarrow \pi^*$-transitions above 400 nm (Table 8, Figure 13) are observed in the spectra of most diazo compounds. With the exception of o-chlorophenyldiazosulfonate Ia, the long-wavelength $\pi \rightarrow \pi^*$-transition of the E-isomers is bathochromically shifted and highly intense, as compared with the Z-isomers of the compounds mentioned above.

The difference in wavelength between the absorption maximum of the long-wavelength $\pi \rightarrow \pi$-transition of the Z- and E-isomers of diazo compounds varies widely. For instance, for o-chlorophenyldiazosulfonate Ia it is 8 nm, for variously substituted aryldiazocyanides II — from 2 to 20 nm, for disubstituted aryldiazosulfides III — from 17 to 55 nm, and for substituted 1-aryl-3-benzothiazolinylidenetriazenes IV — from 8 to 100 nm.

The difference in wavelength of the absorption maxima of the $n \rightarrow \pi^*$-transitions is more pronounced, with the E-form being also absorbed, as a rule, in a longer-wavelength region than the Z-isomer. In the case of aryldiazocyanides, ϵ_{max} of the $n \rightarrow \pi^*$-transition of the Z-isomer is greater than in the E-form. An analogous phenomenon is observed in the case of

Table 8. The spectral data and parameters of the photochemical conversions of aryldiazo compounds

$XC_6H_4—N{=}N—SO_3Na$ [281,285,289]; $XC_6H_4—N{=}N—CN$ [288,355]; $XC_6H_4—N{=}N—SR$ [288,290]; $A{=}N—N{=}NC_6H_4X$, where A equals

I II III

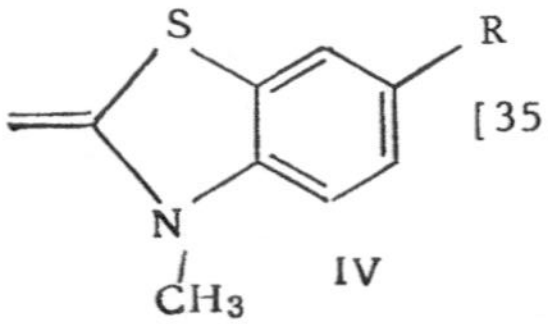

[35,328,329]; IV

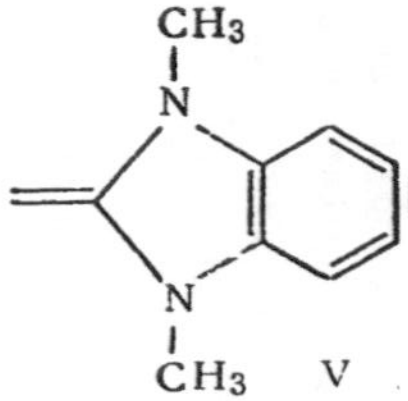

[356]; V

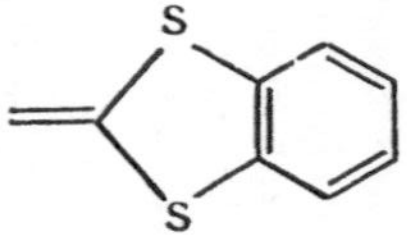

[357]; VI

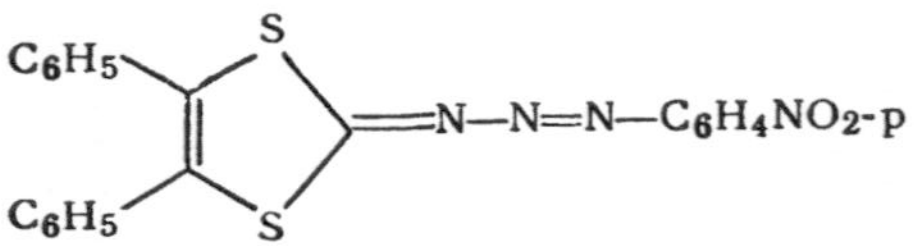

[357]; VII

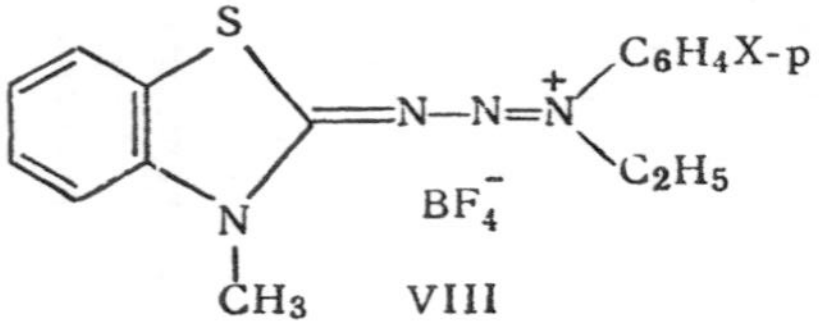

VIII [325]; $CH_3O—N{=}N—C_6H_4NO_2$-p [302]; $XC_6H_4—N{=}N—O^-$ [303-305]

IX X

No. of compound	At X, R>1 the number of hydrogen atoms in the nuclei is less than 4	λ_{max}, nm (log ε) long-wavelength π→π*-transition Z-isomer	λ_{max}, nm (log ε) long-wavelength π→π*-transition E-isomer	λ_{max}, nm (log ε) n→π*-transition Z-isomer	λ_{max}, nm (log ε) n→π*-transition E-isomer	Solvent	$\phi_{Z\to E}$	$\phi_{E\to Z}$	Note
Ia	2-Cl	300(3.2)	292(3.9)	448(2.3)	428(2.3)	water			Z-isomer cannot be isolated at room temperature.
Ib	4-$N(CH_3)_2$		~390(4.40)			ethanol			They are thermally unstable and quickly dissociate to produce
Ic	4-$N(CH_3)_2$-3-Cl		~360(4.05)			"			
Id	4-$N(CH_3)_2$-3,5-Cl_2		~400(3.74)			"			
Ie	4-Morpholino-3-Cl		~360(4.00)			"			

Compound	Substituents	λ, nm (log ε)						Solvent	Remarks
If	2-OCH₃					0.29 (254 nm)	0.37 (366 nm) pH 4		diazonium salts and sulfite.
IIa	4-Cl	330(4.12)	338(4.29)	432(3.07)	438(2.48)			diethyl ether	Z-isomers can be isolated. Upon irradiation with light with λ 405 nm photoisomerization occurs, with λ 300 nm – both photoisomerization and photodecomposition.
IIb	4-Br	335(3.96)	341(4.05)	430(3.08)	425(2.77)			"	
IIc	2,4,6-tri-Br	338(3.42)	342(4.00)	426(2.76)	481(2.67)			"	
IId	2-NO₂	310(3.23)	330(4.79)	410(2.75)	388(3.92)			"	
IIe	3-NO₂	300(3.71)	305(4.13)	425(2.82)	443(2.19)			"	
IIf	4-NO₂	306(3.92)	308(4.22)	405(2.94)	443(2.38)			"	
IIIa	H; CPh₃	312(3.37)	335(4.12)	383(2.78)	shoulder			benzene	The same as for compounds II. The isomerization is sensitized by triplet sensitizers with E$_T$ ≥ 163 kJ/mole.
IIIb	4-OCH₃; CPh₃	326(3.70)	343(4.24)	392(3.13)	"			"	
IIIc	4-NO₂; CPh₃	304(3.94)	359(4.24)	shoulder				"	
IIId	2,4,6-(CH₃)₂; CPh₃	shoulder	337(4.09)	368(2.65)	408(2.99)			"	
IIIe	H; C(CH₃)₃	305(3.40)	329(4.11)	380(2.75)	390(2.90)			"	
IIIf	2,4,6-(CH₃)₃; C(CH₃)₃	shoulder	322(3.90)	372(2.49)	403(2.92)			"	
IIIg	4-NO₂; C(CH₃)₃	~350(4.30)							
IVa	2,4-di-NO₂; H	~350	436					dioxane	Photoisomerization is sensitized by triplet sensitizers; for IVo, IVe and IV1 E$_T$(Z) about 167–209 kJ/mole, E$_T$(E) about 250 kJ/mole; φ is independent of the wavelength and solvent.
IVb	4-NO₂; H	346(4.25)	413(4.44)			0.52	0.34		
IVc	3-NO₂; H	341(4.29)	385(4.38)			0.58	0.37	"	
IVd	2-NO₂; H	333(4.25)	389(4.33)			0.19	0.23	"	
IVe	4-CN; H	338(4.31)	394(4.48)			0.55	0.40	"	
IVf	4-Cl; H	338(4.30)	379(4.39)			0.57	0.43	"	
IVg	3-Cl; H	339(4.28)	383(4.43)						
IVh	4-Br; H	339(4.32)	382(4.44)			0.50	0.47	tetrahydrofuran:H₂O (9:1) dioxane	
IVi	H; H	335(4.32)	370(4.46)			0.50	0.48	"	

Table 8. (continued)

No. of compound	At X, R>1 the number of hydrogen atoms in the nuclei is less than 4	λ$_{max}$, nm (log ε) long-wavelength π→π*-transition Z-isomer	E-isomer	n→π*-transition Z-isomer	E-isomer	Solvent	φ$_{Z→E}$	φ$_{E→Z}$	Note
IVj	4-CH$_3$; H	338(4.33)	370(4.44)			dioxane	0.49	0.45	
IVk	4-OCH$_3$; H	343(4.22)	379(4.43)			"	0.50	0.48	
IVl	4-NO$_2$-2-CH$_3$; H	343(4.33)	413(4.41)			"	0.49	0.30	
IVm	4-NO$_2$-2,6-(CH$_3$)$_2$; H	344(4.38)	370(4.18)			"	0.41	0.30	
IVn	4-Cl; OCH$_3$	343(4.30)	386(4.45)			"	0.61	0.42	
IVo	4-Cl; Cl	337(4.28)	377(4.47)			"	0.60	0.42	
IVp	4-Cl; NO$_2$	367(4.33)	395(4.45)			"			
IVq	4-nitro-1-naphthyl; H	∿340	440(4.27)			"			
IVr	4-NO$_2$; NO$_2$	373(4.37)	409(4.47)			"			
IVs	4-NO$_2$; Cl	344(4.27)	408(4.37)			"	0.50	0.36	
IVt	4-NO$_2$; OCH$_3$	365(4.23)	429(4.40)			"	0.41	0.28	
Va	4-NO$_2$	∿328	426(4.22)			"			Z-isomers are thermally unstable; they change into the E-form at room temperature.
Vb	3-NO$_2$		372(4.42)			"			
Vc	4-CN	∿326	387(4.46)			"			
Vd	4-Cl	∿329	373(4.47)			"			
Ve	H	∿330	368(4.41)			"			
Vf	4-OCH$_3$	∿333	372(4.42)			"			
VIa	4-NO$_2$	357(4.19)	404(4.22)			"			
VIb	4-CN	355(4.15)	398(4.17)			"			
VIc	3-NO$_2$	354(4.13)	386(4.14)			"			
VId	4-Cl	352(4.08)	385(4.09)			"			

				$\epsilon_E - \epsilon_Z$		
VII	389(4.11)	440(4.23)	dioxane			Acichromism as in the case of compounds IV.
VIIIa OCH$_3$	406(4.27)	440(4.34)	1,2-dichloro-	0.40	0.022	
VIIIb H	389(4.33)	414(4.39)	ethane	0.18	0.06	
VIIIc Cl	390(4.35)	416(4.40)	"	0.20	0.095	
VIIId NO$_2$	405(4.30)	420(4.39)	"			
IX		277(4.1)	methanol			Other ethers cannot be isolated because of their instability. Z-isomers cannot be isolated.
Xa H		273($\sim$4.1)	aqueous	7000		The Z-form exhibits increasing extinction in the 310 to 230 nm region with a more or less pronounced shoulder at $\sim\lambda_{max}$ of the E-form. The Z-Form is unknown, no E→Z-isomerization can be detected upon irradiation.
Xb p-Cl		277($\sim$4.1)	alkaline	12000		
Xc p-Br		284($\sim$4.1)	solution	7000		
Xd p-(SO$_2$NH$_2$)		279($\sim$4.1)	"	15000 (280 nm)		
Xe p-NO$_2$ [183,184]		330($\sim$4.1)	"			
Xf m-NO$_2$		268	"			
Xg o-NO$_2$		255				

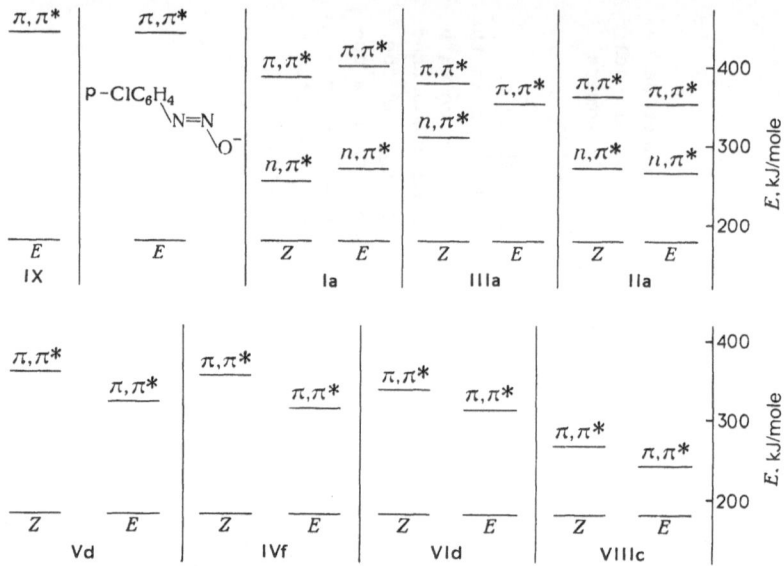

Fig. 13. Energy levels of the n,π*- and π,π*-states of different diazo compounds.

azobenzenes, where in contrast to the E-isomer the n → π*-transition in the Z-isomer is allowed, which is likewise more intense [51]. For 1-aryl-3-heterocyclidenetriazines IV-VIII, no n → π*-transitions were found. Possibly the n → π*-transition is overlapped by the π → π*-transition, which takes place also in the case of heavily substituent-polarized azobenzenes [23].

Relatively substantial differences in the position of the absorption maxima of both isomers of 1-aryl-3-heterocyclidenetriazenes can be explained, in an analogy with stilbenes and azobenzenes, by the disturbed mesomerism of the sterically hindered aryl residue of the Z-isomer. The influence of ortho-substituents in the aryl residue on the spectra of both isomers supports this explanation [323] (Figure 14). In the case of the E-isomer the aryl residue, because of steric hindrances created by the methyl groups, exits from the molecule's plane, the mesomerism is upset, and a hypsochromic shift results. Because of such a rotation, violent torsional vibrations about the N—Ar bond are to be expected. One consequence of this is a widening of the bands and a lowering of the intensity in the absorption maximum. The spectra of the E-isomers of azobenzenes heavily polarized by substituents [121] change in an analogous way upon the introduction of ortho-substituents.

In Z-isomers, even without ortho-substituents, the aryl fragment is twisted out of the molecule's plane for steric reasons. The wide, low-intensity band of the Z-form indicates torsional vibrations around the N—Ar bond. The introduction of ortho-substituents greatly increases the steric hindrances, thereby fixing the aryl fragment more rigidly in a severely twisted form. As a result, the torsional vibrations decrease, which is manifested in a steeper form and a higher intensity of the absorption band. Quantum-chemical calculations performed by the PPP method confirm the twisted arrangement of the aryl fragment in the Z-isomer and correctly register differences in the wavelength of the absorption maxima of both isomers. The results of investigations by means of ^{15}N and ^{13}C NMR spectroscopy are in agreement with these ideas on the steric structure of the isomers [325-327]. The relatively long-wavelength absorption of the E-isomer of p-nitro-benzeneazotriphenylsulfide can be attributed to mesomerism.

142

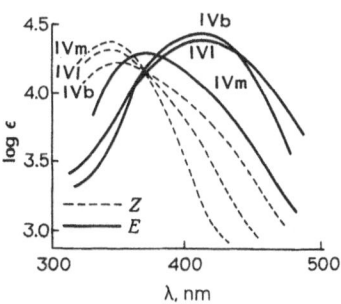

Fig. 14. The electronic spectra of the Z- and E-isomers of compounds IVb,
IVl, and IVm (see Table 8) in dioxane.

Because of the steric hindrances between the H atom in position 2 of
the phenyl residue and the SCPh$_3$ group, the aryl fragment in the Z-form
should be rotated out of the molecule's plane. This disrupts the mesomerism
and results in a rather pronounced (by 55 nm) hypsochromic shift of the
absorption maximum as compared with the E-form. In the E-isomers of aryldiazocyanides, aryldiazosulfonates, and aryldiazosulfides (the latter with
donor substituents in the nucleus) the mesomerism is hindered. That is why
the difference, in the case of these compounds, between the maxima of the
long-wavelength bands of the E- and Z-isomers is small. The donors and
acceptors in the aryl residue induce a weak bathochromic shift in the absorption maximum of the Z-isomer of type IV compounds. This shift is more
pronounced in the E-isomer because of the better conjugation (cf. compounds
IVa and IVo in Table 8).

The simultaneous presence of electron donors and acceptors in the
benzothiazolinylidene residue causes a bathochromic shift of the longest-
wavelength $\pi \rightarrow \pi^*$-transition; electron donors exert a roughly equal influence on the Z- and E-isomer band; the acceptors bathochromically shift
the band of the Z-isomer more than the analogous band of the E-form [328]
(see Table 8). The nature of the heterocycle influences the position of the
long-wavelength $\pi \rightarrow \pi^*$-transition (cf. compounds IV-VI in Table 8) [328].
With the replacement of the N—CH$_3$ group by sulfur atoms the long-wavelength
absorption maxima of the Z- and E-isomers are bathochromically shifted, with
the effects in the Z-isomer being more strongly pronounced, similar to the
influence of substituents in the benzothiazolinylidene residue. Because of
the differing influence of substituents on the position of the absorption
maxima of the Z- and E-isomers, the difference in wavelength between them
also changes. The greatest difference in the position of the spectral bands
for the E- and Z-isomers is observed in the case of triazenes IV (Table 8);
an example of the spectra is given in Figures 14 and 15.

Quantum-chemical calculations showed [324] that the excited state of
triazenes is more polar than the ground state. That is why heterocyclic
triazenes exhibit positive solvatochromism [328].

In contrast to benzoid triazenes, the E-isomers of the above-mentioned
heterocyclic compounds are stable in acids and their color is heavily dependent on the pH [329], which was also observed for azobenzenes. As the concentration of the acid increases, a new band on the long-wavelength edge of
the visible band is formed. In most cases its intensity is higher, which
indicates a protonated form (Table 9). The protonated salt of the E-isomer
can be isolated from solution.

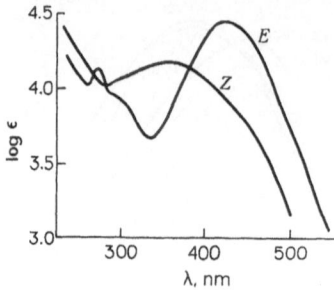

Fig. 15. The electronic spectra of the Z- and E-isomers of compound IVa (see Table 8) in dioxane.

No salts were detected in the Z-isomers of 1-aryl-3-heterocyclidene-triazenes at room temperature since they disintegrate in acid. The proton-ated form of the Z-isomer was detected only at a temperature of −70°C (Table 10). Compared with the E-isomer, the bathochromic shift $\Delta\nu_Z$ of the long-wavelength band is distinctly larger upon protonation. The same relation-ships are observed upon the protonation of Z-azobenzenes [329]. Since the alkylation of triazenes at N^1 results in a similar shift of the long-wave-length band [325], it can be assumed that protonation also proceeds at N^1. The polymethine dye-like structure of the protonated triazene, in contrast to the more polyene-like character of an unprotonated base, accounts for the longer-wavelength absorption and, in the majority of cases, the higher ex-tinction coefficient of the salt.

Investigations of unprotonated and protonated triazenes by means of ^{15}N NMR spectroscopy confirm the validity of the current ideas on the forma-tion of a polymethine system upon protonation [326,329]. The high stability of the E-isomers to acid compared with other triazenes may be attributable to the formation of a polymethine structure since other azapolymethines are also stable in acid [131,330].

Table 9. The maxima of the long-wavelength absorption bands of the unprotonated (λ) and protonated (λ') E-form of certain triazenes [329,336] in mixtures of tetrahydrofuran with water or with HCl [9:1]

Number of compound in Table 8	λ, nm (log ϵ)	λ', nm (log ϵ)	$\Delta\nu_E$, cm^{-1}
IVa	441(4.44)	444(4.43)	154
IVb	422(4.44)	436(4.55)	761
IVc	386(4.38)	426(4.44)	2433
IVd	398(4.43)	434(4.52)	2085
IVe	383(4.45)	437(4.50)	3227
IVf	374(4.40)	433(4.44)	3643
IVg	377(4.41)	443(4.51)	3952
IVh	385(4.46)	462(4.47)	4329

Table 10. The maxima of the long-wavelength absorption bands of the unprotonated (λ) and protonated (λ') Z-form of certain triazenes [329,336] in mixtures of tetrahydrofuran with water or with HCl [9:1]

Number of compound in Table 8	λ, nm (log ϵ)	λ', nm (log ϵ)	$\Delta\nu_Z$, cm^{-1}
IVb	360(4.14)	422(4.49)	4081
IVc	343(4.25)	407(4.37)	4584
IVf	340(4.28)	414(4.50)	5257
IVi	337(4.30)	412(4.40)	5404
IVj	340(4.30)	420(4.48)	5602
IVk	345(4.27)	435(4.36)	5998

The bathochromic shift $\Delta\nu$ upon protonation is a measure of the change in the electron distribution in the molecule. It is smaller, the stronger the polarization of the unprotonated compound. The larger bathochromic shift upon the protonation of the Z-isomer shows that the electronic structure of the unprotonated base, as compared with the E-isomer, is more like that of a polyene.

The influence of substituents on the position of the absorption maxima of triazene salts differs from the effects in bases. For asymmetric cyanine dyes it has been shown [331] that the smaller the difference in the basicity of the terminal groups, the longer the wavelength region in which the dye in question absorbs. This difference in basicity should be the least (among the known compounds of type IV) in the case of the salt of p-methoxy derivative IVk, which results in the longest-wavelength absorption (see Table 10).

The Z- and E-isomers of 1-aryl-3-(3-methylbenzothiazolin-2-ylidene)-triazenes IV can be alkylated at N^1 with triethyloxonium tetrafluoroborate. The resulting 1-aryl-1-ethyl-3-(3-methylbenzothiazolin-2-ylidene)triazenium tetrafluoroborates VIII are a new photochromic system [325,327]. Alkylation in the absence of light results in individual isomers which can be mutually converted photochemically (cf. Table 8 compounds VIIIa-VIIId).

As in protonation, bathochromic shifts of the long-wavelength absorption bands in the Z- and E-isomers are observed in alkylation. The smaller difference, as compared with bases, in the position of bands in the spectra of salts of triazene isomers makes them less suitable for use in data-recording processes.

In contrast to the Z-isomers, the E-isomers of 1-aryl-3-(3-methyl-benzothiazolin-2-ylidene)triazenes form Cu(II) complexes with halides. These can be isolated in the solid state, and they have the following composition: $[Cu(triazene)_2]Hal_2$. The absorption maxima of the complexes are bathochromically shifted compared with the E-triazenes, and almost irrespective of the nature of the substituents in the aryl fragments they are located at 420 nm. For instance [332], for triazene IVk the maxima of the bands in the spectrum of the complex with $CuCl_2$ or $CuBr_2$ lie at 419 nm; for triazenes IVj and IVb with the same halides — at 408 and 426 nm. If irradiation centers on the absorption band of the complex, then E → Z-isomerization does not occur; instead, the triazene is split into a diazonium salt and amine [218].

The Mechanisms of Z/E-Photo- and Thermal Isomerization and Photo- and Thermal Decomposition

The mechanisms of the photochemical and thermal Z/E-isomerization of aryldiazo compounds, unlike that of azobenzenes, have been investigated by only some authors. One reason for this might have been the lower stability of the Z-isomers. The mechanisms of the photo- and thermal isomerization of compounds II, III, and IV have been studied in detail; the Z-isomers of these compounds are generally thermally stable at room temperature.

The quantum yields of the direct photoisomerization of triazenes IV are independent of the nature of the substituents in the para- and meta-positions in the aryl fragment (cf. compounds IVb-IVk in Table 8), of the wavelength of the actinic light, of the polarity of the solvent, and also of the presence or absence of triplet quenchers such as oxygen and guaiazulene [323]. The mean values are as follows: $\phi_{Z \to E}$ 0.53; $\phi_{E \to Z}$ 0.43. Ortho-substituents hinder isomerization and reduce its quantum yields (see compounds IVb, IVl, and IVm in Table 8). As the data presented in Table 11 show, at low temperatures $\phi_{E \to Z}$ decreases to a greater extent than $\phi_{Z \to E}$.

At room temperature no short-lived intermediate states have been detected by means of flash photolysis in the microsecond and nanosecond regions.

Table 11. The temperature dependence of quantum yields of the photo-isomerization of triazenes IV. Ethyl ether:ethanol 1:2, λ 365 nm, measurement error ± 15% [323]

Number of compound in Table 8	t °C	$\phi_{Z \to E}$	$\phi_{E \to Z}$	$\Sigma\phi$	$\phi_{Z \to E}/\phi_{E \to Z}$
IVb	20	0.56	0.32	0.88	1.75
	-115	0.47	0.20	0.67	2.35
IVc	20	0.57	0.44	1.01	1.33
	-115	0.46	0.24	0.70	1.92
IVh	20	0.53	0.47	1.00	1.13
	-115	0.36	0.31	0.67	1.16
IVj	20	0.47	0.45	0.92	1.04
	-115	0.36	0.25	0.61	1.44
IVk	20	0.49	0.50	0.99	0.98
	-115	0.41	0.36	0.77	1.14

The photoisomerization of triazenes IV is sensitized by triplet sensitizers. From the dependence of the position of the photostationary state of
E/Z-photoisomerization on sensitizer energy, $E_T \sim 251$ kJ/mole was obtained
for the E-isomers of the three triazenes shown in Figure 16. Since few measurements were made for the Z-isomers, their E_T could be estimated only
roughly at 167-209 kJ/mole. The sum of $\phi_{Z \to E}$ and $\phi_{E \to Z}$ for sensitized photoisomerization upon a quantitative energy transfer corresponds to ϕ_{isc} of the
sensitizer [219]. Thus, for compound IVf and phenanthrene sensitization,
ϕ_{isc} 0.70-0.88, $\phi_{Z \to E}$ 0.51, $\phi_{E \to Z}$ 0.34, and $\phi_{Z \to E} + \phi_{E \to Z}$ 0.85; if biphenyl is
the sensitizer, then ϕ_{isc} 0.51-0.81, $\phi_{Z \to E}$ 0.39, $\phi_{E \to Z}$ 0.25, and $\phi_{Z \to E} +$
$\phi_{E \to Z}$ 0.64.

The ratio of the deactivation of the T_1^X state calculated from the quantum yields or from the position of the photostationary state is in good
agreement with the ratio of the quantum yields of direct photoisomerization
(corresponds to the ratio of deactivation of the excited intermediate
state). For instance, for compound IVb, $k_{X^\bullet \to E}/k_{X^\bullet \to Z}$ upon direct photoisomerization is 1.35 ± 0.27, while upon sensitized photoisomerization this ratio
is 1.30 ± 0.23. The same values for compounds IVf and IVk are, respectively, 1.25 ± 0.25, 1.45 ± 0.24 and 1.05 ± 0.21, 1.15 ± 0.16 [333].

When the temperature is lowered, the $\phi_{Z \to E}/\phi_{E \to Z}$ ratios of sensitized and
direct photoisomerization change in a similar fashion in the same direction. For instance, for triazene IVf in a mixture of diethyl ether and
ethanol (1:2) at 20°C direct isomerization gives the ratio 1.33, naphtha-
lene-sensitized isomerization — 1.47, biphenyl-sensitized isomerization —
1.50. The same values at −110°C are 1.92, 2.45, and 2.00, respectively. It
follows from this that isomerization probably proceeds through the triplet
state [219].

Figure 17 reflects the isomerization path. In the S_2-state, triazenes
may disintegrate with low quantum yields. From this the conclusion is drawn
regarding the presence of an excited intermediate state T_1^X. The direct
radiationless deactivation of the S_1- or T-states of the Z- and E-isomers
into initial ground states can be neglected since the total of the quantum
yields of the forward and back reactions is approximately unity. Since the
sums of the quantum yields of sensitized photoisomerization correspond
roughly to ϕ_{isc} of the sensitizers, the deactivation constants k_1 and k_1' are
negligibly small compared with the constants k_2 and k_2'. The transitions of
the T_1^Z- and T_1^E-states into the T_1^X-state are responsible for the temperature
dependence of direct and sensitized photoisomerization. With falling k_2

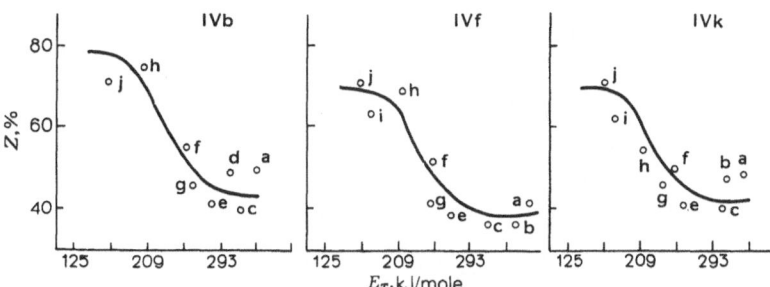

Fig. 16.   The dependence of the position of the photostationary state of
E/Z-photoisomerization on the energy of the sensitizer triplet
for compounds IVb, IVf, and IVk (see Table 8). Sensitizers: a)
benzene; b) anisole; c) aniline; d) diphenylamine; e) biphenyl;
f) phenanthrene; g) naphthalene; h) benzyl; i) eosine; j) anthracene; k) methylene blue.

and k_2' at low temperatures, the relation k_2, $k_2' \gg k_1$, k_1' is not obeyed and the direct radiationless deactivation of the T_1^Z- and T_1^E-states competes with the transition into the T_1^X state. That is why the sums of the quantum yields of direct photoisomerization at low temperatures are less than unity. The mechanism described explains adequately all the properties of the triazenes studied to date. However, in the forward reaction, the inter-mediate triplet states were not proven definitely.

The quantum yields of the photoisomerization of N-ethyltriazenium salts VIIIa-VIIIc (see Table 8), especially in the E $\rightarrow$ Z direction, are smaller compared to the bases. Triazenium salts VIII are highly photostable. This photostability is apparently due to photochromism since by means of photo-isomerization the energy of the exciting light is efficiently converted into thermal energy. That is why salts VIII are used as dyes for polyacrylo-nitrile fibers [131,334,335].

The photoisomerization of triazenes IV is iodine-sensitized [42]. No complex formed between the triazene and iodine has been detected in a cyclo-hexanone solution of components. Irradiation in the iodine absorption band results exclusively in Z $\rightarrow$ E-isomerization, and the E $\rightarrow$ Z-transition is not initiated by iodine. The study of the dependence of $\phi_{Z \rightarrow E}$ on the quantity of absorbed light proves unequivocally the splitting of the iodine molecule into atoms, which then initiate isomerization. The rate constant of the reaction between iodine atoms and Z-triazene is over 2.5×10^5 liters/ (mole $\cdot$ s); this is more than what was found in the case of azobenzenes and stilbenes. Radical initiators such as azobisisobutyronitrile and dibenzoyl peroxide are unsuitable for initiating isomerization.

The dependence of the rate of thermal Z $\rightarrow$ E-isomerization of substi-tuted triazenes IV on the nature of the substituents in the aryl or con-densed nucleus and also on the temperature and nature of the solvent was studied [5]. The isomerization rate goes up upon the introduction of elec-tron-accepting substituents into the ortho- and para-positions of the aryl residue (ρ 4.4) and donor substituents into position 6 of the condensed nu-cleus (ρ about -1.9); donors in the aryl fragments and acceptors in the meta-position hinder thermal isomerization in solution. For instance, for compounds IVb,d,e,l,m,s,t the rate constants ($k \times 10^6$) for E/Z-isomerization in dimethylformamide at 50°C are 493, 103.4, 6.2, 346, 38, 191, and 1960 sec^{-1}, respectively [43].

The influence of the nitro group from the ortho-position is much less than from the para-position. One or two additional ortho-methyl groups noticeably slow the reaction rate with respect to the para-nitro group. This points to steric hindrances of thermal isomerization. The replacement of the benzothiazolinylidene fragment with the more basic N,N'-dimethyl-

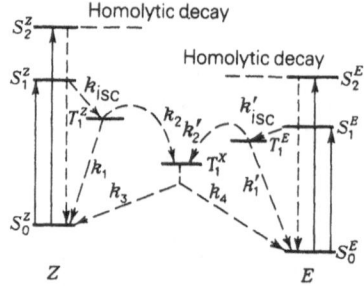

Fig. 17. Yablonskii's diagram for the photochemical Z/E-isomerization of 1-aryl-3-(3-methylbenzothiazolin-2-ylidene)triazenes [333].

benzimidazolinylidene fragment increases the isomerization rate, while the substitution of the dithiolylidene fragment produces the reverse effect.

The rate of the isomerization of polarized triazenes increases dramatically with rising polarity of the solvent (Table 12). Within the protic solvent group the isomerization rate increases with an increase in their acidity. There is a linear dependence between the logarithm of the rate constants and the pK_a of alcohols [5]. Weak acids such as tri-tert-butylphenol accelerate isomerization in an aprotic-dipolar solvent, e.g., in dimethylformamide.

In the quaternary salts of triazenes VIII thermal isomerization is facilitated compared with bases IV. The influence of the electronic effects of substituents on the rate of isomerization of N'-alkylated triazenes is similar to their influence on bases [325]. The values of log k correlate linearly with the donor number of the corresponding solvent. It follows from this correlation that the solvent, coordinating around the cation of the dye because of donor—acceptor interactions, solvates the transition state more efficiently than the ground state.

The parameters of the activation of thermal isomerization in the case of triazenes IVs and IVt are as follows:

In the monomethyl ether of ethylene glycol at 40-72°C for triazene IVt, E_a 90.4 kJ/mole and $\Delta S^{\neq}$ −54.4 J/(mole · K); in dimethyl sulfoxide at 30-45°C these values are 80.8 and −71.2, respectively. For triazene IVs in o-dichlorobenzene at 40-60°C, E_a 83.7 kJ/mole and $\Delta S^{\neq}$ −71.2 J/(mole · K); in CH_3CN at 30-50°C these values are 69.9 and −92.1, respectively; in dimethylformamide at 30-50°C, 69.9 and −92.1; and in dimethyl sulfoxide at 20-40°C and 30-45°C, 73.3 and −62.8 [328].

The activation energy is influenced by the solvent; the activation energies of triazenes are lower than those for the isomerization of many substituted azobenzenes but are far greater than in the case of substituent-polarized azobenzenes. The high negative values of activation entropy indicate a stronger solvation of the transition state compared with the starting state and thus to a higher polarity.

Table 12. The influence of solvents on the rate constants of thermal Z → E-isomerization of triazenes IVs and IVt (see compounds in Table 8)[24]

Solvent	$k_{IVs} \cdot 10^7$, s^{-1} (40°C)	$k_{IVt} \cdot 10^7$, s^{-1} (30°C)
Dioxane	7.3	18.3
2-Isopropyl alcohol	58.8	
Methanol	224.5	
Ethanol	145.2	
Monomethyl ether of ethylene glycol	274.0	
Pyridine	204.5	
Dichlorobenzene	51.0	84.0
Acetonitrile	314.5	1614.0
Dimethylformamide	638.5	3660.0
Dimethyl sulfoxide	2155.0	8650.0
Sulfolane	1764.0	
Hexametapol	1366.0	

If we take into account the criteria mentioned earlier for distinguishing between the mechanisms of rotation and inversion, we can conclude that the results described suggest a polar rotational mechanism of the thermal isomerization of 1-aryl-3-heterocyclidenetriazenes [24].

The photolytic degradation of triazenes IV as compared with photoisomerization plays a subordinate role; it is a side reaction especially in the case of triazenes with donor substituents in the aryl fragment. In this process the ϕ of photodegradation is at least 1-2 orders of magnitude lower than the ϕ of isomerization. The following products of the photodegradation of 1-(4-chlorophenyl)-3-(3-methylbenzothiazolin-2-ylidene)triazene in dioxane were identified: nitrogen (92.4%), chlorobenzene (45.0%), benzene (10.6%), 2-imino-3-methylbenzothiazoline (24.7%), 2-(4-chlorophenylimino)- or 2-phenylimino-3-methylbenzothiazoline (11.6%), and the hydrochloride of 2-imino-3-methylbenzothiazoline (15.9%) [328]. This ratio of products suggested a radical mechanism of photodegradation. The thermolysis of triazenes, owing to the high stability of both isomers, does not contribute to photodegradation. However, the difference in reactivity of the E- and Z-isomers upon triazene protolysis is noteworthy: the Z-isomer dissociates in acid medium with rate constants 4-5 orders of magnitude higher than in the case of the E-isomer. In this process the hydrochloride of 1-imino-3-methylbenzothiazoline and the corresponding diazonium salt are formed [329, 336].

The following are the rates of protolysis of the E- and Z-isomers of triazenes IV [329] in a THF medium with HCl (9:1) at c_{HCl} 1.29 moles/liter. The data are presented in the following order — the number of the compound in Table 8; $k_Z \cdot 10$, s^{-1}; $k_E \cdot 10^5$, s^{-1}; k_Z/k_E. IVb; 1.4; 5.0; 2800. IVc; 2.3; 1.4; 16,000. IVe; 2.2; 2.5; 8800. IVf; 8.0; 0.86; 93,000. IVg; 5.4; 0.58; 93,000. IVh; 8.3; 0.75; 111,000. IVi; 20.4; 2.2; 93,000. IVj; 39.2; 6.8; 58,000. IVk; 71.3; 10.0; 37,500.

The Z- and E-isomers of $Ph-N=N-S-C_nH_{2n+1}$ compounds were synthesized in the aryldiazosulfide series [291,292]. The Z-isomers are sufficiently stable to enable the investigation of the photochemistry involved without interference from thermal reactions. Upon irradiation in methanol solution with 405 nm wavelength light (n → π*-excitation) a photostationary state is established. Upon irradiation with light of 300 nm wavelength (π → π*-excitation) no photostationary state was detected because photodissociation occurs alongside isomerization [283]. Quantitative analysis of the absorption in the ultraviolet and visible regions shows that photodissociation products are formed only from Z-isomers upon excitation into the S_2-state (π → π*-excitation) [294]. For instance, the data of EPR and CIDNP in the NMR spectra indicate that decay products are formed, at any rate some of them are, from the singlet of the tert-butylthiyl—diazenyl radical pair, which corresponds to dissociation from the S_2-state. Heating the Z-isomer gives the same decay products that it yields upon photolysis [292]:

An analysis of the quantum yields and experiments involving the use of triplet sensitizers show that photoisomerization apparently proceeds through triplet states [288,337]. A polar rotational mechanism [290] and an inversion mechanism [338] are proposed for the thermal isomerization of the Z-isomers of aryldiazosulfides into E-isomers.

Aryldiazocyanides exhibit a behavior similar to aryldiazosulfates. The ratio of photoisomerization and photodissociation in this class of compounds also depends on the wavelength of the actinic light, and photodissociation proceeds in the S_2-state ($\pi \to \pi^*$-excitation) of the Z-isomer [288].

The influence of substituents on the rates of ionization, dissociation, and Z $\to$ E-isomerization was investigated in the case of Z-O-ethylaryldiazotates. The results support the ionization—recombination mechanism for thermal Z $\to$ E-conversion [10].

Given below is a scheme of thermal and photoreactions of aryldiazo compounds which also describes the photolysis of azoalkanes:

The E-form is relatively stable and may photochemically change initially only into the Z-form. The Z-isomer is often unstable and may photochemically and thermally isomerize into the E-form, and it may also disintegrate homolytically or heterolytically.

The contribution of the various paths to the overall scheme of Z-isomer conversion depends on the stability of the compound involved. For instance, in compounds (IV) degradation, because of the high stability of the Z-isomer, plays but a secondary role; in compounds I, II, and III the Z-isomers are less stable and degradation becomes dominant. The Z-isomers of aryldiazosulfones, ethers, and carboxylic acids and amines appear to be so unstable as to be undetectable at room temperature, at any rate. The irradiation of these compounds, which usually exist in the form of the E-isomers, should have resulted initially in the formation of Z-isomers, which immediately undergo thermal disintegration. The nature of bond cleavage upon photolysis and thermolysis is strongly dependent on the kind of reaction environment: for instance, in polar solvents heterolytic cleavage of aryldiazosulfones is predominant, while in nonpolar solvents — homolytic cleavage [322].

The Uses of the Photochromic Properties of Aryldiazo Compounds

The narrow differences between the positions of the absorption maxima of both isomers coupled with the instability of the Z-isomers of most aryldiazo compounds are unfavorable for a direct use of their photochromism in data-recording materials. Triazenes IV, however, are suitable for this purpose since the difference in wavelength between the absorption maxima of the E- and the relatively stable Z-isomers in them is as much as 100 nm.

A careful selection of substituents or the right heterocyclic imine makes it possible to put together photochromic systems with the thermal back reaction of the Z-isomer into the E-form $\left(E \underset{h\nu'\ \Delta}{\overset{h\nu}{\rightleftarrows}} Z\right)$ and photochromic systems without a thermal back reaction $\left(E \underset{h\nu'}{\overset{h\nu}{\rightleftarrows}} Z\right)$. Triazenes which at room temperature do not undergo Z $\to$ E-conversion can be used as reversible data-recording materials. Depending on the nature of substituents and by selecting a suitable wavelength of the actinic light, the content of the desired isomer in solution can be brought up to 70-90%. The change in optical density achieved in the process could, in principle, be used for data recording [339]. 1-Aryl-3-heterocyclidenetriazenes can be deposited on F-cellite layers, where photoisomerization also occurs. The quantum yields in the solid layer are lower as compared with those in solution and amount to $\phi_{Z \to E}$ 0.1 (333, 350 nm) and $\phi_{E \to Z}$ 0.4 ± 0.2 for IVf. The 30 experimentally determined cycles achieved in the case of this particular triazene can be extrapolated to 150 cycles. Apparently, by optimizing the layer's composition in a suitable way, degradation could be suppressed and the number of cycles increased still further.

The photochromic properties of aryldiazo compounds can be used for data recording indirectly since the two isomers have different stabilities and reactivities. The use of aryldiazo compounds as a depot of diazonium salts offers a number of advantages compared to diazonium salts because of their higher stability and lower polarity; other advantages include safer processing and handling, easier introduction into the nonpolar layer-forming media, and absorption in a longer-wavelength region. A number of data-recording processes incorporating the use of diazo compounds have therefore been developed and patented. Thus, aryldiazosulfonates are suitable for obtaining negative diazo materials which upon irradiation isomerize and then dissociate into a diazo cation and a sulfite anion. The diazonium ion can enter into a diazo reaction to form a dye [25]. On the other hand, the sulfite may create a latent image [340-344] from the metal salt present in the layer which may be capable of a physical manifestation:

$$Ar \overset{N=N}{\diagdown} SO_3^- \overset{h\nu}{\rightleftharpoons} Ar \overset{N=N}{\diagdown} SO_3^- \rightleftharpoons SO_3^{2-} + ArN_2^+ \longrightarrow azo\ dye$$

$$SO_3^{2-} + Hg_2^{2+} \longrightarrow (HgSO_3)\ complex + Hg \to physical\ manifestation$$

It is recommended to use such systems in the production of printing plates and ultramicrofilms [344].

Similarly, diazosulfides may also be used for developing recording materials with a physical manifestation [341,344,345].

$$Ar \overset{N=N}{\diagdown} SR + \frac{y+1}{2} Hg_2^{2+} + zAg^+ \longrightarrow 2Ar-N_2^+ + Hg(SR)_2 + Hg_y Ag_z$$

$$y = 1 \longrightarrow z = 0; \qquad y = 0 \longrightarrow z = 1$$

The resulting Z-form, unlike the E-form, is capable of giving a metallic latent image from the Hg(I) salt or from mixtures of Hg(I) and silver salts. Such benzenediazosulfide films are easy to produce and are recommended for use in reprography (document reproduction, etc.) and for the production of masks in microelectronics [229].

Aryldiazosulfones disintegrate upon irradiation (apparently, this is preceded also by E → Z-isomerization), while the initial compound, persisting in areas not exposed to irradiation, or the products of its conversions form, together with the azo component, a positive image [346-348]. The creation of negative materials is based on the photochemical disintegration of diazosulfone, whose radical cleavage products catalyze the polymerization in the layer of the material developed for data recording [349]. Diazosulfone-containing materials can be sensitized up to the red region of the spectrum [350].

Diazosulfonates have found application in negative photopolymerization processes to produce printing plates [357] and to photochemically develop colored prints on fabrics [352] ("photo rapid"). Triazene derivatives have been described as light-sensitive components of diazonegative materials [350,353].

In compounds IV the rates of isomer dissociation in acid media can be used for data recording [354]. E-Triazene is introduced into the layer along with a weak acid, which does not result in the disintegration of the E-form, and the azo component (e.g., phloroglucinol). Later, in the irradiated areas the Z-isomer forms, which is more readily split by acid. The resulting diazonium salt combines with the azo component to form a dye, and eventually a negative image appears. Towards the end, the unirradiated E-triazene can be removed from the layer by dissolution or by photodissociation upon ultraviolet irradiation (after the layer had been acidified by gaseous HCl) [329,354]. The advantages of this method include the good preservation of the materials, thanks to the high thermal stability of the triazenes, as well as the convenient region of the materials' photosensitivity, which lies at 350-600 nm; the sensitivity is the same as in the case of ordinary normal diazo materials.

THE PHOTOCHROMISM OF COMPOUNDS CONTAINING THE C=N CHROMOPHORE

Many groups of compounds with the C=N chromophore are photochromic due also to E/Z-isomerization. Compared with systems featuring C=C and N=N bonds, the body of available data on the mechanisms of the photochromic conversions of these compounds is smaller, even though their photochromism has long been a recognized fact (see surveys [358-360]). This is largely due to the spectroscopic properties of both stereoisomers, thermochromism strongly pronounced for certain types of these compounds, and the catalytic action of acids on thermal E/Z-isomerization. Photoinduced E/Z-isomerization has been detected for azomethines, oximes, oxime esters, hydrazines, semicarbazones, and azines.

Electronic Spectra

The authors of many earlier works dealing with the spectra of compounds with the C=N bond completely ignored the existence of configurational and conformational isomers, and this led them to err in estimating the results. Serious work in this field was not published until recently. However, the authors made no systematic study of the impact of stereoisomerization on the spectra of such substances.

A few examples of the spectra of isomers are presented in Table 13. Data are available on azomethine dyes [240,245] which, however, do not enable us to draw conclusions regarding the C=N chromophore since it forms part of the azapolymethine system. Data are also available on azomethines [238,239], but they are of a qualitative character. In all the groups of compounds with the C=N chromophore the $n \rightarrow \pi^*$- and $\pi \rightarrow \pi^*$-transitions are

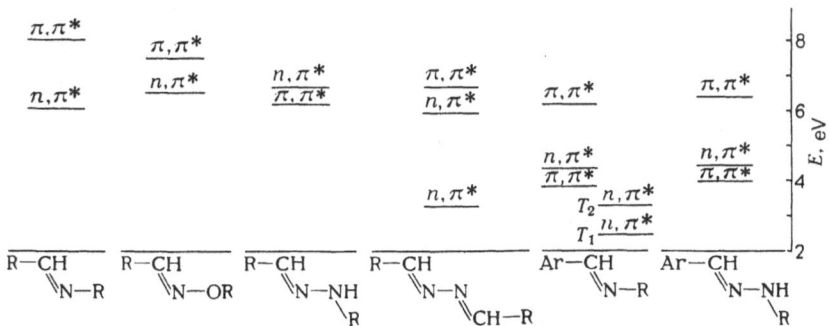

Fig. 18. Energy levels of the n,π^*- and π,π^*-states of the E-isomers of various compounds with the C=N chromophore.

Table 13. Data on the photochromic properties of compounds with the C=N chromophore in different solvents

Compound	λ_{max}, nm (log ϵ)		$\phi_{Z \to E}$; $\phi_{E \to Z}$	Note
	Z-isomer	E-isomer		
2-Naphthyl-CH=NNHPh [442]	340(4.05), cyclohexane 345(4.12), benzene	348(4.45), cyclohexane 353(4.51), benzene	0.22; 0.084, cyclohexane 0.15; 0.015, ethanol	ϕ is independent of wavelength; solvent influence is strong - different in both directions
PhCH=NNHR [443]		330(4.30), ethanol	0.14; 0.025, ethanol	Strong solvent influence
ArCH-NNHC$_6$H$_4$NO$_2$-p [379]	370(4.43), benzene	388(4.54), benzene	0.28; 0.28, benzene, 313 nm, Ar=C$_6$H$_5$	ϕ is dependent on wavelength and substituents. Isomerization sensitized
2-Pyridyl-CH=NNHPh [444]		347, benzene; 355, ethanol	0.007; 0.02 benzene	ϕ is independent of wavelength; photosteady state* 0.68
PhCH(OH)C(Ph)=N-NHCOC$_6$H$_4$Cl-p [385]	294(4.27), methanol	265(4.18), methanol	0.11; 0.16, methanol, 254 nm	ϕ is dependent on wavelength, little dependent on solvent; photosteady state 7 (303 nm)
PhCH=NOH [430]			0.40; 0.38 (benzene, 290 nm)	Isomerization sensitized, photosteady state 1 (290 nm)
PhC(CH$_3$)=NOCH$_3$ [430]^{3*}	238, cyclohexane	248(4.04), cyclohexane	0.29; 0.37, pentane	Photosteady state 2.2 (254 nm)
2-Naphthyl-C(CH$_3$)=NOCH$_3$ [383,439]	337(2.04), pentane 279(3.93)	338(2.14), pentane 295(4.18)	0.40; 0.40 (pentane, 313 nm) 0.27; 0.28 (pentane, 254 nm)	ϕ is dependent on concentration and similarly depends on photosteady state

Compound			Quantum yield	Notes
$CH_3COC(CH_3)=NOC_2H_5$ [390][4*]	244(3.51), cyclohexane 328(1.79)	234(4.03), cyclohexane 318(1.43)	0.53; 0.18 (acetonitrile, 366 nm)	Isomerization sensitized; photosteady state 6 (366 nm)
$PhCOC(CH_3)=N-OC_2H_5$ [392][5*]	246(4.04), cyclohexane 322(1.60)	251(4.09), cyclohexane 328(1.51)		Isomerization sensitized; photosteady state 1.3 (366 nm). Photosteady state 3.2 (254 nm)
$(Ph)_2C=N-N=CH-9$-anthryl [449]	391(3.81), toluene	416(4.15), toluene	0.21; 0.005, toluene	ϕ is independent of wavelength; $Z \rightarrow E$-isomerization sensitized
$PhC(CH_3)=NNHCONH_2$ [394]	240, ethanol	273, ethanol		
$PhCH(OH)C(Ph)=NNHCONH_2$ [385]	275(4.10), methanol	238(4.25), methanol	0.25; 0.24 (methanol, 254 nm)	ϕ is dependent on wavelength; photo-steady state 5.7 (254 nm)

* Photosteady state given [E]/[Z].

** E_S^E 405, E_T^E = 220 kJ/mole.

3* E_S^Z 410, E_S^E 402, E_T 300, E_T^E 145 kJ/mole.

4* E_S^Z 310, E_S^E 320, E_T^E 235, E_T^E 285 kJ/mole.

5* E_T^E 310, E_T^Z 265 kJ/mole.

possible, while below 190 nm, Rydberg transitions are also possible. The positions of these electronic transitions feature a complex dependence on the substituents in the C=N group, and especially on the nature of residues at the N atom, on the type of stereoisomer, and on the nature of the sol- vents (see Table 13). A typical position of the levels of E-isomers of compounds with the C=N chromophore is given in Figure 18.

The electron shell of the atoms of the isomethine C=N bond corresponds to sp^2-hybridization. It follows from this that the unshared electron pair of the nitrogen atom lies on the sp^2-orbital and the substituent X of the C=N—X group is rotated about 115-120° in the plane:

| E-isomer | Z-isomer |

| 1E, 3Z | 1E, 3E | 1Z, 3Z | |
| s-trans-conformation | | | s-cis-conformation |

The resulting stronger s-character of the unshared pair leads to an increased excitation energy of these electrons. Since in the process the π^*-orbital of isoelectronic C=N and C=O chromophores corresponds to an atomic orbital of the 2p type, the frequency of the n → π^*-transition of the C=N chromophore is hypsochromically shifted relative to the C=O band. The n → π^*-transition of the C=N chromophore belongs to the C_s point group and thus is symmetry-allowed. This leads to relatively high values of ϵ (Table 14).

The substituent X in the C=N—X group may interact with its n- and π-electrons. This interaction was investigated experimentally by means of photoelectronic spectroscopy [364,369,370] and is strongly influenced by the angle of rotation about the N—X bond. In E-azomethines this angle is be- tween 40 and 55° [369,371-375]; in hydrazones, between 25 and 35° [369, 370]. This rotation makes possible interactions between the π-electron system of the substituent X and the C=N chromophore, which was proven by the study of the spectra of benzalanilines [376]. If the angle of rotation θ_{N-Ar} is increased by the introduction of an ortho-substituent into the nitrogen aryl, such compounds have the same spectrum that the N-alkylbenzal- anilines have; i.e., no π-interactions with substituents at the nitrogen occur in this case.

Similar influences are observed in the case of hydrazones (X = NR_2), where n,n- and n,π-conjugations are also heavily dependent on steric rela- tions involving the participation of the substituent R [366]. If the n-electron of the NR_2 group is blocked by means of protonation or quater- nization, n,π-conjugation disappears, and these compounds absorb in the shorter-wavelength region. The angle of rotation θ_{N-X} in Z-isomers is about 90°. Therefore, conjugation with the π- and n-electrons of the chromophore in them is hindered or stopped, and the Z-isomers of these groups of com- pounds absorb usually in a shorter-wavelength region than do the E-iso- mers. Typical examples are presented in Table 13.

Table 14. Band maxima in the electronic spectra of compounds with
          the C=N chromophore in different media. In the case of possible
          E/Z-isomerism the data refer to a pure E-isomer

Compound	λ, nm (log ϵ)	
	$\pi \to \pi^*$	$n \to \pi^*$
$C_2H_5CH=NC_2H$ [361]	168(4.03) gas phase	232(2.30) gas phase
$(CH_3)_2C=NCH_3$ [362]	181(3.78) cyclohexane	244(2.20) cyclohexane
Ph-CH-NCH$_3$ [363]	240(f0.32) hexane	278(f0.01) hexane
$CH_3CH=CHCH=NC_2H_5$ [361]	211(4.23) gas phase	270(2.42) gas phase
$CH_2=NOH$ [364]	180(3.75) gas phase	213(2.52) gas phase
$CH_2=C(CH_3)CH=NOH$ [365]	217(4.19), 229(4.23) ethanol	
$(CH_3)_2C=NNH_2$ [366]	212(3.45) ethanol	
$(CH_3)_2C=NNHCH_3$ [366]	227(3.65) ethanol	
$(CH_3)_2C=NNHPh$ [367]	270(4.23) ethanol	
$PhCH=NNH_2$ [368]	212(4.07), 271(4.16) methanol	
$PhCH=NNHPh$ [366]	237(4.16), 302(4.05), 344(4.38) ethanol	
$(CH_3)_2C=NNHCSNH_2$ [365]	229(3.85), 271(4.33) ethanol	

     Substituents at the carbon atom of the C=N group also exert a strong
influence on the energy of electronic transitions due to electronic and
steric effects (see Table 13). The electronic influence of the C-substitu-
ents on the n $\to$ π^*- and $\pi \to \pi^*$-transitions (see Figure 18) is considerable
in the absence of hindrances to a coplanar arrangement of substituents,
especially aromatic ones, with the C=N chromophore; this is observed in the
case of aldehyde derivatives having E- and Z-configurations. However, the
authors of some papers advocate the opposite point of view for Z-isomers,
viz., that the substituent at the carbon atom is twisted and does not lie in
the plane of the C=N chromophore [279,380]. At the same time, it is obvious
that the interaction with substituents at the C=N chromophore differs from
stilbenes and azobenzenes, where the substituents at both ends of the chromophore in the Z-isomers are twisted.

     In ketone derivatives the interaction of π-electron systems bonded to
the carbon atom in the C=N group may be disturbed, owing to unfavorable
steric relationships [366,382,383]. In extreme cases, it results in a total
cessation of conjugation, with the appearance of the spectra of individual
chromophores as, for instance, in the oxime esters of alkyl naphthyl ketones
[383].

     The relationships change if an intramolecular hydrogen bond arises in
one of the isomers [379,381]. Typical examples are provided by the Z-isomers of 2-pyridinaldehyde arylhydrazones [384] and benzoinhydrazone derivatives [385]. A six-membered chelate is formed in them, thanks to which the
n-electrons of the substituted nitrogen atom effectively interact with the
C=N group. As a result, the long-wavelength band of this isomer is bathochromically shifted compared with the E-isomer (see Table 13).

     The relationships in azines are complex, and this makes interpretation
of the spectra difficult. Because of the presence of two chromophore
groups, the existence of three steroisomers in them is possible, from which,
moreover, additional s-trans- and s-cis-conformers may arise [386,387].
Added to this is rotation about the C-R bonds. Quantum-chemical calculations showed [388] that the ground state of many azines is severely twisted.

This is also observed in the case of α-acyloximes [389-393] and certain azo-methine dyes [25]. Apart from various configurational isomers, the existence of numerous conformational isomers is also possible.

In the spectra of aldimines and ketimines from aliphatic carbonyl compounds, the n → π*- and π → π*-transitions display wide differences in energy and well-defined bands [361,362,395,396]. The lower excited state has an unmistakable n,π*-nature. In the case of oximes and hydrazones the frequency of the n → π*-transition increases. In oximes this results in a converging of the excitation energies of both transitions, which explains why there are contradictory data in the literature on the position of the n → π*- and π → π*-bands in the spectra of this class of compounds [364, 397,398]. In hydrazones, because of decreased energy levels, the π,π*-state is the lower excited singlet state [366,369,370]. Because of the narrow energy difference, the π,π*-transition is overlapped by the intense n → π*-band and is not identified by means of routine ultraviolet spectroscopy.

Aliphatic azines feature two low-energy n → π*-transitions. In azines with aromatic substituents the energy of the n → π*- and π → π*-transitions declines sharply because of conjugation, with the result that the π,π*-state becomes the lower excited singlet state [363,367-369,373,376].

The data reported in the literature [399,400] on the n → π*-nature of the S_1-state are possibly based on an error (see Figure 18).

The data presented in Table 13 indicate that differences in the absorption maxima of the E- and Z-isomers are not considerable, and this important property of photochromic systems is but poorly expressed. In most cases, the absorption bands, moreover, are extraordinarily heavily overlapped, so much so that for all practical purposes no region exists where only one of the isomers of any of the substances studied to date will absorb. Somewhat more favorable are the relations within those groups of compounds in which the Z-isomer possesses an intramolecular hydrogen bond. In this case, the values of Δλ are greater and the absorption bands are overlapped to a lesser extent and, in addition, the Z-isomer absorbs in the longest-wavelength region. Thus, the spectroscopic properties of the E- and Z-isomers of C═N chromophore systems make their practical application more difficult and hamper a detailed investigation of the mechanism of the photochromic transition.

Attempts to measure the luminescence spectra of the azomethines of benzaldehyde under a wide variety of experimental conditions were not successful [363,399], although N-salicylidene-β-naphthylamine [413] fluoresces well, as do both isomers of the methyl ether of β-naphthyl methyl ketone oxime [383]. However, the E- and Z-isomers of benzophenone-9-acridinal-dehyde azine display differences in luminescence: the E-isomer lacks fluorescence, while the Z-isomer fluoresces at 77°K in an EPA (ether—pentane—alcohol) matrix with a quantum yield of 1×10^{-3}. This is attributed to the differing nature of the lower excited states: in the case of the E-isomer we are talking of an n,π*-nature, while in the case of the Z-isomer — of a π,π*-nature [412].

The Mechanisms of the E/Z-Photo- and Thermal Isomerization of Compounds with the C═N Chromophore

There is scant and often contradictory information in the literature on the mechanisms of the E/Z-photoisomerization of compounds with the C═N chromophore. We have already discussed the reason why. Apart from geometric isomerization, almost all of the compounds in the class described also enter into a variety of photoreactions. Apparently, the direction of the

158

latter is influenced by the nature (n,π or π,π*) of the lower excited state in terms of energy. Because of the slight energy differences, it is difficult to populate selectively both these states upon direct excitation (see Figure 18), which is why several parallel reactions often occur.

Aldimines and ketimines undergo photoreduction in solvents, which can easily act as hydrogen donors (reactions involving n,π*-states) [405-416]; E → Z-isomerization takes place concurrently. For many derivatives of aromatic carbonyl compounds, because of the low activation barrier of reverse thermal isomerization, it is possible to study Z → E-photoisomerization only at low temperatures [376,404,405] or, alternatively, by pulse techniques [377,378,406-411]. It is not known whether Z → E-isomerization is induced by light. The amount of published quantitative data on photoisomerizations is limited. For instance, for benzalanilide $\phi_{E\to Z}$ is 0.09 [378]; the thermal back reaction at room temperature occurs so quickly that it is impossible to detect any change in the system using conventional macroscopic photolysis techniques. The activation barriers of thermal Z → E-isomerization are between 55 and 80 kJ/mole [25,404,407,409]. A great many factors influence the mechanism of this reaction [382,419-424]; in solutions the mechanism is particularly subject to the influence of traces of acid [382,419-424].

E/Z-Photoisomerization about the C=N bond of azomethines has been studied in some detail (Figure 19) [26,402,418]. This photoisomerization may proceed either by direct excitation or by means of triplet sensitization.

The study of the influence of oxygen, as well as of external and internal heavy-atom effects, revealed that upon direct excitation isomerization proceeds through the S_1-state. Internal heavy atoms or a solvent with a heavy atom and triplet sensitizers channel the reaction through triplet states [418]. In dyes containing electron-donor substituents, isomerization proceeds by a rotational mechanism, while in dyes with acceptor substituents it proceeds by means of inversion.

In compounds with the C—H group in the α-position, thermal isomerization proceeds by the mechanism of imine—enamine tautomerism [426], which was confirmed by deuterium—hydrogen exchange experiments:

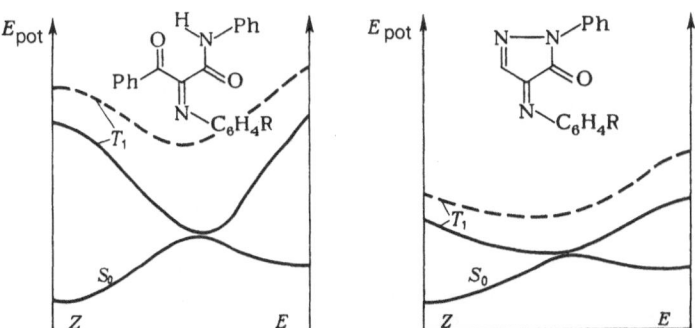

Fig. 19. Potential curves for the photoisomerization of azomethine derivatives with donor (——) and acceptor (- - -) substituents.

159

For azomethines an inversion mechanism of the thermal Z → E-transition [374,408,410,411] is accepted; this mechanism has been substantiated by means of quantum-chemical calculations [12]. A similar mechanism in compounds with acceptor substituents is discussed for photochromic azomethine dyes. The introduction of donor substituents modifies the reaction mechanism on account of the better stabilization of the transition state [244]. A rotational mechanism is realized in these compounds.

Thus, a characteristic feature of aldimines and ketimines is that light energy in the $E \xrightarrow{h\nu} Z \xrightarrow{\Delta} E$ cycle is converted into heat without causing any substantial chemical change. That is why the quantum yields of other photoreactions of this class of compounds are generally low, which reflects the considerable photostability of the system involved. For azomethine dyes, which play an important part in the production of color photographic and cine film images, this effect is highly important [427]. Azomethines could also be used as photostabilizers of polymers. Unfortunately, no such application of azomethines has yet been reported in the literature.

The thermal E/Z-isomerization of oximes and their ethers under normal conditions does not occur; because of the high energy barriers (170-210 kJ/mole), the rate constants of the thermal reaction are 10^{-13} s^{-1} even at 333°K [428]. This slowdown is apparently due to the acceptor influence of the OR or OH groups. The energy barriers calculated by quantum-chemical techniques and determined experimentally exhibit a good linear dependence on the electronegativity of the X substituents of the C=N—X group (Figure 20) [429].

The thermal stability of the stereoisomers of oximes makes it possible to investigate photochemical reactions far more thoroughly in this series than in the case of azomethines. Upon the photolysis of these substances the following events occur: E/Z-isomerization, Beckmann's photorearrangement [398,430,432], dissociation of the N—O bond [380,394,398,433,434,435], and the formation of an aldehyde or a corresponding ketone [398,436]. Naturally, other reactions are also possible due to other chromophores [437, 438]. Along with structural prerequisites, the decisive influence on the reaction's direction is exerted by the solvent. Solvent effects are at this time difficult to link to the change in relative position of the n,π*- or π,π*-levels or, for that matter, to the specific impact of the medium

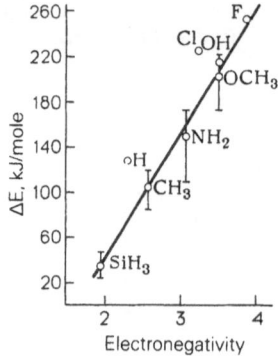

Fig. 20. The dependence of the energy barriers of thermal E/Z-isomerization on the electronegativity of substituents X in the C=N—X group. The points indicate values calculated by the CNDO/2 method for models; vertical lines show the energy barriers for the corresponding classes of compounds found experimentally.

on the reactivity of the corresponding states. E/Z-Isomerizations proceed for the most part in nonpolar solvents (see the literature survey in the paper [358]).

This isomerization occurs in both directions E → Z and Z → E with great efficiency both upon direct excitation and upon triplet sensitization. Examples for which quantitative data are available are summarized in Table 13.

Opinions differ regarding the multiplicity of the reacting state upon direct excitation. For benzaldoxime a triplet state isomerization has been postulated [431] since the composition of the photostationary mixture is not in agreement with the calculations obtained for isomerization in a singlet state. However, more detailed investigations of the photochromism of oxime ethers [428] and oxime α,β-unsaturated carbonyl compounds [398] showed that isomerization proceeds in a singlet state. These reactions are not quenched by piperylene, and the composition of the photostationary mixture corresponds with a priori calculations. Both isomerization directions should proceed through one and the same intermediate state. Radiationless transitions are postulated from inverted excited states, for which reason the sum $\phi_{E \to Z} + \phi_{Z \to E}$ is less than unity.

The nonsensitized isomerization of the oxime ether of β-naphthyl methyl ketone [383,439] features a far more complex mechanism. Although a singlet state isomerization would appear to be more preferable in this case, it turned out that the composition of the photostationary state was dependent on the concentration of oxime ether (E:Z 1.8 at 3×10^{-3} mole/liter and 0.7 at 1.35 moles/liter), temperature (the fraction of the Z-isomer drops at elevated temperatures), and on the nature of the solvents. These dependences can be accounted for with the help of concepts of the partial formation of exciplexes identifiable by a recognizable fluorescence band at higher concentrations of the starting material. The deactivation of the exciplex into the ground state of the E- or Z-isomer should proceed with varying efficiency, which explains the concentration dependence of the ratio of the resulting isomers. The experiments involving quenching by means of a singlet quencher − 1,3-cyclohexadiene − established that fluorescence of oxime ether isomers is subject to quenching to a greater extent than isomerization. This is evidence for the formation of excimers upon this type of quenching.

For α-ketooximes it is assumed that the direct and sensitized E/Z-isomerizations proceed through the triplet π,π^*-state. This conclusion was drawn on the basis of the discrepancy between the relationship of the photostationary state and preliminary calculations for the reaction path through the singlet and the lack of isomerization quenching with Z-1,3-pentadiene. However, the photodissociation products obtained seem to indicate a reaction in the S_1 (n, π^*)-state [390,393].

Regarding the potential practical applications of E/Z-photoisomerization of oximes, we should note only slight differences between the spectra of isomers, short-wavelength absorption, and numerous side reactions upon photolysis. For this reason, it is difficult to expect the direct use of these compounds as bases for photochromic materials. It has been suggested instead to use oxime isomerization for the photostabilization of polymers [440], as well as for synthesizing the Z-isomers of isonicotinaldoxime, with their valuable pharmacological effects [441]. So far, the observed different reactivity of E- and Z-isomers of oximes has not been put to practical use in the chemical industry [389]. Irradiation of the oxime would probably change the reaction's direction.

Upon irradiation, hydrazones either undergo E/Z-isomerization [385,379, 381,442-444,445] and form dissociation products of the N—N bond [445] (nitriles from aldehyde hydrazones, alternatively — amines and imines) or undergo reduction into hydrocarbons [436,445] (Wolff—Kishner photoreduction). The last two reactions proved to be preferable upon the population of the higher excited states [442]. As for other classes of compounds containing the C—N chromophore, the dependence of the direction of photochemical reactions on the $\pi,\pi*$- or n,$\pi*$-nature of the excited state has not been studied. This makes the formulation of any broad generalizations difficult. In a series of these structures, E/Z-isomerization is light-induced in both directions and proceeds upon direct irradiation and upon triplet sensitization (Table 13). The data for the series of hydrazones of aryl- and 2-naphthaldehydes show that the value of $\phi_{Z \to E}$ is greater than $\phi_{E \to Z}$. This is attributable primarily to the steric interactions between the NH group and aryl rings in Z-isomers. The quantum yields of both isomerization directions within the 254-365 nm wavelength range are independent of the wavelength of the actinic light. This suggests a general transition state in isomerizations. The solvent exerts a far greater influence on E → Z-isomerization since the formation of associates through N—H bridges with the solvent is possible above all in the E-isomer. It is assumed that the isomerization proceeds via a rotary mechanism in the triplet state, even though it is quenched by azulene or ferrocene. Both isomerization directions are sensitized by fluorene, with the fluorescences of the sensitizer not being subject to quenching. Energy transfer occurs in the phenylhydrazone of 2-naphthaldehyde in the exciplex and results in a tenfold increase in $\phi_{E \to Z}$. The $\phi_{E \to Z}/\phi_{Z \to E}$ ratios of the sensitized and nonsensitized reactions differ widely, which provides evidence against the triplet mechanism of the nonsensitized isomerization. This is supported by the findings of a study of the E/Z-isomerization of benzoinacyl- and benzoinarylhydrazones [385], as well as of the phenylhydrazone of α-pyridinaldehyde [444,445], for which a singlet path was established on the basis of the absence of an effect of oxygen, internal and external heavy-atom effects, and the influence of triplet quenchers.

By means of laser photolysis the singlet mechanism of the photoisomerization of α-ketohydrazones was found; the photoproduct is unstable and at room temperature changes (k about 1 sec^{-1}) into a more stable isomer. The ratio of isomers in this process was not ascertained [446].

The dependence of the quantum yields and the composition of isomers in the stationary state of benzophene-sensitized isomerization of p-nitrophenylhydrazones of aromatic aldehydes on the nature of the substituents [379] was investigated. It turned out that in the case of unsubstituted ($\phi_{E \to Z}$ 0.65, $\phi_{Z \to E}$ 0.27) and p-methoxy-substituted derivatives ($\phi_{E \to Z}$ 0.56, $\phi_{Z \to E}$ 0.35), the twisted (rotated) $\pi,\pi*$-triplet serves as a common intermediate particle. Conversely, the T-states of the E- and Z-isomers of p-nitrophenylhydrazone of p-nitrobenzaldehyde appear to have different characters: for the E-isomer a $\pi,\pi*$-nature, and for the Z-isomer an n,$\pi*$-nature. It becomes clear why the ϕ of both isomerization directions differ so widely: $\phi_{E \to Z}$ 0.45, $\phi_{Z \to E}$ 0.053. Nonsensitized photoisomerization apparently proceeds through the short-lived T_1-state; isomerization proper (by one of the paths considered above) should proceed in a hot ground state. Evidence of this is provided by the dependence of ϕ on the wavelength of the actinic light. Different conformers become excited in the process. In general, substituents reduce the quantum yields in both directions of the isomerization (see Table 13).

Z → E-Isomerization of different hydrazones may also proceed thermally [385,442,443,447]. An inversion mechanism was discussed for this isomerization [240]. Small activation enthalpies (106 kJ/mole for the phenylhydrazone of 2-naphthaldehyde [442,443], 62 kJ/mole for the p-chlorobenzoylhydrazones of benzoin) are at variance with the calculations performed for such a path. Apparently, the relationships are far more complex, and allowance has to be made for the possibility of inversion on the azomethine nitrogen atom by means of rotation about the N–N or C–N bonds. Such variants have been discussed for benzoin derivatives in the first instance. For these compounds a thermal E → Z-isomerization was also established at 301°K ($\Delta H^{\neq}$ 51 kJ/mole); an intramolecular hydrogen bond which may be present in the Z-isomer facilitates this isomerization.

Only a few instances of the photochromism of semicarbazones have been described in the literature [380,385,394]. In the semicarbazone of benzoin, substitution of the carbonyl group [379] may compete with E/Z-photoisomerization upon prolonged irradiation; both isomerization directions are light-induced. Oxygen has no effect on the efficiency of photoisomerization, which can be sensitized by aromatic hydrocarbons with an energy of $E_S > 400$ kJ/mole; no quenching by triplet quenchers was observed, which suggests isomerization in the singlet state.

The geometrical isomerization of hydrazones and semicarbazones is noted for the small values of $\Delta\lambda$ and for the accompanying irreversible side reactions. The available data on these systems do not offer much hope for the direct practical application of these systems in photochromic materials. As for other areas of the potential application of the photoisomerization of hydrazones and semicarbazones, the ideas set out for oximes also apply to it.

Depending on their photochromic properties, azines can be divided into four groups [449]. Certain compounds (e.g., benzalazine) do not isomerize on exposure to light, but instead they undergo cleavage at the N–N bond [450,451]. If in azines (see scheme on p. 156) R = R' and R" = R''', then E → Z-isomerization is possible. Azines in which R = R''' ≠ R" = R' belong to the third group. In these cases the following isomerizations occur: 1E, 3E → 1E, 3Z → 1Z, 3Z. The fourth group includes azines possessing various R–R''' substituents which makes the formation of four stereoisomers conceivable. No studies of the photoisomerization of compounds in this group have yet been reported.

In the case of benzophenonaldazines (aldehyde — 9-formylanthracene or 9-formylacridine) both stereoisomers can be isolated from the photolysis product upon direct irradiation in an air-saturated solution [449]. The mutual conversion of these isomers is also light-induced. In the process $\phi_{E \to Z}$ is always smaller than $\phi_{Z \to E}$ (for the diphenylmethylidenehydrazone of 9-formylanthracene: 0.0053 and 0.21; for the diphenylmethylidenehydrazone of 9-formylacridine: 0.003 and 0.15, respectively).

These reactions cannot be quenched by a triplet quencher with an E_T between 270 and 154 kJ/mole; the E → Z-isomerization cannot be initiated by triplet sensitizers in the same energy region; upon direct excitation neither oxygen nor solvents with a heavy atom have any influence on it. Therefore, it is generally believed that the S_1-state of the E-isomer reacts, while the cause of the low quantum yields is the radiationless $S_1 \to T_n \to S_0$ process. For the Z → E-isomerization a singlet mechanism is also postulated. Radiationless deactivation into the ground state is a competing process to this isomerization. However, the Z → E-isomerization can be initiated by sensitizers with an E_T of 240-165 kJ/mole.

9-Anthracenaldazine as a representative of the third group of azines has been studied in detail, with all three possible stereoisomers of the s-trans-conformation being isolated and described. The photochromism of this compound can be described by the following scheme (quantum yields are indicated):

$$1E, 3E \underset{\substack{hv, \Delta. \\ 0.015}}{\overset{\substack{0.024 \\ hv}}{\rightleftarrows}} 1E, 3Z \underset{\substack{hv, \Delta \\ 0.23}}{\overset{\substack{0.010 \\ hv}}{\rightleftarrows}} 1Z, 3Z$$

Direct conversion of the 1E, 3E-isomer into the 1Z, 3Z-isomer was not proved [449]. Analogous results were obtained for other azines [452].

The photoinduced E → Z-isomerization of the azines of the second group is completely thermally irreversible [448]. Evidence against the simple inversion or rotational mechanism is provided by the activation parameters found for the diphenylmethylidenehydrazone of 9-formylanthracene (in methanol, T 298°K, $\Delta H^{\neq}$ 100.1 kJ/mole, $\Delta S^{\neq}$ −12.6 kJ/mole · K) [386,387]. These findings are in very good agreement with inversion about the nitrogen atom and the simultaneous rotation about the N—N or C—N bond. Evidence for inversion is provided by the dependence of the activation parameters on the nature of the solvents. It was proved in this instance that thermal E → Z-isomerization is possible. At room temperature 97% of the E- and 3% of the Z-isomer correspond to the thermal equilibrium in methanol [448]. In the case of azines, as in other compounds with the C—N chromophore, changes in optical density following E/Z-isomerization are small since the geometrical isomers feature similar spectra.

REFERENCES

1. Techniques of Chemistry. Vol. III. Photochromism, ed. by G. H. Brown, Wiley-Interscience, New York—London (1971).
2. J. Epperlein, B. Hofmann, and K. Stopperka, J. Signal AM, 4(3):155-186 (1976).
3. S. Dähne, Z. Wiss. Photogr., Photophys. Photochem., 62:183 (1968).
4. N. Wiberg and H. J. Pracht, Chem. Ber., 105:1392 (1972).
5. D. Gegiou, K. A. Muszkat, and E. Fischer, J. Am. Chem. Soc., 90:3907 (1968).
6. P. S. Engel and C. Steel, Acc. Chem. Res., 6:275 (1973).
7. H. Kessler, Tetrahedron, 30:1861 (1974).
8. T.A.J.W. Wajer and T. J. de Boer, Recl. Trav. Chim. Pay-Bas, 91:565 (1972).
9. E. J. Grubbs and J. A. Villarreal, Tetrahedron Lett., 1841 (1969).
10. T. J. Broxton, Austr. J. Chem., 32:1031 (1979).
11. N. A. Porter and L. Y. Marett, J. Am. Chem. Soc., 95:4361 (1973).
12. T. Sueyoshi, Sh. Nishimura, Sh. Yamamoto, and Sh. Hasegawa, Chem. Lett., 1131 (1974).
13. D. Wurmb-Gerlich, F. Vögtle, A. Mannschreck, and H. A. Staab, Liebigs Ann. Chem., 708:36 (1967).
14. H. Kessler and D. Leibfritz, Chem. Ber., 104:2143 (1971).
15. E. R. Talaty and J. C. Fargo, Chem. Commun., 65 (1967).
16. P. D. Wildes, J. G. Pacifici, G. Irick, and D. G. Whitten, J. Am. Chem. Soc., 93:2004 (1971).
17. H.-O. Kalinowski, H. Kessler, and A. Walter, Tetrahedron, 30:1137 (1974).
18. J. Binenboym, A. Burcat, A. Lifshitz, and J. Shamir, J. Am. Chem. Soc., 88:5039 (1966).
19. D. R. Kearns, J. Phys. Chem., 69:1062 (1965).
20. N. C. Baird and J. R. Swenson, Can. J. Chem., 51:3097 (1973).
21. J. M. Howell and L. J. Kirschenbaum, J. Am. Chem. Soc., 98:877 (1976).
22. S. Ljunggren and G. Wettermark, Acta Chem. Scand., 25:1599 (1971).
23. G. Gabor and E. Fischer, J. Phys. Chem., 75:581 (1971).
23a. R. Arnaud and P. Jacques, C. R. Acad. Sci. Paris, Ser. C, 282:1097 (1976).
24. E. Fanghänel, R. Hänsel, and J. Hohlfeld, J. Prakt. Chem., 319:485 (1977).

25. W. V. Herkstroeter, J. Am. Chem. Soc., **95**:8686 (1973).
26. W. V. Herkstroeter, J. Am. Chem. Soc., **98**:330 (1976).
27. P. S. Engel, Chem. Rev., **80**:99-150 (1980).
28. N. R. Camp, I. R. Epstein, and C. Steel, J. Am. Chem. Soc., **99**:2453 (1977).
29. N. J. Turro, J. McVey, V. Ramamurthy, and P. Lechtken, Angew. Chem., **91**:597 (1979).
30. G. S. Hammond, J. Saltiel, et al., J. Am. Chem. Soc., **86**:3197 (1964).
31. D. V. Bent and D. Schulte-Frohlinde, J. Phys. Chem., **78**:446 (1978); **78**:451 (1978).
32. G. Zimmermann, L.-Y. Chow, and U. J. Paik, J. Am. Chem. Soc., **80**:3528 (1958).
33. J. Blanc, J. Phys. Chem., **74**:4037 (1970).
34. H. J. Niemann and H. Mauser, J. Phys. Chem. N.F., **82**:295 (1972).
35. W. Ortmann, Dissertation, TH Leuna-Merseburg (1977).
35a. A. M. Thompson, P. C. Goswami, and G. L. Zimmermann, J. Phys. Chem., **83**:314 (1979).
36. J. Saltiel, J. Am. Chem. Soc., **89**:1037 (1967).
37. J. Saltiel and E. D. Megarity, J. Am. Chem. Soc., **94**:2742 (1972).
38. L. M. Stephenson and G. S. Hammond, Angew. Chem., **81**:279 (1969).
39. E. Fischer, Ber. Bunsenges. Phys. Chem., **73**:758 (1969).
40. R. Kretschmer, Dissertation, Universität Tübingen (1972).
41. S. Yamashita, D. P. Cosgrave, H. Ono, and O. Toyama, Bull. Chem. Soc. Jpn., **36**:688 (1963).
42. W. Ortmann and E. Fanghänel, J. Prakt. Chem., **322**:401 (1980).
43. J. Ronayette, R. Arnaud, P. Lebourgis, and J. Lemaire, Can. J. Chem., **52**:1868 (1974).
44. H. B. Gray, M. Wrighton, and G. S. Hammond, J. Am. Chem. Soc., **93**:3285 (1971).
45. D. R. Arnold and P. C. Wong, J. Am. Chem. Soc., **101**:1894 (1979).
46. A. J. Bard, V. J. Puglisi, J. V. Kenkel, and A. Lomax, Faraday Discuss. Chem. Soc., **56**:353 (1973).
47. J. Fabian and H. Hartmann, Light Absorption of Organic Colorants, Vol. VII, Springer-Verlag, Berlin—Heidelberg—New York (1980).
48. W. Adam and O. De Lucci, Angew. Chem., **92**:815-832 (1980).
49. H. Dürr and B. Ruge, Topics in Current Chemistry, Vol. 66, Springer-Verlag, Berlin—Heidelberg (1976), pp. 53-87.
50. J. Munder, "Diazotypie und verwandte Prozesse," IV. Internationaler Kongress für Reprographie und Information, Hannover (1975); Plenarvorträge S. 120 ff.
51. H. Rau, Angew. Chem., **85**:248-258 (1973).
52. S. Patai (ed.), The Chemistry of Hydrazo, Azo, and Azoxy Groups, Wiley, New York (1975).
53. R. J. Boyd, J. C. G. Bunzli, and J. P. Snyder, J. Am. Chem. Soc., **98**:2398 (1976).
53a. M. B. Robin, R. R. Hart, and N. A. Kuebler, J. Am. Chem. Soc., **89**:1564 (1967).
54. D. P. Wong, W. H. Fink, and L. C. Allen, J. Chem. Phys., **52**:6291 (1970).
55. H. Hsu, Chem. Phys., **7**:187 (1975).
56. N. C. Baird, P. De Mayo, J. R. Swenson, and M. C. Usselmann, J. Chem. Soc. Chem. Commun., 314 (1973).
57. B. Tinland, Spectrosc. Lett., **3**:51 (1970).
58. N. W. Winter and R. M. Pitzer, J. Chem. Phys., **62**:1269 (1975).
59. J. R. Lombardi, W. Klemperer, M. B. Robin, H. Bosch, and N. A. Kuebler, J. Chem. Phys., **51**:33 (1969).
60. R. Dichfield, J. E. Del Bene, and J. A. Pople, J. Am. Chem. Soc., **94**:703 (1972).
61. E. Haselbach and A. Schmelzer, Helv. Chim. Acta, **54**:1575 (1971).
62. E. Hasselbach and E. Heilbronner, Helv. Chim. Acta, **53**:684 (1970).

63. K. N. Houk, Y.-M. Chang, and P. S. Engel, J. Am. Chem. Soc., **97**:1824 (1975).

64. J. Kroner, W. Schneid, N. Wiberg, B. Wrackmeyer, and G. Ziegenleder, J. Chem. Soc. Faraday Trans. II, **74**:1909 (1978).

65. M. J. Mirbach, K. C. Liu, M. F. Mirbach, W. R. Cherry, N. J. Turro, and P. S. Engel, J. Am. Chem. Soc., **100**:5122 (1978).

66. F. Brogli, W. Eberbach, E. Haselbach, E. Heilbronner, V. Hornung, and D. Lemal, Helv. Chim. Acta, **56**:1933 (1973).

67. H. Bisle, M. Römer, and H. Rau, Ber. Bunsenges. Phys. Chem.,**80**:301 305 (1976).

68. M. Gisin and J. Wirz, Helv. Chim. Acta, **59**:2273-77 (1976).

69. H. Bisle and H. Rau, Chem. Phys. Lett., **31**:264 (1975).

70. S. Yamamoto, N. Nishimura, and S. Hasegawa, Bull. Chem. Soc. Jpn., **44**:2018 (1971).

71. J. Kroner and H. Bock, Chem. Ber., **101**:1922 (1968).

72. H. Bock, K. Wittel, M. Veith, and N. Wiberg, J. Am. Chem. Soc., **98**:109 (1976).

73. H. Bock, Angew. Chem., **77**:469 (1965).

74. H. Rau and G. Kortüm, Ber. Bunsenges. Phys. Chem., **71**:664 (1967).

75. G. Kortüm and H. Rau, Ber. Bunsenges. Phys. Chem., **68**:973 (1964).

76. P. S. Engel, L. D. Fogel, and C. Steel, J. Am. Chem. Soc., **96**:327 (1974).

77. B. S. Solomon, T. F. Thomas, and C. Steel, J. Am. Chem. Soc., **90**:2249 (1968).

78. S. G. Cohen, R. Zand, and C. Steel, J. Am. Chem. Soc., **83**:2895 (1961).

79. D. Grasso, S. Millefiori, and S. Fasone, Spectrochim. Acta, **31A**:187 (1975).

80. J. Fabian and J. Sühnel, J. Prakt. Chem., **315**:307 (1973).

81. M. L. Heyman and J. P. Snyder, J. Am. Chem. Soc., **97**:4416 (1975).

82. L. D. Fogel and C. Steel, J. Am. Chem. Soc., **98**:4859 (1976).

83. E. V. Brown and G. R. Granneman, J. Am. Chem. Soc., **97**:621 (1975).

84. H.-J. Timpe and Th. Merseburg, Unveroffentlichte Wiss. Ergebnisse.

85. H.-J. Timpe, U. Müller, and J. Franze, Z. Chem., **20**:440 (1980).

86. M. J. Namking, N. K. Naimy, C.-A. Cole, N. Ishikawa, and T. L. Fletcher, J. Org. Chem., **35**:728 (1970).

87. C. J. Brown, Acta Crystallogr., **21**:146 (1966).

88. H. Hope and D. Victor, Acta Crystallogr., **B25**:1849 (1969).

89. H. Suzuki, Electronic Absorption Spectra and Geometries of Organic Molecules, Academic Press, New York (1967).

90. M. Komeyama, S. Yamamoto, N. Nisnimura, and S. Hasegawa, Bull. Chem. Soc. Jpn., **46**:2606 (1973).

91. M. Traetteberg, I. Hilmo, and K. Hagen, J. Mol. Struct., **39**:231 (1977).

92. J. Favrot, J.-M. Leclerq, R. Roberge, C. Sandorfy, and D. Vocelle, Chem. Phys. Lett., **53**:433 (1978).

93. H.-J. Hoffmann and P. Birner, J. Mol. Struct., **39**:145 (1977).

94. A. Mostad and C. Romming, Acta Chem. Scand., **25**:3561 (1971).

95. G. C. Hampson and J. M. Robertson, J. Chem. Soc., 409 (1941).

96. D. L. Beveridge and H. H. Joffe, J. Am. Chem. Soc., **88**:1948 (1966).

97. D. Bontschev and E. Ratschin, Monatsh. Chem., **101**:1454 (1970).

98. D. Bontschev and E. Ratschin, Monatsh. Chem., **103**:1000-1010 (1972).

99. G. Hohlneicher and W. Sänger, in: Quantum Aspects of Heterocyclic Compounds in Chemistry and Biochemistry, Vol. 2, ed. by E. D. Bermann and P. Pullman, Academic Press, New York (1970), p. 193.

100. A. Goursot, P. Jacques, and J. Faure, J. Chim. Phys. Physicochim. Biol., **73**:694 (1976).

101. G. M. Badger, R. J. Drewer, and G. E. Lewis, Austr. J. Chem., **19**:643 (1966).

102. V. Rehak, F. Novak, and I. Cepciansky, Coll. Czech. Chem. Commun., **38**:697 (1973).

103. P. Tomasik, Rocz. Chem., **44**:1369 (1970).

104. I. Skulski and T. Urbanski, Rocz. Chem., 34:141 (1960).
105. P. P. Birnbaum, J. H. Linford, and D.W.G. Style, Trans. Faraday Soc., 49:735 (1953).
106. S. Skulski, W. Waclawck, and B. Müller, Bull. Acad. Pol. Sci., 19:329 (1971).
107. G. S. Kikot, M. G. Chauser, and M. I. Cherkashin, Izv. Akad. Nauk SSSR, Ser. Khim., 2150 (1976).
108. R. W. Castelino and G. Hallas, J. Chem. Soc. B., 793 (1971).
109. K. Kokkinos and R. Wizinger, Helv. Chim. Acta, 54:330 (1971).
110. K. Kokkinos and R. Wizinger, Helv. Chim. Acta, 54:335, 338 (1971).
111. G. Hallas, L. W. Ho, and R. Todd, J. Soc. Dyers Colour, 90:121 (1974).
112. L. M. Yagupolski and L. S. Gandelsman, Zh. Obshch. Khim., 35:362 (1965).
113. L. M. Yagupolski and L. S. Gandelsman, Zh. Obshch. Khim., 37:2101 (1967).
114. J. Fabian, Chimia, 32:328 (1978).
115. J. Griffiths, J. Soc. Dyers Colour, 88:106 (1972).
116. J. Griffiths and B. Roozpeikar, J. Chem. Soc., Perkin Trans. I, 42 (1976).
117. W. T. Hanson, J. Photogr. Sci., 25:189 (1977).
118. J. Griffiths, J. Chem. Soc. B, 801 (1971).
119. J. P. Dix and F. Vogtle, Angew. Chem., 90:893 (1978).
120. P. Gregory and D. Thorp, J. Chem. Soc., Perkin Trans. I, 1990 (1979).
121. E. Hoyer, R. Schickfluss, and W. Steckelberg, Angew. Chem., 85:984 (1973).
122. G. Hallas and K. L. Ng, J. Soc. Dyers Colour, 93:284 (1977).
123. H. Mustroh, J. Epperlein, and D. Ebert, J. Signal AM, 5:439 (1976).
124. A. S. Bailey and J. J. Merer, J. Chem. Soc. C, 1345 (1966).
125. S. Hünig, G. Kiesslich, P. Linhart, and H. Schlaf, Liebigs Ann. Chem., 752:196 (1971).
126. S. Hünig and H.-Chr. Steinmetz, Liebigs Ann. Chem., 1060 (1976).
127. H. Quast and E. Schmidt, Liebigs Ann. Chem., 732:43 (1970).
128. H. Baumann and J. Dehnert, Chimia, 15:163 (1961).
129. P. A. Mikhailenko and L. I. Shevchuk, Khim. Geterotsikl. Soedin., 135 (1974).
130. S. Hünig, G. Kiesslich, K. M. Otto, and H. Quast, Liebigs Ann. Chem., 754:46 (1971).
131. J. Voltz, Chimia, 15:168 (1961).
132. H. Dahn and H. V. Castelmur, Helv. Chim. Acta, 36:638 (1953).
133. A. I. Kiprianov and V. Yu. Buryak, Ukr. Khim. Zh., 34:1016 (1968).
134. A. I. Kiprianov and V. Yu. Buryak, Ukr. Khim. Zh., 179 (1969).
135. R. J. Morris and W. R. Brode, J. Am. Chem. Soc., 70:2485 (1948).
136. I. Ya. Bershtein and O. F. Ginsburg, Usp. Khim., 41:177 (1972).
137. R. A. Cox and E. Buncel, in: Chemistry of the Hydrazo, Azo, and Azoxy Groups, Part II, ed. by S. Patai, Wiley, London (1975), Vol. 18, p. 775.
138. T. Kobayashi, E. O. Degenkolb, and P. M. Rentzepis, J. Phys. Chem., 83(19):2431 (1979).
139. G. Gabor and K. H. Bar-Eli, J. Phys. Chem., 72:153 (1968).
140. A. E. Lutskii, V. V. Bocharova, and M. R. Kreslawskii, Zh. Obshch. Khim., 45:2276 (1975).
141. G. Irick and J. G. Pacifici, Text. Res. J., 42:391 (1972).
142. S. Hünig, G. Bernhard, W. Liptay, and W. Brenninger, Liebigs Ann. Chem., 690:9 (1965).
143. S. Hünig and H. Herrmann, Liebigs Ann. Chem., 636:32 (1960).
144. B. A. Korolev and S. P. Titova, Zh. Org. Khim., 7:1188 (1971).
145. B. A. Korolev, S. P. Titova, and V. N. Ufimzev, Zh. Org. Khim., 7:1191 (1971).
146. A. Steiniger and V. Gutmann, Mh. Chem., 96:1173 (1965).
147. L. M. Yagupolski and L. Z. Candelsman, Ukr. Khim. Zh., 45:145 (1979).

148. F. Gerson, E. Heilbronner, A. van Veen, and B. M. Webster, Helv. Chim. Acta, 43:1889 (1960).
149. E. Haselbach, Helv. Chim. Acta, 53:1526 (1970).
150. M. A. Hofnagel, A. van Veen, and B. M. Webster, Recl. Trav. Chim. Pays-Bas, 88:562 (1969).
151. H. Haselbach, A. Henriksson, A. Schmelzer, and H. Berthou, Helv. Chim. Acta, 56:705 (1973).
152. V. A. Kogan, N. N. Kharabaev, O. A. Osipov, and E. N. Yurchenko, Zh. Obshch. Khim., 45:362 (1975).
153. Sh. Yamamoto, N. Nishimura, and Sh. Hasegawa, Bull. Chem. Soc. Jpn., 46:194-198 (1973).
154. P. D. Bolton, J. Ellis, K. A. Fleming, and I. R. Lantzke, Austr. J. Chem., 26:1005-1014 (1973).
155. K. Y. Chu and J. Griffiths, Tetrahedron Lett., 405 (1976).
156. P. A. Mikhailenko, L. I. Shevchuk, and A. I. Kiprianov, Khim. Getero-tsikl. Soedin., 923 (1973).
157. J. Sühnel, H. Hartmann, and J. Fabian, Z. Chem., 18:183 (1978).
158. J. Sühnel and J. Fabian, Z. Chem., 14:275 (1974).
159. E. Lippert and W. Voss, Z. Phys. Chem., 31:321 (1962).
160. B. J. Baba and L. Godman, J. Chem. Phys., 43:2902 (1965).
161. M. Chowohyry and L. Goodman, J. Chem. Phys., 36:548 (1962).
162. D. S. Andrews and A. C. Day, J. Chem. Soc., Chem. Commun., 477 (1967).
163. H. Rau, J. Luminescence, 112:191 (1971).
164. M. J. Mirbach, M. F. Mirbach, W. R. Cherry, N. J. Turro, and P. S. Engel, Chem. Phys. Lett., 53:266 (1978).
165. H. Rau, Ber. Bunsenges. Phys. Chem., 72:637 (1968).
166. G. Gabor, Y. Frei, D. Gegiou, M. Koyanowitsch, and E. Fischer, J. Chem., 5:193 (1967).
167. H. Hartmann, D. Leupold, and St. Stephan, DDR-Patent WP 133500.
168. K. Hiraki, Bull. Chem. Soc. Jpn., 46:2438 (1973).
169. H. Rau, Ber. Bunsenges. Phys. Chem., 71:48 (1967).
170. F. Pragst, Z. Phys. Chem., 256:312-318 (1975).
171. R. J. Boyd, J. C. Snyder, J. P. Bunzli, and M. L. Heymann, J. Am. Chem. Soc., 95:6479 (1974).
172. T. Kobayashi, K. Yokota, and S. Nagahura, J. Electron Spectrosc. Rel. Phen., 6:167 (1975).
173. F. Brogli, Helv. Chim. Acta, 56:1933 (1974).
174. H. Schmidt, A. Schweig, B. M. Trost, H. B. Neubold, and P. H. Soudder, J. Am. Chem. Soc., 96:622 (1971).
175. K. E. Gilbert, J. Org. Chem., 42:609 (1977).
176. L. N. Domelsmith, K. N. Houk, J. W. Timberlake, and S. Szilagyi, Chem. Phys. Lett., 48:471 (1977).
177. P. S. Engel, VIII IUPAC Symposium on Photochemistry, Seefeld, Austria (1980).
178. N. J. Turro, J.-M. Liu, H. D. Martin, and M. Kunze, Tetrahedron Lett., 21:1299-1302 (1980).
179. N. J. Turro, K.-C. Liu, W. Cherry, J.-M. Liu, and B. Jacobson, Tetrahedron Lett., 555 (1978).
180. P. S. Engel, R. A. Hayes, L. Keifer, S. Szilagyi, and J. W. Timber-lake, J. Am. Chem. Soc., 100:1876 (1978).
181. L. B. Jones and G. S. Hammond, J. Am. Chem. Soc., 87:1219 (1965).
182. C. C. Wamser, R. T. Medary, I. E. Kochevar, N. J. Turro, and P. L. Chang, J. Am. Chem. Soc., 97:4864 (1975).
183. O. A. Mosher, M. S. Foster, W. M. Flicker, I. I. Beauchamp, and A. J. Kupermann, Chem. Phys., 62:3424 (1975).
184. J. Metcalfe, S. Chervinsky, and I. Oref, Chem. Phys. Lett., 42:190 (1976).
185. J. Ronayette, R. Arnaud, P. Lebourgis, and J. Lemaire, Can. J. Chem., 52:1848 (1974).
186. S. Malkin and E. Fischer, J. Phys. Chem., 66:2482 (1962).
187. E. Fischer, J. Am. Chem. Soc., 82:3249 (1960).

188. D. Gegiou, K. A. Muszkat, and E. Fischer, J. Am. Chem. Soc.,90:12 (1968).
189. J. L. Gardette, G. Guyot, R. Arnaud, and J. Lemaire, Nouv. J. Chim., 1:287 (1977).
190. E. Fischer, J. Am. Chem. Soc., 90:796 (1968).
191. R. Arnaud, J. Ronayette, and J. Lemaire, C. R. Acad. Sci., Ser. C., 274:2144 (1972).
192. R. Ronayette, R. Arnaud, P. Lebourgis, and J. Lemaire, Can. J. Chem., 52:1858 (1974).
193. S. K. Lower and M. A. El-Sayed, Chem. Rev., 66:199 (1966).
194. S. Hashimoto and K. Kano, Bull. Chem. Soc. Jpn., 45:852 (1972).
195. G. Irick and J. G. Pacifici, Tetrahedron Lett., 1303, 2207 (1969); J. Am. Chem. Soc., 91:5654 (1969).
196. P. Hugelshofer, J. Kalvoda, and K. Schaffner, Helv. Chim. Acta, 43:1322 (1960).
197. K. H. Grellmann, G. M. Shermann, and H. Linschitz, J. Am. Chem. Soc., 86:303 (1964).
198. F. B. Mallory and C. S. Wood, Tetrahedron Lett., 2643 (1965).
199. C. P. Joshua and V.N.R. Pillai, Tetrahedron Lett., 3559 (1973).
200. G. E. Lewis, Tetrahedron Lett., 12 (1960).
201. C. P. Joshua and V.N.R. Pillai, Tetrahedron, 30:3333 (1974).
202. G. Gabor and E. Fischer, J. Phys. Chem., 66:2478 (1962).
203. P. Haberfield, P. H. Block, and M. S. Lux, J. Am. Chem. Soc., 97:5804 (1975).
204. H. J. Hofmann, Z. Chem., 19:424 (1979).
205. C. D. Eisenbach, Makromol. Chem., 179:2489 (1978).
206. V. I. Pergushov, O. M. Mikhailik, A. K. Vorob'er, and V. S. Gurman, Khim. Vys. Energ., 12:53 (1978).
207. S.-B. Rhee and H. H. Jaffe, J. Am. Chem. Soc., 95:5518 (1973).
208. N. N. Ackermann, J. Am. Chem. Soc., 99:1661 (1977).
209. T. Mill and R. Stringham, Tetrahedron Lett., 1853 (1969).
210. G. Olbrich, Chem. Phys., 27:117 (1978).
211. V. I. Pergushov, O. N. Bormot'ko, and V. S. Gurman, Chem. Phys. Lett., 51:269 (1977).
212. M. B. Robin, R. R. Hart, and N. A. Kuebler, J. Am. Chem. Soc., 89:1564 (1967).
213. G. Wagniere, Theor. Chim. Acta., 31:269 (1973).
214. K. Vasudevan, S. D. Peyerimhoff, R. J. Buenker, W. E. Kamner, and H. L. Hsu, Chem. Phys., 7:187 (1975).
215. P. S. Engel, J. Am. Chem. Soc., 91:6903 (1961).
216. A. Halgren and W. M. Lipscomb, J. Chem. Phys., 58:1569 (1973).
217. R. Cimivaglia, J. M. Rieva, and J. Tomasi, Theor. Chim. Acta, 46:223 (1977).
218. H. Cerfontain and K. O. Kutschke, Can. J. Chem., 36:344 (1958).
219. R. H. Riem and K. O. Kutschke, Can. J. Chem., 38:2332 (1960).
220. J. O. Terry and J. H. Futrell, Can. J. Chem., 45:2327 (1967).
221. R. W. Durham and E. W. R. Steachie, Can. J. Chem., 31:377 (1953).
222. J. R. Dacey, W. C. Kent, and G. O. Pritchard, Can. J. Chem., 44:969 (1966).
223. W. C. Worsham and O. K. Rice, J. Chem. Phys., 46:2021 (1967).
224. G. O. Pritchard, W. A. Mattinen, and J. R. Dacey, Int. J. Chem. Kinet., 2:191 (1970).
225. S. S. Collier, D. H. Slatter, and J. G. Calvert, Photochem. Photobiol., 7:737 (1968).
226. E. C. Wu and O. K. Rice, J. Phys. Chem., 72:542 (1968).
227. G. O. Pritchard, F. M. Servedio, and P. E. Marchant, Int. J. Chem. Kinet., 8:959 (1976).
228. F. M. Servedio and G. O. Pritchard, Int. J. Chem. Kinet., 7:99 (1975).
229. G. O. Pritchard, P. E. Marchant, and C. Steel, Int. J. Chem. Kinet., 11:951 (1979).
230. S. Chervinsky and I. Oref, J. Phys. Chem., 81:1967 (1977).

231. S. Yamashita, K. Okumuva, and T. Hayagawa, Bull. Chem. Soc. Jpn., 46:2744 (1973).
232. R. E. Rebbert and P. Ausloos, J. Am. Chem. Soc., 87:1847 (1965).
233. P. S. Engel and B. M. Monroe, Adv. Photochem., 8:245 (1971).
234. J. A. Den Hollander, J. Chem. Soc. Chem. Commun., 403 (1976).
235. P. S. Engel, D. J. Bishop, and M. A. Page, J. Am. Chem. Soc., 100:7009 (1978).
236. G. S. Hammond and J. R. Fox, J. Am. Chem. Soc., 86:4031 (1964).
237. W. A. Pryor and K. Smith, J. Am. Chem. Soc., 92:5403 (1970).
238. S. Seltzer and F. T. Dunne, J. Am. Chem. Soc., 87:2628 (1965).
239. T. Tagaki and R. J. Crawford, J. Am. Chem. Soc., 93:5910 (1971).
240. W. Duismann, R. Hertel, J. Meister, and C. Rüchardt, Liebigs Ann. Chem., 1820 (1976).
241. R. F. Hutton and C. Steel, J. Am. Chem. Soc., 86:745 (1964).
242. P. S. Engel and P. D. Bortlett, J. Am. Chem. Soc., 92:5883 (1970).
243. S. Kodama, S. Fujita, J. Takeishi, and O. Toyama, Bull. Chem. Soc. Jpn., 39:1009 (1966).
244. Y. Y. Abram, G. S. Milwe, B. S. Solomon, and C. Steel, J. Am. Chem. Soc., 91:1220 (1969).
245. P. Smith and A. Rosenberg, J. Am. Chem. Soc., 81:2037 (1959).
246. P. S. Engel and D. J. Bishop, J. Am. Chem. Soc., 94:2148 (1972).
247. R. Hempel, H. Viola, J. Morgenstern, and R. Mayer, Z. Chem., 19:298 (1979).
248. R. Hempel, H. Viola, J. Morgenstern, and R. Mayer, Faserforsch. Textiltech., 29:190 (1978).
249. D. Schulte-Frohlinde, Liebigs Ann. Chem., 612:131, 138 (1958).
250. D. P. Fisher, V. Piermattie, and J. C. Dabrowiak, J. Am. Chem. Soc., 99:2811 (1977).
251. A. W. Adamson, A. Vogler, H. Kunkely, and R. Wachter, J. Am. Chem. Soc., 100:1298 (1978).
252. G. Gauglitz, J. Photochem., 5:41 (1976).
253. R. Frank and G. Gauglitz, J. Photochem., 7:355-357 (1977).
254. R. Lovrien and J.C.B. Waddington, J. Am. Chem. Soc., 86:2315 (1964).
255. H. Kamogawa, M. Kato, and H. Sugiyama, J. Polym. Sci., A6:2967 (1968).
256. R. Lovrien, Proc. Natl. Acad. Sci., 57:236 (1967).
257. C. S. Paik and H. Morawetz, Macromolecules, 5:171 (1972).
258. D. Tabek and H. Morawetz, Macromolecules, 3:403 (1970).
259. C. D. Eisenbach, Makromol. Chem., 180:565-571 (1979).
260. C. D. Eisenbach, Polym. Bull., 1(7):517 (1979); C.A., 91:75006.
261. E. V. Bystritskaya, T. S. Karpovich, and O. N. Karpukhin, Dokl. Akad. Nauk SSSR, 228(3):632 (1976).
262. H. Fink, Habilitationsschrift, TU Dresden (1969).
263. G. Pelzl, Z. Chem., 17:294 (1977).
264. C. Leier and G. Pelzl, J. Prakt. Chem., 321:197 (1979).
265. B. Schnuriger and J. Bourdon, J. Chim. Phys. Chim. Biol., 73(7-8): 795-798 (1976).
266. W. H. Nutting, R. A. Jewell, and H. Rapoport, J. Org. Chem., 35:505 (1970).
267. Sh. Yamashita, H. Ono, and O. Toyama, Bull. Chem. Soc. Jpn., 35:1849 (1962).
268. P. T. Birnbaum and D.W.G. Style, Trans. Faraday Soc., 50:1192 (1954).
269. P. Roth, Dissertation, Humboldt-Universität, Berlin (1976).
270. R. Arnaud and P. Jaques, C. R. Hebd. Seances Acad. Sci., Ser. C., 282:1097 (1976).
271. E. Hofer, Chem. Ber., 112:2925 (1979).
272. P. S. Engel, R. A. Melaugh, and M. A. Page, J. Am. Chem. Soc., 98:1971 (1976).
273. N. A. Porter, G. R. Dabay, and J. G. Green, J. Am. Chem. Soc., 100:920 (1978).
274. P. S. Engel and D. J. Bishop, J. Am. Chem. Soc., 97:6754 (1975).

275. C. G. Overberger, M. S. Chi, D. G. Pucci, and J. A. Barry, Tetrahedron Lett., 4565 (1972).
276. G. Vitt, E. Hadicke, and G. Quinkert, Chem. Ber., 109:518-530 (1976).
277. D. L. Webb and H. H. Jaffe, J. Am. Chem. Soc., 86:2419 (1964).
278. J. Swigert and K. G. Taylor, J. Am. Chem. Soc., 93:7337 (1971).
279. A. Hantzsch, Chem. Ber., 27:1726 (1894).
280. A. Hantzsch and D. W. Schulze, Chem. Ber., 28:666 (1895).
281. H. C. Freeman and R.J.W. Lefevre, J. Chem. Soc., 415 (1951).
282. E. S. Lewis and H. Suhr, Chem. Ber., 92:3031, 3043 (1959).
283. J. van der Veen, J. Helferich, and L.K.H. van Beek, Recl. Trav. Chim. Pays-Bas, 85:895 (1966).
284. H. Jonker, T.P.G.W. Thijssens, and L.K.H. van Beek, Recl. Trav. Chim. Pays-Bas, 87:397 (1968).
285. H. Jonker et al., J. Photograph. Sci., 19:187 (1971).
286. R.J.W. Lefevre and J. Northcott, J. Chem. Soc., 333 (1949).
287. T. Ignasiak, J. Suszko, and B. Ignasiak, J. Chem. Soc., Perkin I, 2122 (1975).
288. J. Brokken-Zijp, V IUPAC Symposium on Photochemistry, Contributed Papers, Aix-en-Provence, France, Paper 16, July 19-23 (1976).
289. H. Jonker et al., Photogr. Sci. Eng., 13:1 (1969).
290. H. van Zwet and E. C. Kooyman, Recl. Trav. Chim. Pays-Bas, 86:993 (1967).
291. J. Brokken-Zijp and H. van de Bogaert, Tetrahedron, 29:4169 (1973).
292. J. Brokken-Zijp and H. van de Bogaert, Tetrahedron Lett., 249 (1974).
293. J. Brokken-Zijp, Tetrahedron Lett., 2673 (1974).
294. J. Brokken-Zijp, Mol. Photochem., 7:399 (1976).
295. A. B. Sakla, B. A. Riad, S. Abdou, and B. N. Barsoum, Acta Chim. Acad. Sci. Hung., 98:479 (1979).
296. E. Fanghänel, R. Hänsel, W. Ortmann, and J. Hohlfeld, J. Prakt. Chem., 317:631 (1975).
297. C. D. Ritchie and P.O.I. Virtanen, J. Am. Chem. Soc., 94:1589 (1972).
298. W. J. Boyle, T. J. Broxton, and J. F. Bunnett, Chem. Commun., 1469 (1971).
299. A. Hantzsch, Chem. Ber., 36:3097 (1903).
300. T. J. Broxton and D. L. Roper, J. Org. Chem., 41:2157 (1976).
301. C. S. Anderson and T. J. Broxton, J. Org. Chem., 42:2454 (1977).
302. J. F. Bunnett and H. Takayama, J. Org. Chem., 33:1924 (1968).
303. C. D. Ritchie and D. J. Wright, J. Am. Chem. Soc., 93:2425, 6574 (1971).
304. E. S. Lewis and H. Suhr, Chem. Ber., 91:2350 (1958).
305. R.J.W. LeFevre and J. B. Sousa, J. Chem. Soc., 745 (1957).
306. R. Naef and H. Balli, EPA Newsletters, S. 27, April (1980).
307. J. Reichel and R. Vidac, Rev. Roum. Chim., 15:1107 (1970).
308. J. Reichel and R. Vidac, Rev. Roum. Chim., 15:1227 (1970).
309. H. Zollinger, Acc. Chem. Res., 6:335 (1973).
310. B. Lang, Diplomarbeit, TH Leuna-Merseburg (1974).
311. R.J.W. Lefevre and T. H. Liddicoet, J. Chem. Soc., 2743 (1951).
312. H. C. Freeman and R.J.W. Lefevre, J. Chem. Soc., 2932 (1952).
313. DMS-UV-Atlas of Organic Compounds, Butterworths, London; Verlag Chemie, Weinheim, Vols. 1-5 (1966-1971).
314. M. Kobayashi, S. Fugii, and H. Minato, Bull. Chem. Soc. Jpn., 45:2039 (1972).
315. N. D. Trufanova, I. L. Bagal, I. S. Ifros, and B. A. Porai-Koshits, Zh. Org. Khim., 7:1208 (1971).
316. A. Hantzsch and M. Singer, Chem. Ber., 30:312 (1897).
317. H. C. Freeman et al., J. Chem. Soc., 3381 (1952).
318. E. Müller, Methoden der Organischen Chemie, 10(3):559 (1965).
319. H. C. Freeman, R.J.W. Lefevre, and I. R. Wilson, J. Chem. Soc., 1977 (1951).
320. J. Majer et al., Wiss. Z. THLM, 16:335 (1974).
321. E. Fanghänel and H. Linck, Z. Chem., 17:178 (1977).

322. H. Linck, Dissertation, TH Lenna-Merseburg (1977).
323. E. Fanghänel, W. Ortmann, B. Tyszkiewics, and M. Tyszkiewics, J. Prakt. Chem., 320:422 (1978).
324. H.-T. Dorsch, H. Hoffmann, R. Hänsel, G. Rasch, and E. Fanghänel, J. Prakt. Chem., 318:671 (1976).
325. E. Fanghänel, H. Poleschner, R. Radeglia, and R. Hänsel, J. Prakt. Chem., 319:813 (1977).
326. E. Fanghänel, R. Radeglia, and G. Hautmann, J. Prakt. Chem., 320:618 (1978).
327. E. Fanghänel, S. Simova, R. Radeglia, J. Prakt. Chem., im Druck.
328. R. Hänsel, Dissertation, TH Leuna-Merseburg (1976).
329. J. Hohlfeld, Dissertation, TH Leuna-Merseburg (1979).
330. J. Voltz, Angew. Chem., 74:680 (1962).
331. L.G.S. Brooker, J. Am. Chem. Soc., 73:5356 (1951).
332. B. Luders, Diplomarbeit, TH Leuna-Merseburg (1977).
333. E. Fanghänel, W. Ortmann, and K. Hirsch, J. Prakt. Chem. 320:607 (1978).
334. J. Voltz, Angew. Chem., 74:610 (1962).
335. Y. Tanaka, Y. Takeda, T. Kinoshita, and K. Hirabayaschi, Jpn. Patent 7211346, C.A., 77:90032 (1972).
336. E. Fanghänel and J. Hohlfeld, J. Prakt. Chem., im Druck.
337. J. Brokken-Zijp and H.J.L. Bressers, Mol. Photochem., 7:359 (1976).
338. L.K.H. van Beek, J.R.G.C.M. van Beek, J. Boven, and C. J. Schoot, J. Org. Chem., 36:2194 (1971).
339. E. Fanghänel, R. Hänsel, H.G.O. Becker, and J. Epperlein, DD-PS 114 692.
340. N. Baumann, Nachr. Chem. Tech., 22:477 (1974).
341. N. V. Philips, Gloeilampenfabrieken, Eindhofen, Ndld., Patent OS 1622295 and OS 1622296.
342. H. Jonker et al., J. Photogr. Sci., 20:53 (1972).
343. E. Inoue, H. Kokado, and T. Ikawa, C.A., 82:148520h (1975).
344. E. Brinckman, G. Delzenne, A. Poot, and J. Willems, Unconventional Imaging Processes, The Focal Press, London—New York (1978), p. 113.
345. J. A. Sprung and W. A. Schmidt, C.A., 43:7846d (1949).
346. E. Inoue, H. Okado, and N. Yamase, C.A., 84:172167k (1976).
347. F. J. Rauner and R. H. Engebrecht, C.A., 78:78148j (1973).
348. T. Saito and T. Kazami, C.A., 79:151657b (1973).
349. H. Mustacchi, GAF Corp., DE-OS 2609565 (1976).
350. GB PS 1147896.
351. S. Levinos, Oji Paper Co. Ltd., US-PS 3856528; C.A., 82:92045n (1975).
352. Farbwerke Hoechst AG, DE-PS 953787 (1953).
353. H. Mustacchi, GAF Corp., DE-OS 2609565 (1976); C.A., 86:99048g (1976).
354. E. Fanghänel and J. Hohlfeld, DDR-Patent 137157 (1978).
355. R.J.W. Lefevre and I. R. Wilson, J. Chem. Soc., 1106 (1949).
356. W. Schäfer, Diplomarbeit, TH Leuna-Merseburg (1974).
357. R. Borchardt, Diplomarbeit, TH Leuna-Merseburg (1974).
358. A. Padwa, Chem. Rev., 77:37 (1977).
359. A. C. Pratt, Chem. Soc. Rev., 6:63 (1977).
360. R. Exelby and R. Grinter, Chem. Rev., 65:247 (1965).
361. D. Vocelle, A. Dargelas, R. Pottier, and C. Sandorfy, J. Chem. Phys., 66:2860 (1977).
362. D. A. Nelson and J. J. Worman, Tetrahedron Lett., 507 (1966).
363. A. E. Lubarskaya, M. I. Knyazhanskii, and M. B. Stryukov, Zh. Prikl. Spektrosk., 25:938 (1976).
364. A. Dargelas and C. Sandorfy, J. Chem. Phys., 67:3011 (1977).
365. K. L. Evans and A. E. Gilman, J. Chem. Soc., 565 (1943).
366. Don Iffland, M. P. McAneny, and D. J. Weber, J. Chem. Soc. C, 1703 (1969).
367. A. J. Bellamy and J. Hunter, J. Chem. Soc., Perkin Trans. I, 456 (1976).

368. G. Adembri, P. Sarti-Fantoni, and E. Belgodere, Tetrahedron, 22:3149 (1966).
369. R. Dreckschmidt, H. Kessel, and F. Marschner, Tetrahedron, 33:101 (1977).
370. V. V. Zverev, V. I. Vouna, M. S. Elman, Yu. P. Kitaev, and F. I. Visolov, Dokl. Akad. Nauk SSSR, 213:1319 (1973).
371. V. I. Minkin, Yu. A. Zhdanov, E. A. Medyantzeva, and Yu. A. Ostronov, Tetrahedron, 23:3651 (1967).
372. H. B. Bürgi and J. D. Dunitz, Helv. Chim. Acta, 53:1747 (1970); 54:1225 (1971).
373. T. Bally, E. Haselbach, S. Lanyiova, F. Marschner, and M. Rossi, Helv. Chim. Acta, 59:486 (1976).
374. K. Weiss, C. Warren, and G. Wettermark, J. Am. Chem. Soc., 93:4658 (1971).
375. D. Pitea, G. Favini, and G. Capietti, J. Chem. Soc., Perkin Trans. II, 2038 (1977).
376. M. Kobayashi, M. Yoshida, and H. Minato, Chem. Lett., 185, 1100 (1976); J. Org. Chem., 41:3322 (1976).
377. C. Dietrich and W. Jaenicke, Ber. Bunsenges. Phys. Chem., 74:25 (1970).
378. W. Gajewski, H. Wendt, and R. Wolfbauer, Ber. Bunsenges. Phys. Chem., 76:450 (1972).
379. G. Condorelli, L. L. Costanzo, G. Giuffrida, and S. Pistara, J. Phys. Chem. Neue Folge., 96:97 (1975).
380. G. Just and C. Pace-Asciak, Tetrahedron, 22:1069 (1966).
381. C. Condorelli, L. L. Costanzo, and S. Pistara, Ann. Chim., 61:381 (1971).
382. J. Bjargo, D. R. Boyd, C. C. Watson, and W. B. Jennings, J. Chem. Soc., Perkin Trans. II, 757 (1974).
383. A. Padwa and F. Albrechts, J. Org. Chem., 39:2361 (1974).
384. D. Schulte-Frohlinde, R. Kuhn, W. Münzing, and W. Otting, Liebigs Ann. Chem., 622:43 (1959).
385. H.-J. Timpe and U. Müller, Unveröffentlichte Versuche.
386. J. Elguero, R. Jacquieur, and C. Marzin, Bull. Soc. Chim., 1374 (1969).
387. D. S. Malamet and J. M. McBride, J. Am. Chem. Soc., 92:4586 (1970).
388. G. Gustav et al., Z. Chem., 18:26, 456 (1978); 19:259, 347 (1979); J. Prakt. Chem., 321:395 (1979).
389. J. C. Danilcewicz, J. Chem. Soc. C, 1049 (1970).
390. P. Baas and H. Cerfontain, J. Chem. Soc., Perkin Trans. II, 1351 (1977).
391. P. Baas and H. Cerfontain, Tetrahedron Lett., 1501 (1978).
392. P. Baas and H. Cerfontain, J. Chem. Soc., Perkin Trans. II, 151 (1979).
393. P. Baas and H. Cerfontain, J. Chem. Soc., Perkin Trans. II, 156 (1979).
394. V. I. Stenberg, P. A. Barks, D. Bays, D. D. Hammargren, and D. V. Rao, J. Org. Chem., 33:4402 (1968).
395. J. J. Worman, G. L. Pool, and W. P. Jensen, J. Chem. Educ., 47:709 (1970).
396. R. Bonett, J. Chem., 2313 (1965).
397. P. J. Orenski and W. D. Closson, Tetrahedron Lett., 3629 (1967).
398. T. Sato et al., Bull. Chem. Soc. Jpn., 45:1176 (1972).
399. M. Ottolenghi and D. S. McClure, J. Chem. Phys., 46:4613 (1967).
400. J. Meisenheimer and O. Dorner, Liebigs Ann. Chem., 502:156 (1933).
401. W. F. Smith, Jr., J. Phys. Chem., 68:1501 (1964).
402. W. G. Herkstroeter, Mol. Photochem., 3:181 (1971).
403. W. G. Herkstroeter, J. Am. Chem. Soc., 97:3090 (1975).
404. E. Fischer and Y. Frei, J. Chem. Phys., 27:808 (1957).
405. E. Fischer and M. Kaganowich, Bull. Res. Counc. Isr. Sect., A10:138 (1961).

406. G. Wettermark and A. King, Photochem. Photobiol., 4:417 (1965).
407. D. G. Anderson and G. Wettermark, J. Am. Chem. Soc., 87:1433 (1965).
408. G. Wettermark, Ark. Kemi, 27:159 (1967).
409. G. Wettermark, J. Weinstein, J. Sousa, and L. Dogliotti, J. Phys. Chem., 69:1584 (1965).
410. G. Wettermark, Sven. Kem. Tidskr., 79:249 (1967).
411. G. Wettermark and E. Wallström, Acta Chem. Scand., 22:675 (1968).
412. K. Appenroth, M. Reichenbächer, and R. Paetzold, J. Photochem., 14:51 (1980).
413. G. Condorelli and L. L. Costanzo, Boll. Sedute Acc. Gioenia, 9:126 (1967); C.A., 70:15755u (1970).
414. D. A. Nelson, R. L. Atkins, and G. L. Clifton, J. Chem. Soc., Chem. Commun., 399 (1968).
415. A. Padwa, W. Bergmark, and D. Pashayan, J. Am. Chem. Soc., 91:2653 (1969).
416. K. N. Mehrota and B. P. Giri, Indian J.Chem. B, 15:1106 (1977).
417. K. Kanamaru and K. Kimura, Mol. Photochem., 5:427 (1973).
418. W. G. Herkstroeter, J. Am. Chem. Soc., 98:6210 (1976).
419. H. Kessler, Angew. Chem., Vol. 82 (1970).
420. H. Kessler and D. Leibfritz, Tetrahedron, 26:1805 (1970).
421. C. G. McCarthy, in: The Chemistry of the Carbon Nitrogen Double Bond, ed. by S. Patai, Wiley-Interscience, New York (1970), p. 405.
422. J. M. Lehn, Fortschr. Chem. Forsch., 15:311 (1970).
423. H.J.C. Yeh, H. Ziffer, D. M. Jerina, and D. R. Boyd, J. Am. Chem. Soc., 95:2741 (1973).
424. D. R. Boyd, C. G. Watson, W. B. Jennings, and D. M. Jerina, J. Chem. Soc., Chem. Commun., 183 (1972).
425. W. B. Jennings, S. Al-Showimann, M. S. Tolley, and D. R. Boyd, J. Chem. Soc., Perkin Trans. II, 1575 (1975).
426. W. B. Jennings and D. R. Boyd, J. Am. Chem. Soc., 94:7187 (1972).
427. W. F. Smith, Jr., W. G. Herkstroeter, and K. L. Eddy, Photogr. Sci. Eng., 20:140 (1976).
428. A. Padwa and F. Albrecht, J. Am. Chem. Soc., 96:4849 (1974).
429. R. J. Cook and K. Mislow, J. Am. Chem. Soc., 93:6703 (1971).
430. H. Izawa, P. De Mayo, and T. Tabata, Can. J. Chem., 47:51 (1969).
431. H. Suginome, Kagaku No Ryoiki, 30:578 (1976); C.A., 86:5221h (1977).
432. H. Suginome and F. Yagihashi, J. Chem. Soc., Perkin Trans. I, 2488 (1977).
433. A. Stojiljkovic and R. Tasovacs, Tetrahedron Lett., 1405 (1970).
434. G. A. Delzenne, U. Laridon, and H. Peeters, European Polym. J., 6:933 (1970).
435. S. I. Hong, Soul Taehakkyo Noumunjip, 25:119 (1975); C.A., 87:85457y (1977).
436. T. Tezuka and N. Narita, Nippon Kagaku Kaishi, 1097 (1976); C.A., 85:159045v (1976).
437. A. Mustafa, A. K. Mansour, and H.A.A. Zaher, J. Org. Chem., 25:946 (1966).
438. R. Calas, R. Lolande, F. Moulines, and J. G. Fangers, Bull. Soc. Chim. France, 121 (1965).
439. A. Padwa and F. Albrecht, Tetrahedron Lett., 1083 (1974).
440. M. Tsunooka, T. Nishino, and M. Tanaka, Chem. Lett., 1107 (1977).
441. E. J. Poziomek, J. Pharm. Sci., 54:333 (1965).
442. C. Condorelli, S. Pistara, and S. Giuffrida, J. Phys. Chem. Neue Folge, 90:58 (1974).
443. C. Condorelli and L. L. Costanzo, Boll. Sedute Accad. Gioenia, 8:753, 755 (1966); C.A., 70:3120v, 10815d (1969).
444. D. Schulte-Frohlinde, Liebigs Ann. Chem., 622:47 (1959).
445. G. Condorelli, L. L. Costanzo, L. Alicata, and A. Giuffrida, Chem. Lett., 227 (1975).
446. J. McVie, A. D. Mitchell, R. S. Sinclair, and T. G. Truscott, J. Chem. Soc., Perkin Trans. II, 286 (1980).

447. C. I. Stassinopoulou, C. Zioudron, and G. J. Karabatsos, Tetrahedron, 32:1147 (1976).
448. K. Appenroth, M. Reichenbächer, and R. Paetzold, Tetrahedron, im Druck.
449. R. Paetzold, K. Appenroth, and M. Reichenbächer, Z. Chem., 16:447 (1976); J. Photochem., 14:39 (1980).
450. R. W. Binkley, J. Org. Chem., 33:2311 (1968); 34:931, 2072 (1969).
451. J. Corse and R. W. Binkley, J. Org. Chem., 37:575 (1972).
452. M. Naulet, M. Leymari-Beljean, and G. J. Martin, Org. Magn. Reson., 11:16 (1978).

Chapter 4

PHOTOCHROMIC TAUTOMERIC SYSTEMS

A. V. El'tsov,* A. I. Ponyaev,* É. R. Zakhs,*
D. Klemm,** and E. Klemm**

*Lensovet Institute of Technology, Leningrad, USSR
**Wilhelm Pieck University, Jena, GDR

In pentad tautomeric systems the shift of Z groups between the ends X and Y of the chain is accompanied by a bond rearrangement which may produce a marked change in the electronic spectrum (color); if in this case the Z shift is light-induced, the system is photochromic:

$$ Z\diagup \overset{X}{\underset{Y}{\diagdown}}{}^{1}_{2}{}_{3} \quad \underset{h\nu',\,\Delta}{\overset{h\nu,\,\Delta}{\rightleftarrows}} \quad Z\diagdown \overset{X}{\underset{Y}{}}{}^{1}_{2}{}_{3} $$

Until recently, photochromic o-alkylarylcarbonyl compounds (X = C, Y = O, Z = H), salicylalanils (X = O, Y = NR, Z = H), and o-hydroxyazoaromatic compounds (X = O, Y = NAr, Z = H) were subjected to a detailed study. In all of them C^1 and C^2 are included in the aromatic nucleus; in azo compounds Y is bonded with nitrogen [1]. The action of light produces unstable, deeply colored quinoid tautomers — quinonedimethanes, quinonemethanes, and quinonimines, respectively. These quickly, thermally reverse back to the starting substances. As a rule, it is not possible to observe quinoid tautomers without irradiation because of the impossibility of overcoming the high activation barrier of their formation, e.g., for o-alkylaromatic compounds:

$$ \underset{C-H}{\overset{C=O}{\bigcirc}} \quad \underset{\Delta}{\overset{h\nu}{\rightleftarrows}} \quad \underset{C}{\overset{C-OH}{\bigcirc}} $$

Comprehensive studies of many hypothetical compounds of this type with different X, Y, and Z, with carbon atoms or heteroatoms in the principal chain links which may be included in various rings, will undoubtedly identify photochromic substances with unexpected properties suitable for a variety of practical purposes.

The authors of this chapter, using the example of (nitroaryl)(hetaryl)-alkanes (X = CH$_2$-2-hetaryl, C^1–C^2 bonds of the aromatic ring, Y = oxygen of the nitro group, Z = H), revealed a system of acid–base equilibria involving the participation of a thermodynamically unstable, quinoid photoinduced tautomeric form; the state of this equilibrium system determines, at each given point in time, the light-generated coloration.

In their study of aryloxyquinones the authors for the first time observed light-induced pentad tautomerism with aryl migration (X = Y = O, Z = Ar, C^1–C^2–C^3 bond of the aromatic rings). In the series of benzothiophenone derivatives, acyl photomigration was discovered which proceeds intramolecularly by the nonadiabatic path (X = O, Y = N, Z = Ac). This photoacylotropy can be of interest as a method of accumulating solar energy.

THE PHOTOCHROMISM OF (NITROARYL)(HETARYL)ALKANES

Structural Features of Photochromic (Nitroaryl)(hetaryl)alkanes

Nitroaromatic compounds with the nitro group in the ortho-position with respect to the methyl, methylene, or methine groups possess photochromic properties [1,2]. We now know of the existence of a series of polynitro-arylalkanes and colorless nitrobenzyl and nitronaphthylmethyl derivatives of nitrogenic heterocycles which, upon irradiation with ultraviolet light, induce absorption in the visible region of the spectrum (Table 1).

The following are additional spectral characteristics for a number of compounds that have been ommitted from the table:

4, 460 in water/12 M H_2SO_4; 500, 540, and 630 in methanol and acetonitrile. 10, 600 at 23°C in heptane [13,14]; 600 and 700 in ethanol. 11, 12, 16, 500 at 23°C in heptane [14]. 15, 550 at 23°C in heptane [14]. 26, 564 in methanol [20]; in crystals 545, 620 [23], and 575 [17]; 600 adsorbed on NaCl [24]; 567 in sorbite–fructose glass [17]; 547 at −20°C in super-cooled melt [25]; 566 in polymethyl methacrylate [26]; at 23°C 400-420 and 550-590 in ethanol, 400-450 and 510-560 in propanol, 390-410 and 530-580 in heptane [14,27]. 48, 580 in crystals [37]; 583 and 571 in polymethyl meth-acrylate with a molar mass M of 23,600 and 3900 g/mole, respectively; 561 in polystyrene with M of 47,000 g/mole and 571 in polyethyl acrylate with M of 35,000 g/mole [26]. 52, at 20°C 430 and 550 in ethanol and 430 and 505 in carbon tetrachloride [38]. 54, 557 at −100°C in ethanol [11,39] and at 20°C in isopropyl alcohol acidified with $HClO_4$ [41]; 613 at −27°C in supercooled melt [25]. 59, 590 or 517, 625 sh. at 20°C in ethanol (50%) with pH 5 or 10 [42]. 64, 515 or 480, 620 sh. at 20°C in ethanol (50%) with pH 2.75 or 9.06 [45]. 65, 580 or 440, 580 sh. at 20°C in ethanol (50%) with pH 4.9 or 9.1 [40]. 66, 520 at −75°C in ethanol [44]. 67, 520 at pH 2.7; 517 at pH 6.5; 495 and 620 at pH 10 in ethanol (50%) at 20°C [45]. 75, 503 or 503, 630 sh. at 20°C in ethanol (50%) with pH 2.86 or 9.92 [47]. 77, 535 or 512, 630 sh. at 20°C in ethanol (50%) with pH 2.68 or 10.07 [47].

The study of the photochromism of o-nitrobenzyl compounds was pioneered by Chichibabin and his associates who discovered, as early as 1925, that crystals of 2-(2,4-dinitrobenzyl)pyridine (2-DNBP) 26 possess photochromic properties [53]. Under the influence of sunlight, pale-yellow crystals of 2-DNBP assumed a blue coloration, and after several hours in the dark became decolorized. Heating accelerated the process, while upon dissolution it occurred instantly.

After the discovery of the photochromism of 2-DNBP crystals, investigators were confronted with a series of questions: a) which light-induced structural changes were behind the process? b) further, which structural factors of the starting molecule determine these changes? c) whether the photochromism was characteristic of the crystals only or whether it could be observed in solution as well, and whether it was characteristic of other aggregation states of the material? etc.

Chichibabin synthesized the model compound 1-methyl-2-(2,4-dinitro-benzylidene)-1,2-dihydropyridine 80, whose color coincided with the color of the photoinduced form of 2-DNBP [53]:

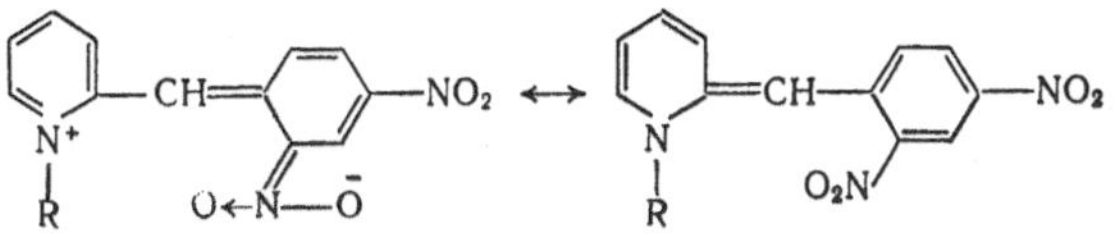

R = CH_3, **80**; R = H, **81**

It was this that enabled Chichibabin et al. [53] to adopt the structure of azamerocyanine **81** for the photocolored form.

In the early 1960s, when there was a demand for photochromic materials, researchers in different countries took a fresh look at o-nitrobenzyl compounds. The work of a number of research teams in the United States and subsequently in the USSR and the GDR demonstrated that under different conditions photochromism was a quality of substituted nitroarylalkanes; 2- and 4-(2-nitrobenzyl)pyridines having substituents in the benzene ring, in the methylene group, and in the heteronucleus; analogous derivatives of other heterocycles and some complexes of transition metals on a 2-DNBP base (see Table 1).

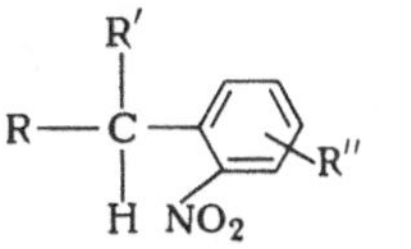

R = H, COOH, aryl, hetaryl;
R′ = H, OH, halogen, aryl;
R′′ = various substituents.

Concerning compounds **82-90** given below, data are available in the literature to the effect that they are photochromic, but no characteristics of their photoinduced forms are given (the compound number, its name, temperature, and solvent are indicated):

82, ammonium 2,4-dinitrophenylacetate, −100°C, ethanol [11];
83, 2,4-dinitrobenzyl alcohol, −111, −42, 20°C, ethanol [18];
84, 2,4-dinitrobenzyl acetate, ethanol (cooled) [31];
85, 2,4-dinitrobenzyl acrylate, ethanol (cooled) [31];
86, 2,4-dinitrobenzyl methacrylate, ethanol (cooled) [31];
87, salt of bis(2-nitro-4-trimethylaminophenyl)methane, −25°C, DMFA [49];
88, 2,4-bis(2,4-dinitrophenyl)pyridine, −50°C, ethanol [19];
89, 2-[4-(p-nitrobenzylidenamino)-2-nitrobenzyl]pyridine [50];
90, copolymers of 2,4-dinitrobenzyl acrylate, 2,4-dinitrobenzyl methacrylate, and 2-(4-acrylamido-2-nitrobenzyl)pyridine with methacrylic acid [51,52].

The conditions for the observation of the photoinduced coloration of all the photochromic compounds of this type (**1-79**) known before 1979 are presented in Table 1. Data on their photochromism were obtained for cooled alcoholic solutions upon stationary irradiation, for solutions at room temperature (in this instance the flash photolysis method is employed), for crystals, and for supercooled melts.

Compounds **26**, **29-32**, **38-40**, **43**, **44**, and **47-49** are colored upon ultraviolet irradiation in solution and in crystal form, but the crystals of substances **29** and **38**, at liquid nitrogen temperatures only. Compounds **26** and **44** retain photochromic properties on being adsorbed on NaCl. The photochromic properties of compound **41** have been observed on gelatinous film, and compounds **26**, **44**, **47**, **48**, in polymethyl methacrylate, polyethyl acrylate, and polystyrene films. Compounds **36**, **37**, and **84-86**, containing groups capable of polymerization, form photochromic copolymers with methyl methacrylate.

The data summarized in Table 1 indicate that the presence of a nitro group in the ortho-position with respect to the CHRR′ group is apparently a

Table 1. Spectral characteristics of photoinduced forms of nitroarylalkanes; DNB – dinitrobenzyl; DNPh – dinitrophenyl

No.	Compound name	λ_{max}, nm	Registration conditions: t(°C); solvent/pH
1	2-nitrotoluene [3-5]	370 or 420	24; water/0.0 or 13
2	2,4-dinitrotoluene [3,6,7]	420 or 530	30; water/0.3 or 13
3	2,6-dinitrotoluene [8]	410 or 500	30; water/0.0 or 13
4	2,4,6-trinitrotoluene [9]	460	20; cyclohexane, benzene
	diphenylmethanes (substituents in nuclei indicated):		
5	2,2'-dinitro-4,4'-diformyl- [10]	526	-55; ethanol
6	2,2'-dinitro-4,4'-distyryl- [10]	540	
7	2,2'-dinitro-4,4'-bis(β-cyanostyryl)- [10]	625, 740	
8	2,4,2'-trinitro- [11,12]	425, 645	-100; ethanol
9	2,4,4'-trinitro- [11,12]	580	
10	2,4,2',4'-tetranitro- [10-12]	718 or 712	-55 or -100; ethanol
11	-5-methyl- [10,13,14]	699 or 500,600	-55 or 23; ethanol
12	-5,5'-dimethyl- [10,13,14]	684 or 500,600	
13	-5-styryl- [10]	666	
14	-5,5'-distyryl- [10]	740	-55; ethanol
15	2,4,2',4',6-pentanitro-5-methyl-[10] [13,14]	625	
16	2,4,2',4',6-pentanitro-5,5'-dimethyl- [10,13,14]	550,600	23; ethanol
17	2,4,2',4',2'',4''-hexanitrotriphenylmethane [11]	619 or 500,600	-55 or 23; ethanol
		715	-7; methanol
18	alkaline DNPh-acetates [11]	425 shoulder, 650	-72; ethanol
19	methylbis(DNPh) acetate [11]	470 shoulder, 665	
20	ethylbis(DNPh) acetate [11]	470 shoulder, 665	
21	methyl(2,4,4'-trinitrodiphenyl) acetate [12]	482 shoulder, 645	-103.5; ethanol
22	ethyl(2,4,4'-trinitrodiphenyl) acetate [12]	475 shoulder, 650	
23	alkaline 4-(2-nitrobenzyl) benzoates [15]	412	water/10
24	alkaline 2,2'-dinitrodiphenylmethane-4,4'-dicarboxylates [15]	515	1,5; water/12
25	2-(2-nitrobenzyl)pyridine [16]	400	25; ethanol
26	2-DNB-pyridine [3,11,17-21,22]	567.5	-100; ethanol
		460,630 shoulder	20; ethanol (50%)/9.15
	2-(2-nitro-4-X-benzyl)pyridines (reduced X):		
27	chloro [16]	580	25; ethanol

#	Compound	λ (nm)	Conditions
28	cyano [18]*	405, 455, 476, 488	-70; ethanol
	[28]		-196; ether-pentane-alcohol
29	carboxy [29]*	450, 590	
30	methoxycarbonyl [29]*	405, 450, 470, 580	-91; ethanol
31	ethoxycarbonyl [29]*	405, 460, 485, 575	
32	carbamyl [29]*	410, 460, 485, 585	
33	hydroxy [16]*	403, 450, 485, 580	
34	amino [16]*	580	25; ethanol
35	acetyl [30]	500	
36	methacryloylamino [31]	blue-violet	ethanol (cooled)
37	isobutyrylamino [31]	blue-green	
38	3,3'-dinitro-4,4'-di(2-pyridylmethyl)azoxybenzene [29,32,33]*	blue-green	
	2-(2,4-dinitro-X-benzyl)pyridines (reduced X):	610	-91; ethanol
39	5-methyl [19]*	575 or 567	-50; ethanol or
40	5-chloro [19,25]*	550 or 558	-20; supercooled melt
41	5-sodiumsulfonate [19]	580	gelatin film
42	α-hydroxy [29]	600	-91; ethanol
43	1-DNPh-1-(2-pyridyl)ethane [25,34]*	625	-20; supercooled melt and
			-55; ethanol
44	DNPh(2-pyridyl)bromomethane [19,25]*	610 or 625	-50; ethanol or -20; super-
			cooled melt
45	2,4,4'-trinitrodiphenyl(2-pyridyl)methane [35]*	565	20; ethanol
46	1-DNPh-1-(2-pyridyl)ethylene [19]	465	dioxane
47	meso-1,2-bis(DNPh)-1,2-bis(2-pyridyl)ethane [19,26,36]*	550	-50; ethanol
	pyridines (substituents indicated):	589	polyethyl acrylate
48	2-DNB-4-methyl-[19,25]	565 or 555	-50; ethanol or -20; super-
			cooled melt
49	2-(2,4-dinitro-5-methylbenzyl)-4-methyl-[19,25]*	564	-50; ethanol
		568	-20; supercooled melt
		590	-50; ethanol
50	2-DNB-4-carboxy- [19]	410, 555	20; ethanol
51	hydrochloride 2-DNB- [38]	501, 601 shoulder	-80; ethanol
52	1-oxide 2-DNB- [11]	430, 550	20; ethanol
53	1-methyl-2-DNB-pyridinium nitrate	430	20; carbon tetrachloride
54	4-DNB-pyridine [40]	575	20; ethanol (50%)/4.1
		465, 650 shoulder	20; ethanol (50%)/9.8
55	4-(2-nitro-4-cyanobenzyl)pyridine [29]	448, 555, 590	-91; ethanol

Table 1. (Continued)

No.	Compound name	λ_{max}, nm	Registration conditions: t(°C); solvent/pH
56	4-DNB-pyridine-1-oxide [11]	527, 650 shoulder	-80; ethanol
57	1-benzyl-4-DNB-pyridinium [42]**	580	20; ethanol (50%)/2.7
58	1-methyl-4-(2,4-dinitronaphthylmethyl)pyridinium [42]**	607	
59	2-DNB-quinoline [42]	600	-80; ethanol
60	1-methyl-4-DNB-quinolinium [42]**	598	20; ethanol (50%)/2.7
61	1-methyl-4-(2,4-dinitronaphthylmethyl)quinolinium [42]**	635	
62	1-benzyl-4-DNB-quinolinium [42]**	595	
63	2-(2-nitrobenzyl)benzimidazole [43]	620	20; ethanol
64	2-DNB-benzimidazole [44]	515	-75; ethanol
65	1-DNPh-1-(2-benzimidazolyl)ethane [44]	580	
66	1-DNPh-1-(1-methyl-2-benzimidazolyl)ethane [40]	507	20; ethanol (50%)/4.25
		440, 580 shoulder	20; ethanol (50%)/8.95
67	1-methyl-2-DNB-benzimidazole [44]	520	-75; ethanol
68	1,3-dimethyl-2-DNB-benzimidazolium perchlorate [46]	495	20; ethanol (50%)/2.75
		495	20; ethanol (50%)/5.78
	2-DNB-5(6)-X-benzimidazoles [44] (reduced X):		
69	methyl	516	-75; ethanol
70	methoxy	520	
71	chloro	522	
72	bromo	523	
73	nitro	546	
74	2-(2-nitrobenzyl)benzoxazole [43]	500	20; ethanol
75	2-DNB-benzoxazole	500	
76	1-DNPh-1-(2-benzoxazolyl)ethane [47]	533	-75; ethanol
77	2-DNB-benzothiazole [47]	526	
78	chloro-cis-dicarbonyl(2-DNB-pyridine)rhodium(I) [48]	567	20; ethanol (50%)/2.75
79	chloro-cis-dicarbonyl(1-methyl-2-DNB-benzimidazole)rhodium(I) [48]	520	

* Photochromism was also recorded in crystals.

** The salts were obtained by dissolving the corresponding azamerocyanines in an acetate-phosphate-borate buffer solution with pH 2.7 in 50% ethanol.

necessary condition for the appearance of photochromism in nitroarylalkanes. Indeed, without these structural elements, obtained for the express
purpose of photochromic studies, in a large group of substances 93, 97,
99-106, 108-110, and 112-114 (see below) no reversible light-induced color
changes were observed.

The following are structurally related nonphotochromic compounds with
data cited in the following order: compound number, name, and test conditions (designated with letters — a: −196°C, ethanol, 94%, 400-W mercury lamp
[11]; b: −55°C, ethanol, HBO200 mercury lamp [10]; c: −50°C, ethanol,
HBO200 mercury lamp [19]; d: crystals, −20°C, ethanol, HBO200 mercury lamp
[35]; e: 25°C, ethanol, IFP-5000 pulse xenon lamp [54]):

91, 2,2'-dinitrodiphenylmethane, a;
92, 2,4'-dinitrodiphenylmethane, a;
93, 4,4'-dinitrodiphenylmethane, a;
94, 2,2'-dinitro-4,4'-diamino-5,5'-dimethyldiphenylmethane, b;
95, 2,2'-dinitro-4,4'-bis(benzylidenamino)diphenylmethane, b;
96, 2,2'-dinitro-5,5'-dimethyl-4,4'-bis(salicylidenamino)diphenylmethane,
    b;
97, 4,4',4"-trinitrotriphenylmethane, a;
98, 2,4'-dinitrophenylacetic acid, a;
99, 2-benzylisonicotinic acid, b;
100, 2-(3-methylbenzyl)pyridine, b;
101, 2-(3-chlorobenzyl)pyridine, b;
102, 2-(4-nitrobenzyl)pyridine, a;
103, 2-(5-sodiumsulfonatebenzyl)pyridine, b;
104, 2-(4-chloro-3-nitrobenzyl)pyridine, b;
105, 2-(4-chloro-3,5-dinitrobenzyl)pyridine, a;
106, 2-(2,4-dinitrobenzyl)pyridine, a;
107, ethyl ester of 1-ethoxycarbonyl-3-(2,4-dinitrophenyl)-3-(2-pyridyl)-
    butyric acid, b;
108, diphenyl(2-pyridyl)methane, d;
109, 2-pyridyl-4-nitrodiphenylmethane, d;
110, 2-pyridyl-di(4-nitrophenyl)methane, d;
110, 1-(2-pyridyl)-1-(2,4-dinitrophenyl)-2-piperidinoethane, c;
112, 4-(4-nitrobenzyl)pyridine, a;
113, 2-benzyl-4-methylpyridine, c;
114, 2-(3-methylbenzyl)-4-methylpyridine, c;
115, 1-methyl-2-(3,5-dinitro-2-pyridylmethylene)benzimidazoline, e.

In these compounds nitro or CHRR' groups either are absent (99-101,
103, 106, 108, 113, 114) or are in positions other than the ortho-position
with respect to one another (93, 97, 102, 104, 105, 109-112).

In the polynitrodi- and polynitrotriphenylmethanes, compounds 5-22 and
85 are photochromic with no fewer than three nitro groups (8-17, 19-22), of
which at least one is always in the ortho-position, or with two nitro groups
(5-7, 87), where both are in the ortho-position with respect to the CHRR'
group and with electron-acceptor substituents in the 4,4'-positions.

There is reason to believe that the ortho-position of the nitro group
and the hydrogen-donor is an essential, if insufficient, condition: in com-
pounds 91, 92, 94-96, 98, and 107, although they do have the $O_2N-\overset{|}{C}=\overset{|}{C}-CHR_2$
group, no photochromism was discovered. It is possible that suitable condi-
tions have not yet been selected for the photochromic effect of these substances to be observed. For instance, it is known that ethanol solutions of
2-(2-nitrobenzyl)pyridine 25 and 2-(2-nitro-4-aminobenzyl)pyridine 34 lack
the photochromic effect during photolysis by the unfiltered light of a
mercury lamp at low temperature [11], but reversible coloration is registered upon pulse photoexcitation at room temperature [16].

The spectra of photochromic (nitroaryl)(hetaryl)alkanes are cited in many papers [11,12,17,18,20,28,32]. In the ultraviolet spectra of the heterocyclic derivatives **26**, **64-67**, and **69-72** in neutral alcohol there is a wide band in the 230-280 nm region (Figure 1), which is due to electron transfers both in the dinitrophenyl ring and in the heterocycle; in benz-imidazole derivatives **64**, **65**, **67**, and **69-73** there are two narrow bands of a vibrational structure at 275 and 280 nm [44]. The hypsochromic shift of the absorption band up to 224 nm in 2-DNBP upon the replacement of the p-nitro group with the CN group is due to the lower electronegativity of the latter [18]. The replacement of solvent leads to a bathochromic shift of the absorption band [18] appearing in the series isooctane—ethanol—water for 2-DNBP.

In a benzimidazole derivative with a nitro group in the condensed ring **73** a vibrational structure of the long-wavelength band is absent, and a wide band appears in the 300-310 nm region. A comparison of the absorption spectra of the model compounds of 2-methylbenzimidazole and 5-nitro-2-meth-ylbenzimidazole shows that this band is caused by the conjugation of the nitro group with a heterocycle in position 5 [44]. In 0.1 N H_2SO_4 the 2-DNBP and benzimidazole derivatives **64**, **65**, **67**, and **69-73** display a hypso-chromic shift of the absorption band, apparently, as a result of the hetero-cycle's protonation [44,71]. For the low-basicity benzoxazole **75** and benzo-thiazole **77** derivatives under the same conditions, protonation is insignifi-cant. For this reason their absorption spectra in neutral and acid media are the same.

The hydrogen atom of the CHRR′ groups in such systems is noticeably acidic, but so far no data are available for which the relationship between the protonic mobility of such a hydrogen atom in the ground and excited states and photochromism parameters could be usefully discussed. What has been established is that in benzimidazole derivatives, upon the replacement of the dinitrobenzyl fragment with 3,5-dinitro-2-pyridyl derivatives, tauto-meric equilibrium in alcoholic media is shifted towards the NH tautomer **115**, which is colored but not photochromic [54].

Margerum et al. [11] showed that alkaline and ammonium **82** salts of 2,4-dinitrophenylacetic acid **98** are photochromic, although the acid itself under the same conditions is not. Later it was established that these salts and nitrophenylacetate and 4-nitrohomophthalate ions (acids **126-130**; see below) generally photodecarboxylate far more efficiently than the acids themselves, producing the corresponding colored nitrobenzyl anions [6,55].

Apart from acids **126-130**, a group of photochromic nitrocarboxyaryl-alkanes **117-120** (see below) is known which, unlike the above-mentioned substances, have a nitro group in the para-position and a carboxylate anion in the ortho- or para-positions with respect to the CHRR′ group [56].

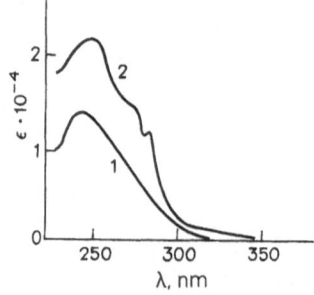

Fig. 1. The electronic absorption spectra of ethanol solutions:
1) 2-(2,4-dinitrobenzyl)pyridine **26**; 2) 2-(2,4-dinitrobenzyl)-benzimidazole **64**.

The following are photosensitive nitrobenzyl compounds having the COOH
group. Data are given in the following order: the compound number, name,
test conditions (designated by letters — a: 25°C, water, pH 10, the flash
photolysis method [56]; b: ammonia (aqueous), stationary photolysis, λ <
400 nm [57]; c: 22°C, water, pH 13, the flash photolysis method [6]; d:
22°C, water, pH 12, the flash photolysis method [6]):

116, 2-methyl-3-nitrobenzoic acid, b;
117, 2-methyl-5-nitrobenzoic acid, a, b;
118, 2-(4-nitrobenzyl)benzoic acid, a;
119, 4-(4-nitrobenzyl)benzoic acid, a;
120, 1,2-di(4-nitro-2-carboxyphenyl)ethane, a;
121, 4-methyl-3-nitrobenzoic acid, b;
122, 4'-nitro-2-carboxydiphenylmethane, b;
123, 2'-nitro-4-carboxydiphenylmethane [57], b;
124, 4'-nitro-4-carboxydiphenylmethane, b;
125, 2',4',5-trinitro-2-carboxydiphenylmethane, b;
126, 2-nitrophenylacetic acid, c;
127, 3-nitrophenylacetic acid, c;
128, 4-nitrophenylacetic acid, c;
129, 4-nitrohomophthalic acid, c;
130, 4,4'-dinitrodiphenylacetic acid, d.

The flash-induced photoexcitation of these compounds in aqueous solutions produces a nonequilibrium concentration of the dianions of these acids
generated by back synthesis in high-basicity media under dark reaction
conditions. In this way the structure of the registered colored particle
was established clearly.

The Nature of Photoinduced Forms of (Nitroaryl)(hetaryl)alkanes and the Mechanism of Their Formation

Until recently, despite its apparent simplicity, the question of the
structure of the photoinduced form of 2-DNBP, its analogues, and other compounds of this type (see p. 177) was not resolved unequivocally.

The reversible appearance of color, light-induced in toluene nitro
derivatives 1-4 (see Table 1) in highly acidic media, discovered fairly
recently [3-5,7-9], is attributed to the appearance of colored acinitro
acids. The pK_a values for compounds 1, 2, and 3 were found to be, respectively, 3.7 [5], 1.1 [7], and 1.8 [8]. The absorption bands of the electronic spectrum of nitronic acids are bathochromically shifted as the number
of nitro groups increases (see Table 1). However, the character of the
spectrum remains. This similarity made it possible to draw a conclusion
regarding the acinitro structure in organic solvents of the photolysis product of compound 4 [9]. In less acidic and neutral media, more deeply colored mesomeric anions are formed, identified by the absorption electronic
spectra with anions generated under ordinary conditions. Similarly, upon
the stationary ultraviolet irradiation of cooled alcoholic solutions of
nitro-substituted forms of di- and triphenylmethane 5-17, a colored anion
appeared which was proved on the basis of the spectrum [10,11].

The example of compound 4 shows that in nonpolar benzene and cyclohexane, laser photolysis gives acinitro acid, and in acetonitrile, alcohols,
and water, an anion of this acid, where the magnitude of the charge is confirmed by the presence of the salt effect in methanol [9].

In substituted forms of polynitrodiphenylmethanes 10-12, 15, and 16 in
heptane right after laser-induced excitation, the absorption of a short-
lived colored form (τ_e 2-1.3 μs) in the 500-600 nm zone is observed, which
is assigned to acinitro acid [13,14]. In alcohol, apart from nitronic acid,

a longer-lived product appears (τ_e 2.5 × 10^{-3} to 0.9 × 10^{-2} s) which absorbs at 600-700 nm and is identified as an anion by comparing its absorption spectrum with the spectrum of the assured potassium salt of the corresponding tetranitrodiphenylmethane [13,58]:

Upon the transition to the hetaryl-substituted forms of 2,4-dinitrotoluene, notably to the pyridine-substituted forms, the possibility arises of forming, as a result of photoreaction, not only acinitro acid and an anion but also an azamerocyanine dye; Chichibabin assigned this very structure to the colored form of 2-DNBP **26** crystals [53]. After Chichibabin's work a number of authors, after repeating experiments involving 2-DNBP crystals, also proposed an azamerocyanine structure for the photoinduced form, on the strength of the agreement of the absorption spectra of the colored form and of the model [23,59,60].

The possibility was discussed in the literature of the appearance of coloration in photochromic (nitroaryl)(hetaryl)alkanes due to the transition of the molecule into a metastable excited state upon ultraviolet irradiation, i.e., a triplet state that absorbs visible light, and in this way gives coloration which disappears as the molecules cross from the excited state to the ground state [17,61]. However, attempts to detect EPR signals in colorless and colored crystals and in cooled solutions of 2-DNBP were unsuccessful [17,21,34].

Upon the irradiation of 2-DNBP crystals other authors detected an EPR signal which is a wide (about 50 Oe) asymmetric doublet with a g-factor of about 2 [62]. It is more likely that the EPR signal is due to the effect of phenyl or nitroxy radicals, the possibility of whose appearance was discussed in the papers [63-67]. In particular, upon ultraviolet irradiation of o-nitrotoluene, apart from reversible isomerization, the formation of relatively stable free phenylnitroxyl radicals [33] is observed.

The studies carried out by Hardwick et al. [17] showed that not only crystals but also 2-DNBP solutions are photochromic at low temperature. 2-DNBP displays a qualitatively similar behavior in different solutions upon irradiation. The absorption spectrum of the photoinduced form of 2-DNBP in solid sorbite—fructose glass, just as in methanol solution, has one band with a maximum at 567 nm, in contrast to two bands in crystals. Since in experiments conducted by the authors the coloration of the 2-DNBP crystals irradiated and dissolved in acetone cooled to −70°C was found to be identical to the color of the solution prepared at room temperature, cooled and later irradiated, it follows that photocolored forms in liquid and crystal states are identical, and the spectral difference is attributed to the special features of the crystal structure.

Despite this, the authors considered the acinitro acid structure to be a "tempting alternative" to the merocyanine structure. They attempted to use infrared spectroscopy to directly establish the structure of the colored form in crystals. However, the positions arising upon the ultraviolet irradiation of the bands (1290, 1120, 1570, 1160 cm^{-1}) and the changes in the starting materials do not allow one to choose between the azamerocyanine and the acinitro acid structure for the colored form. In particular, the appearance of the 3390 cm^{-1} band could be attributed to the absorption of both the NH and the NOH groups in merocyanine and acinitro acid, respectively.

One argument in favor of the acinitro structure of the registered form was the discovery of photochromic properties in 4-(2,4-dinitrobenzyl)pyridine (4-DNBP) 54 [39], whose photochromism was not observed in crystals [53,60]. Mosher et al. [39] believed that the paths of the appearance of coloration in 2- and 4-DNBP were of the same type and consisted of the intramolecular hydrogen transfer to the ortho-located group, which in 4-DNBP cannot be anything but the nitro group. It followed from this that in the photoisomer of 2-DNBP the hydrogen is also localized on the nitro group.

The authors of a number of papers devoted to the study of the photochromism of 2- and 4-DNBP and their derivatives, while regarding the structure of the observed photocolored form, failed to take into account environmental influences and the acid—base properties of the starting and photoinduced forms when analyzing the results. That led to erroneous logical constructs and conclusions.

Thus, without special analysis the acinitro acid structure was assigned* to the photoinduced registered form of a number of benzylpyridine derivatives [16,29,68], with substituents on the bridge carbon and in the para-position of the benzene ring. During the study of the photochromism of 2-(2-nitro-4-cyanobenzyl)pyridine 28, 2-(2-nitro-4-aminobenzyl)pyridine 34, 2,4-dinitrobenzyl alcohol 83, and 2-DNBP 26 in alcoholic media, only acinitro acid was taken into account, and the energy of the transition state upon decoloration was calculated for it; on the basis of the substantial decrease in entropy of activation [−184.4 J/(mole · K) for 2-DNBP in ethanol] this state was considered to be planar [18].

Ben-Hur and Hardwick [169] believed that under the influence of light on a 2-DNBP solution, nitronic acid was formed, which could dissociate to form an anion. They also believed that the equilibrium in 95% 2-propanol was shifted towards the acinitro acid (at least at low temperatures), although the pK_a values of nitronic acids of o-nitrotoluene and its analogs were known in the literature [5,7,8]. Mosher et al. [41] attempted to take into account all possible structures of the photocolored form (nitronic acid, its anion, azamerocyanine) while studying 2- and 4-DNBP by the flash photolysis method in 95% 2-propanol. However, they failed to find the conditions for the existence of each of them, although they did show that in alkaline solutions an anion was registered.

Hiraoka and Hardwick [70] proposed to obtain fixed models of colored structures by acting upon 2- and 4-DNBP with epoxy compounds. In both cases they obtained colored and photoinduced forms and stable alkylation products, but they were unable to establish with certainty the position of the alkyl group (N or O) in the colored compound, and thus failed to choose between either the azamerocyanine or the nitronic acid structure.

---

*The data were obtained for solutions (ethanol, ether, water, isooctane) at low temperature (stationary photolysis) or at room temperature (flash photolysis) and in some instances also for crystals.

Finally, in experiments involving the flash photolysis of 2-DNBP in aqueous solutions, Wettermak [71] recorded an acid–base equilibrium not only between the 2-DNBP base and its pyridinium cation (pK$_a$ 4.1) but also between the colored forms (pK$_a$ about 4-6) to which he assigned the anion and aza-merocyanine structures, depending on the medium's pH. Analogous conclusions were drawn for the products of the photolysis of 2-(2-nitro-4-cyanobenzyl)-pyridine 28 [18].

The synthesis and investigation of the photochromism of 2-DNBP and its derivatives and a series of meso-substituted benzazoles and quinoline formed the basis for developing generalized ideas on the structure, mutual trans-formations of photoinduced forms, and the character of photoinduced color changes, depending on the nature of the solvent, its acid–base properties, and the nature of the heterocycle [13,14,27,42,44-47].

First, using the example of the N-oxides of 2- and 4-DNBP 52, 56 [11] and the protic salts of dinitrobenzylbenzimidazole 64, 67 [45], and later other salts 51, 53, 59-62, and 68 [34,38,46], it was shown that the quater-nary salts are photochromic, just as the bases are.

The electronic absorption spectra of short-lived particles generated by flash photolysis in 50% aqueous ethanol buffer solutions coincided com-pletely with the spectra of the corresponding azamerocyanine dyes obtained by C-deprotonation of salts [46]. This is unequivocal evidence for the azamerocyanine structure of the observed photoinduced particles from these onium salts.

Simultaneous absorption at 430 and 550 nm by laser photolysis (time resolution of 20 ns) in alcohol of the nitrate of N-methyl-2-DNBP 53 and the N-oxide 52 and hydrochloride 51* of 2-DNBP, attributed to acinitro acids and azamerocyanines, respectively (Table 2), was detected [38]. The acinitro acids changed completely into azamerocyanines (Figure 2).

The formation of azamerocyanine in quaternary salt 53 in CCl$_4$ was not observed because of the absence of a proton-acceptor center from the mole-cule of the photochrome itself and from the medium.

---

*Photochromism of the salts of benzimidazole derivatives 64 and 68 was discovered upon the flash photolysis of solutions in 50% alcohol [45,46].

Table 2. The absorption region and decay time (τ_e) of photoinduced forms from 2-DNBP and its N-substituted forms upon irradiation with the second harmonic of a ruby laser (23°C) [27,38]

Compound number in Table 1	Solvent	Nitronic acid		Azamerocyanine	
		absorption region, nm	τ_e, s	absorption region, nm	τ_e, s
26	hexane	390–410	$8 \cdot 10^{-7}$	530–580	$5.8 \cdot 10^{-4}$
	ethanol	400–420	$12 \cdot 10^{-6}$	550–590	0.5
	propanol	400–450	$1.5 \cdot 10^{-7}$	510–580	0.7
51	ethanol	410	$3.0 \cdot 10^{-7}$	555	0.5
52	carbon tetra-chloride	430	$1.0 \cdot 10^{-7}$	505	$2.0 \cdot 10^{-5}$
	ethanol	430	$2.0 \cdot 10^{-7}$	550	$2.0 \cdot 10^{-3}$
53	carbon tetra-chloride	430	$1.5 \cdot 10^{-7}$	no absorption	
	ethanol	430	$2.0 \cdot 10^{-7}$	550	0.8

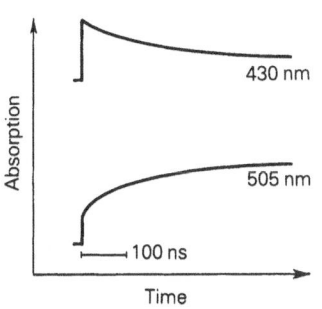

Fig. 2. The change in the absorption of the solution of compound 52 in CCl₄ upon irradiation by the second harmonic of a ruby laser (23°C, E 20 mJ/s).

Azamerocyanine and acinitro acid, which completely transformed into azamerocyanine, were detected by the same method as for 2-DNBP in heptane and in alcohol, which was confirmed (as for compounds 52 and 53) by the coincidence of the accumulation times of azamerocyanine and the disappearance of acinitro acid [27]. Photolysis in alcoholic alkali with pH 7.2 produces, apart from the products mentioned above, also an anion of 2-DNBP with absorption at 450-490 nm. Azamerocyanine is not detected with increasing alkalinity (pH > 7.5).

During the study of heterocyclic derivatives 26, 54, 64-67, 75, and 77 by the flash photolysis method with a time resolution of 50 μs in absolute ethanol, the kinetic curves in the long-wavelength region (λ > 500 nm) showed a fast component of decay, and concurrently with it a build-up of coloration with the same characteristic time (Figure 3) [40] was registered in the short-wavelength region (380-500 nm). 2-DNBP and α-C-methyl derivatives of benzimidazole 65 and 66 displayed the most pronounced effect. The absorption spectra were calculated immediately after the flash and a few milliseconds later, when the contribution of the fast component could be neglected (Figure 4). The coloration in the first instance corresponded to merocyanine, while the accumulating one to the anion.* Therefore, the disappearance of azamerocyanine in pure ethanolic solutions was accompanied by anion accumulation, while the fast process represented the establishment of equilibrium between them. This made it possible to find for 65 the rate constant k_{-1} of the N-protonation of the anion to azamerocyanine from

_____

*Earlier, the fast component was observed only in one instance [41] on the initial section of the decoloration curves for 2-DNBP in 2-propanol. It appeared, in the opinion of the authors, as a result of convective currents caused by the pulse heating during the flash. This effect was observed in gas systems [72], but was practically unknown for flash photolysis in condensed media.

189

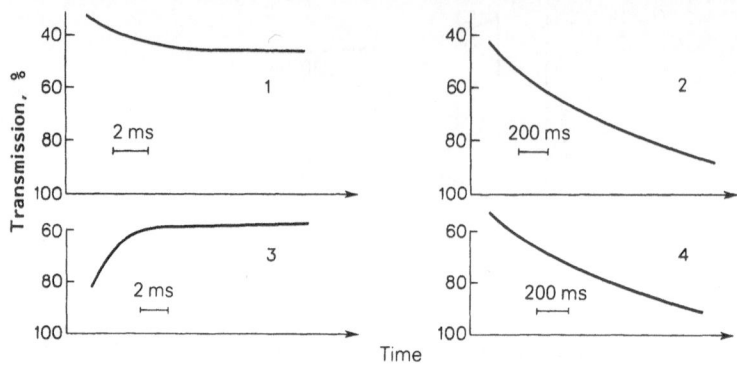

Fig. 3. The change in the absorption of the solution of compound 65 in
absolute alcohol over time under the effect of a flash (20°C,
c_0 2 × 10⁻⁵ mole/liter, tray length 20 cm, E 1000 J): the probing
wavelength, nm: 1, 2) 580; 3, 4) 440.

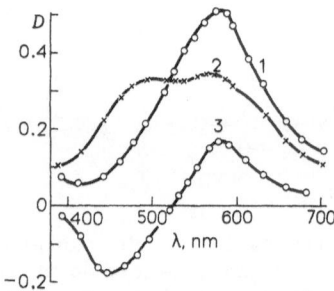

Fig. 4. The absorption spectra of the photoinduced forms of compound 65
in absolute alcohol (20°C, c_0 2 × 10⁻⁵ mole/liter, E 1000 J):
1) right after the flash pulse; 2) 4 ms after the flash pulse;
3) the difference between spectra 1 and 2.

$K_a = k_1/k_{-1}$ of the equilibrium $BH^{\pm} \underset{k_{-1}}{\overset{k_1}{\rightleftharpoons}} B^- + H^+$ and the experimentally obtained (Figure 3) value of k_1, equal to 8.3 × 10³ s⁻¹. If we accept that K_a
does not change substantially upon the transition from 50% (pK$_a$ 6.9) to 95%
ethanol, then the calculated value of k_{-1} amounts to about 10¹⁰ liters/
(mole · s), which is close to the value of the diffusion constants in ethanol [73].

Apparently, the ratio between azamerocyanine and the anion is determined by the acid–base properties of the medium and the substrate in protic
media. This was clearly demonstrated for the first time in the case of the
heterocyclic compounds 26, 64, and 67 and especially for 1-methyl-2-(2,4-di-
nitrobenzyl)benzimidazole 67 [46].

For compounds 26, 54, 59, 64, 67, 75, and 77 the absorption spectra of
colored forms generated by flash photolysis in alkaline buffer solutions
coincide with the spectra of the corresponding mesomer anions which form
without the action of light in much higher-basicity media [41,45,47]. In
the intermediate pH region, which is different for different compounds, the
absorption spectrum of an azamerocyanine and anion mixture is observed. The
equilibrium between them is established in much less than 10⁻⁶ s. Given
below is a scheme for a benzimidazole derivative 67 [46]. It appears as a
result of the combined action of light and environmental factors. Within it
the structure of azamerocyanine BH$^{\pm}$ requires special proof.

190

It was found that the spectrum of the supposed merocyanine had an absorption band maximum at 520 nm, while the model 1,3-dimethyl-2-(2,4-di-nitrobenzylidene)benzimidazoline had a maximum at 495 nm in 50% aqueous ethanol solutions. However, the decoloration rate constant (k_T) of this compound, which was light-induced from salt **68**, numerically coincided with the disappearance rate constant of the BH$^\pm$ form over a wide range of pH values (Figure 5), created by the general-purpose buffer mixture [46]. Besides, in aqueous HCl solutions containing 1% ethanol, linear dependences of log k_T on pH with identical slopes of −1 are observed for the photo-induced forms of compounds **67**, **75**, and **77**, which correspond to the interaction between the base and the proton at a rate proportional to [H$^+$] (Figure 6) [47]. This made it possible to positively identify these structures as azamerocyanines.

The pK$_a$ value for equilibrium 1 (see the diagram) was measured for al-coholic solutions by potentiometric titration (3.95) [44,45]; equilibria 2 for 50% alcoholic solutions, spectrophotometrically (12.60) [44]; equilibria 3, also spectrophotometrically with the generation of colored particles by the flash photolysis method with variation of the solution pH (6.9) [46]; the pK$_a$ of equilibrium 4 is assumed to equal the pK$_a$ of the model system 1,3-dimethyl-2-(2,4-dinitrobenzyl)benzimidazolium−1,3-dimethyl-2-(2,4-dini-trobenzylidene)benzimidazoline (9.0) [46], found by measuring the absorption spectra at different pH values. For the closed system of equilibria 1-4, due to the law of mass action, the following equation must be observed [74]:

$$pK_a^{(1)} + pK_a^{(2)} = pK_a^{(3)} + pK_a^{(4)}$$

This equation is fulfilled for the system under study within an ac-curacy of ±0.5 of a pK$_a$ unit [46], which, allowing for the assumptions made, confirms the validity of the proposed scheme.

These experiments show that the role of the light quantum during photo-chromic conversions of (nitroaryl)(hetaryl)alkanes consists in generating azamerocyanines and isomeric acinitro acids and anions in concentrations ex-ceeding equilibrium ones for these dark conditions. The tautomeric equi-librium constant (K_T) between them can be calculated for dark conditions from the pK$_a$ values of tautomers BH$^\pm$ and BH of compound **67** mentioned above: $K_T = K_a^{(2)}/K_a^{(3)} = 2 \times 10^{-6}$. In the initial stages of the photoreaction aza-merocyanine and nitronic acid are formed (Figure 7) [27]. As is shown in the case of 2-DNBP [14], acinitro acids survive for no longer than 10^{-7} s and, depending on environmental properties, may turn into azamerocyanine and anion. The mechanism of nitronic acid transition into azamerocyanine has

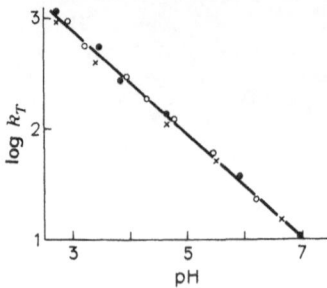

Fig. 5. The pH dependence of the rate constants of decoloration of
azamerocyanines photoinduced from compounds **64** (O), **67** (•),
and **68** (×) in 50% water—ethanol buffer solutions.

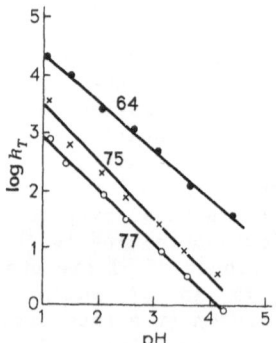

Fig. 6. The pH dependence of the rate constants of decoloration of
azamerocyanines photoinduced from compounds **64**, **75**, and **77**
in diluted hydrochloric acid.

not been studied, but for hydrocarbon media an intramolecular path may be
proposed [14,17]. Under conditions of ordinary flash photolysis, as a rule,
azamerocyanine and the anion are the actually registered forms [41,45,46,
71]. Azamerocyanine in basic media turns into the anion, and their ratio is
determined by environmental properties and the corresponding substrates,
while in protic media — by acid—base properties [27,40,46].

Electron densities and transfer energies were determined by the Hückel
quantum-chemical calculation method for nitro- and aci-structures of
o-nitrotoluene, 2,4-dinitrotoluene, and 2-DNBP [75]. It was established
that upon excitation, the electron density increases on the nitro group and
decreases on the bridge carbon atom. This should lead to a strengthening of
the coulomb attraction of the alkyl hydrogen towards the neighboring nitro
group. It follows that acinitro acid is the most probable structure that
appears following the excitation. In 2-DNBP, as a result of excitation, the
electron density of the pyridine part of the molecule not only does not
increase but decreases, something that cannot contribute to the attraction
of the alkyl hydrogen by the nucleus [75].

Quantum-chemical calculations by the PPP method also led to the conclusion regarding the initial formation of the acinitro acid [76]. It was
shown by the same method that the π-electron density on the alkyl carbon increases as the electron-donor ability of the substituent on the benzyl
fragment of 2-DNBP increases [77]; as is known, the rate constant of the
colored form's transition to the starting molecule [16] increases in the
same sequence; the registered colored form, in reality, corresponds to

azamerocyanine and not to acinitro acid, as Weinstein et al. [16] wrongly believed (see p. 186). Ryaboi and Bazov, using quantum-chemical calculations, established that in the electronic spectra of 2,4-dinitrotoluene and 2-DNBP the most intense bands correspond to electron transfer from the benzene ring to the nitro group.

As a result of these calculations, it was assumed that the intramolecular phototransfer of hydrogen, with the formation of acinitro acid in o-nitrotoluene, proceeds in the lower excited state — π,π^*-triplet (or the n,π^*-triplet, which differs little from it in energy terms). The authors believed that a triplet of some intermediate structure with a "quinoid equilibrium configuration which is very close in terms of geometric parameters to the final reaction product" — acinitro acid — is formed from the reaction triplet; in this way, the adiabatic path of photoisomerization was substantiated [78,79]. This path was not proved experimentally.

The Effect of Structure on the Color of Observed Photochromes

The expected coloration range in photochromes of this class can be predicted on the basis of laws governing the color change in the corresponding nitronic acids, anions, and azamerocyanines with variation in structure.

For the anions — tetranitrodiphenylmethane derivatives 10-12, 15, and 16 — it has been shown that electron-donor and electron-acceptor substituents in positions 6 or 5,5' cause a hypsochromic shift of the absorption maximum because of the disturbed symmetry of π-electron distribution [34].

Given below are the λ_{max} values (nm) of long-wavelength absorption bands for alcoholic solutions of merocyanine dyes, derivatives of 3-ethylrhodanine [80-82], and anions and photoinduced azamerocyanines from (nitroaryl)(hetaryl)alkanes [20,21,40,42,44,45,47]:

Comp. No., A	Mero-cya-nine	Starting compound (see Table 1)	Azamero-cyanine in 50% alcohol	Anion in 50% alcohol
135, 1-ethylbenzoxazolin-2-ylidene	490	75	503	503, 620 sh.
136, 1,3-diethylbenzyl-imidazolin-2-ylidene	512	67	520	495, 645 sh.
137, 1-ethylbenzothiazolin-2-ylidene	526	77	535	512, 630 sh.
138, 1-ethylpyrid-2-ylidene	540	26	567	460, 640 sh.
139, 1-ethylpyrid-4-ylidene	565	54	575	465, 645 sh.
140, 1-ethylquinol-2-ylidene	565	59	580	517, 625 sh.

As these data show, the spectral bands of the azamerocyanines are similar to those of the well-studied ethylrhodanine dyes based on the same heterocyclic systems [42,47]. Thus, the laws governing the chromaticity in the series of well-known merocyanine dyes can be extended to azamerocyanines. In particular, the extension of the conjugation chain in the heterocyclic fragment of the azamerocyanines causes a deepening of the color, as it does in the ethylrhodanine series. Azamerocyanines, like other intraionoid dyes [83], are polarized and thus exhibit pronounced solvato-

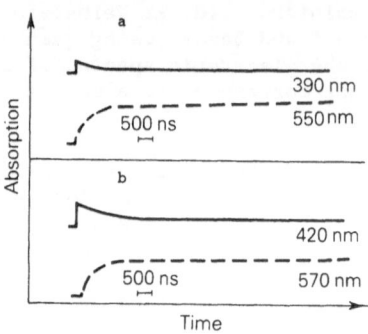

Fig. 7. The change over time in the absorption of the 2-DNBP **26** solution
in heptane (a) and ethanol (b) upon the irradiation by the second
harmonic of a ruby laser (23°C, c_0 10^{-3} mole/liter, E 20 $\mu J/s$);
——) absorption of nitronic acid; ---) absorption of azamerocyanine.

chromic properties: all the dinitroarylidene derivatives based on the
N-methyl derivatives of dihydropyridine systems (an example of fixed azamerocyanines) [34,42] have a positive solvatochromism, while the analogous
benzimidazole derivatives have a negative solvatochromism [44,84]. This
corresponds to the higher polarization of the latter in the ground state.
For instance, for compound **53** the λ_{max} of the band in ethanol is 566 nm
(log ϵ 4.30), in carbon tetrachloride it is 544 nm, and the pK_a in 50% ethanol is 9.90. The same parameters for compound **57** are 584, 548, 9.61; for
58: 607 (4.50), 596, 9.10; for **60**: 598 (4.40), 558, 8.64; for **61**: 637
(4.10), 578, 8.58; and for **68**: 504 (4.52), 515, 9.0. A similar color change
(in the same direction) is observed in a series of fixed azamerocyanine
derivatives of condensed dihydropyridine systems and the corresponding
"unfixed" azamerocyanines light-induced from analogous (nitroaryl)(hetaryl)-
alkanes [42]. Therefore, synthetically more accessible, fixed azamerocyanines can serve as a basis for evaluating the spectral characteristics of
hypothetical "unfixed" analogs.

The spectra of anions of (dinitroaryl)(hetaryl)alkanes consist of two
bands in the wavelength region between 500 and 600 nm [42,41,47]. Using the
example of a series of compounds (Table 3), it has been shown that the
growth of the conjugation chain and steric hindrances in the way of the
coplanar anion position is accompanied by a shift of the band, which is
difficult to predict and is due to the combined effect of all of these
factors [86,87].

The registration of the absorption spectra of anions of benzimidazole
derivatives with the C-methyl group **65**, **66** proved impossible because of
their irreversible dissociation in strong alkaline solutions [88]. The
flash photoexcitation of compounds **65**, **66** in buffer solutions with pH under
11.5 produces short-lived colored forms whose absorption spectrum and lifetime are dependent on the pH of the medium [40]. The nature of the dependence of the decoloration rate constant on the pH of the solution for these
compounds is the same as that of the benzimidazole derivatives without the
C-methyl group **64**, **67**. This leads us to assume that for compounds **65** and **66**
the photoinduced form, depending on the acidity of the medium, is an anion
(pH > 8, λ_{max} about 440 nm) or an azamerocyanine (pH < 5, λ_{max} 580 nm for
compound **65** and 507 nm for compound **66**).

The Decoloration of Colored Forms

Upon photolysis nitronic acids, azamerocyanines, and anions from
(nitroaryl)(hetaryl)alkanes form in concentrations exceeding the equilibrium

Table 3. pK_a Values and absorption band maxima of anions and photoinduced azamerocyanines from (nitroaryl)-(hetaryl)alkanes [42,47]

Number of compound in Table 1	pK_a of the equilibrium* $BH \rightleftarrows B^- + H^+$	Anion		Azamerocyanine** λ, nm (log ϵ)	
		λ, nm (log ϵ)	λ_{sh}, nm	method A	method B
16	17.12	460(4.29)	640	567(4.63)	567(4.63)
59	16.68	517(4.30)	625	590(4.50)	
64	12.58	480(4.45)	645	515(4.54)	515(4.54)
67	15.74 / 12.60	495(4.47)	646	520(4.66)	520(4.65)
69	12.54	485(4.47)	645		516(4.67)
70	12.52	485(4.44)	645		520(4.68)
71	12.43	480(4.48)	640		522(4.74)
72	12.38³*	490(4.45)	650		523(4.70)
73		505(4.45)			546(4.56)⁴*
75	10.93	503(4.55)	620	503(4.62)	500(4.68)⁴*
77	11.54	512(4.53)	630	535(4.59)	526(4.68)⁴*

* For compounds 26, 59, and 67 (pK_a 15.74) – in solutions of sodium methylate in methanol with the use of the H_ function from the paper [85]; for compounds 64, 67 (pK_a 12.60), 69-73, 75, and 77 – in 50% ethanol.

** Method A is based on the constancy of the total concentration of anion and azamerocyanine upon their generation by means of flash photolysis in 50% ethanol at 20°C; method B – spectral-kinetic (ethanol, –78°C).

³* "Apparent" value of pK_a since ionization of the NH group of the imidazole ring occurs concurrently. In N-methyl-substituted compounds the pK_a for ionization of the methylene group is 11.8 [44].

⁴* Mixture of azamerocyanine and anion.

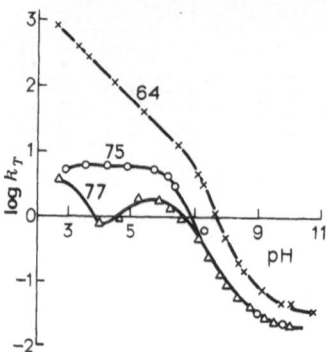

Fig. 8. The dependence of the logarithm of the rate constant of
decoloration on the pH of the acetate–phosphate–borate buffer
solution in 50% ethanol for compounds 64, 75, and 77.

values for dark conditions; the transition to this equilibrium follows different paths, depending on the structure of the colored form and the properties of the medium, in accordance with the laws that govern tautomeric
systems. In nonpolar media the conversion of nitronic acids and azamerocyanines into (nitroaryl)(hetaryl)alkanes proceeds, apparently, intramolecularly [13,14,27]. In protic media, anions and azamerocyanines become protonated, in the limiting stage, at the carbon until the establishment of an
acid–base equilibrium between all the forms (see, for instance, the scheme
for 1-methyl-2-(2,4-dinitrobenzyl)benzimidazole 67).

Rate constants have been calculated for the C-protonation of the azamerocyanine in water and of the anion in 50% ethanol for compounds 64, 65,
67, 75, and 77 on the basis of available data on dependences analogous to
those given in Figures 6 and 8. They amounted to 10^5 and 10^8 liters/
mole·s, respectively, while for the N-protonation of the anion of the α-C-
methyl derivative of benzimidazole 65 the rate constant in alcohol is
10^{10} liters/mole·s. The obtained ratio of rate constants of protonation for
various centers is in good agreement with the available data [89,90] for the
protonation of C- and N-bases. In neutral or weakly alkaline media, colorless or light-colored (nitroaryl)(hetaryl)alkanes, as well as their onium
salts, are not ionized at the alkyl group. Therefore, the establishment of
the dark equilibrium more often than not corresponds to decoloration
(bleaching).

Because decoloration proceeds either intramolecularly or in diluted
solutions, the kinetics of decoloration obeys a first-order or pseudo-first-
order equation. Most of the results on the decoloration kinetics of colored
forms have been obtained for solutions at room temperature by the flash
photolysis technique with a time resolution of at least 20 μs. The following is a summary of the latest available data on the decoloration kinetics. Data are given in the following order: compound number (taken from
Table 1), medium, temperature, k_T, s^{-1}.

1, Water, 1 M HCl, 24°C, 7.6·10 [4]; water, 0.1 N NaOH, 24°C, 1.0 [4]; ethanol (abs.), 24°C, $1.3 \cdot 10^2$ [3]. 2, Water, pH 10-13, 30°C, 1.0 [5,7]; water,
pH 0-1, 30°C, $6.5 \cdot 10^3$ [7]. 3, Water, 0.1 M HClO$_4$, 30°C, $2 \cdot 10^3$ [8].
4, Methanol, 20°C, 1.74 [9]; acetonitrile, 20°C, $4.4 \cdot 10^{-1}$ [9]; cyclohexane,
20°C, $1.04 \cdot 10^3$ [9]; water, 12 M H$_2$SO$_4$, 20°C, $3.7 \cdot 10^2$ [9]. 9, Ethanol
(abs.), −50°C, $6.95 \cdot 10^{-4}$ [12]. 10, Ethanol (abs.), −50°C, $2.84 \cdot 10^{-3}$ [12],
−92°C, $1.9 \cdot 10^{-3}$ [11]; ethanol, −78°C, $6.5 \cdot 10^{-4}$ [21], −55°C, $1.8 \cdot 10^{-4}$ [10],
23°C, 10^6, $1.25 \cdot 10^2$ [14]; heptane, 23°C, 10^6 [14]. 11, Ethanol, −55°C,

$3.9 \cdot 10^{-4}$ [10], 23°C, $7.7 \cdot 10^5$, $2.4 \cdot 10^2$ [14]; heptane, 23°C, $7.7 \cdot 10^5$ [14].
12, Ethanol, −55°C, $8.0 \cdot 10^{-4}$ [10], 23°C, $5.0 \cdot 10^6$ [14]; heptane, 23°C, $5 \cdot 10^6$
[14]. **13**, Ethanol, −55°C, $2.4 \cdot 10^{-4}$ [10], 23°C, $8.3 \cdot 10^5$, $2.1 \cdot 10$ [14]; heptane, 23°C, $7.7 \cdot 10^5$ [14]. **16**, Ethanol, −55°C, $1.6 \cdot 10^{-4}$ [10], 23°C, $5 \cdot 10^6$,
$2.6 \cdot 10$ [14]; heptane, 23°C, $5 \cdot 10^6$ [14]. **19**, Ethanol (abs.), −50°C, $6.03 \cdot 10^{-3}$ [12]. **20**, Ethanol (abs.), −50°C, $7.85 \cdot 10^{-4}$ [12]. **25**, Ethanol, 25°C,
$5.63 \cdot 10$ [16]. **26**, Ethanol, −78°C, $2 \cdot 10^{-4}$ [21], −55°C, $1.8 \cdot 10^{-4}$ [10], 23°C,
$8.3 \cdot 10^5$, 2.0 [14,27]; 24°C, $1.22 \cdot 10^{-1}$ [12,18]; ethanol (abs.), 24°C, $1.2 \cdot 10^{-1}$ [18]; isopropyl alcohol, −78°C, $1.6 \cdot 10^{-4}$ [69]; −46°C, $1.02 \cdot 10^{-2}$ [69];
23°C, $6.7 \cdot 10^6$, 1.43 [14,27], 25°C, $7.8 \cdot 10^{-2}$ [41]; isopropyl alcohol, 10^{-5} M
HCl, −46°C, $1.19 \cdot 10^{-2}$ [69]; isobutyl alcohol, 24°C, $2.77 \cdot 10^{-1}$ [18]; secondary butyl alcohol, 24°C, 1.42 [18]; tert-butyl alcohol, 24°C, $1.53 \cdot 10^{-1}$
[18]; diethyl ether, 24°C, $1.93 \cdot 10^{-1}$ [18]; benzene, 24°C, $1.52 \cdot 10$ [18];
isooctane, 24°C, $1.02 \cdot 10^3$ [18]; water, pH 2, 25°C, $1 \cdot 10^4$ [72], pH 10-12,
25°C, $1 \cdot 10^{-1}$ [71]; 95% isopropyl alcohol, $1.3 \cdot 10^{-10}$ M HClO$_4$, 25°C, $1 \cdot 10^{-1}$
[41], 10^{-7} M HClO$_4$, 25°C, $7.8 \cdot 10^{-2}$ [41], $4.6 \cdot 10^{-1}$ M HClO$_4$, 25°C, $2.2 \cdot 10^4$
[41]; heptane, 23°C, $1.25 \cdot 10^6$, $1.7 \cdot 10^3$ [14,27]; crystals, 20.8°C, $3.5 \cdot 10^{-5}$
[3]; adsorbed on SiO$_2$, 27.5°C, $9.5 \cdot 10^{-5}$ [24], 27.5°C, $5.3 \cdot 10^{-5}$ [24]; supercooled melt, −27°C, $8.2 \cdot 10^{-3}$ [25]. **27**, Ethanol, 25°C, $2.74 \cdot 10$ [16]. **28**,
Ethanol (abs.), 24°C, $3.7 \cdot 10^{-1}$ [3], −88°C, $3.24 \cdot 10^{-3}$ [18]; ethanol, 25°C,
1.17 [16]. **29**, Ethanol, 25°C, $2.63 \cdot 10$ [16]. **30**, Ethanol, 25°C, 2.42 [16].
31, Ethanol, 25°C, 2.45 [16]. **32**, Ethanol, 25°C, 5.94 [16]. **33**, Ethanol,
25°C, $3.29 \cdot 10^2$ [16]. **34**, Ethanol, 25°C, $3.7 \cdot 10^3$ [16]. **45**, Ethanol, −55°C,
$2.4 \cdot 10^{-3}$ [10]. **48**, Ethanol, −55°C, $2.9 \cdot 10^{-4}$ [10]; supercooled melt, −27°C,
$3.8 \cdot 10^{-4}$ [25]. **51**, Ethanol, 20°C, $3.3 \cdot 10^6$, 2.0 [38]. **52**, Carbon tetrachloride, 20°C, 10^7, $5 \cdot 10^4$ [38]. **53**, Ethanol, 20°C, $5 \cdot 10^6$, 1.25 [38];
carbon tetrachloride, 20°C, $6.7 \cdot 10^6$ [38]; ethanol, 20°C, $5 \cdot 10^6$, 1.25 [38].
54, Ethanol (abs.), −100°C, $2 \cdot 10^{-3}$ [41]; ethanol, −78°C, $2 \cdot 10^{-5}$ [21]; isopropyl alcohol 25°C, $4.0 \cdot 10^{-2}$ [41]; isopropyl alcohol, $3.6 \cdot 10^{-7}$ M HClO$_4$,
25°C, 7.0; acetonitrile, 25°C, $5.75 \cdot 10^{-2}$, acetonitrile, $5 \cdot 10^{-4}$ M HClO$_4$,
25°C, 9.2; tert-butyl alcohol, 25°C, $4.6 \cdot 10^{-2}$; tert-butyl alcohol, $5 \cdot 10^{-4}$ M
HClO$_4$, 25°C, $1.25 \cdot 10$; methanol, 25°C, $6.9 \cdot 10^{-1}$; methanol, $5 \cdot 10^{-4}$ M HClO$_4$,
25°C, $1.06 \cdot 10$; acetone, 25°C, 1.5; acetone, $5 \cdot 10^{-4}$ M HClO$_4$, 25°C, 7.67;
isopropyl alcohol, $1.2 \cdot 10^{-10}$ M HClO$_4$, 25°C, $2.9 \cdot 10^{-2}$, $1.5 \cdot 10^{-8}$ M HClO$_4$,
25°C, $2.2 \cdot 10^{-2}$, $3.0 \cdot 10^{-6}$ M HClO$_4$, 25°C, $4.5 \cdot 10^{-2}$, $5.0 \cdot 10^{-3}$ M HClO$_4$, 25°C,
$1.4 \cdot 10^2$, $4.6 \cdot 10^{-1}$ M HClO$_4$, 25°C, $1.5 \cdot 10^4$. **63**, Ethanol (50%), universal
buffer mixture of phosphoric, acetic, and boric acids in a concentration of
0.04 mole/liter each and 0.2 mole/liter NaOH (UBM for short), pH 3.60, 20°C,
$5.5 \cdot 10^3$ [43], pH 5.05, 20°C, $1.58 \cdot 10^3$, pH 5.64, 20°C, $1.03 \cdot 10^3$. **64**, Ethanol, −75°C, $2 \cdot 10^{-5}$ [44]; ethanol (50%), UBM, pH 2.69, 20°C, $9.4 \cdot 10^2$ [45], pH
4.58, 20°C, $1.2 \cdot 10^2$, pH 6.53, 20°C, $1.3 \cdot 10$, pH 8.57, 20°C, $1.9 \cdot 10^{-1}$, pH
10.17, 20°C, $5 \cdot 10^{-2}$. **65**, Ethanol, −75°C, $9 \cdot 10^{-4}$ [44]; ethanol (50%), HCl,
pH 1.32, 20°C, $1.0 \cdot 10^4$ [40], pH 2.32, 20°C, $1.1 \cdot 10^3$, UBM, pH 2.72, 20°C,
$9.15 \cdot 10^2$, pH 4.62, 20°C, $1.55 \cdot 10^2$, pH 6.95, 20°C, $1.26 \cdot 10$, pH 8.47, 20°C,
$1.85 \cdot 10^{-1}$, pH 10.40, 20°C, $1.92 \cdot 10^{-2}$. **66**, Ethanol, −75°C, $5.5 \cdot 10^{-4}$ [44];
ethanol (50%), UBM, pH 2.73, 20°C, $4.48 \cdot 10^2$, pH 4.67, 20°C, $7.41 \cdot 10$, pH
6.56, 20°C, $1.06 \cdot 10$, pH 8.0, 20°C, $3.04 \cdot 10^{-1}$, pH 9.79, 20°C, $1.02 \cdot 10^{-2}$. **67**,
Ethanol, −75°C, $2 \cdot 10^{-5}$ [44], 20°C, $1.5 \cdot 10^{-1}$ [91]; heptane, 20°C, $2.24 \cdot 10$
[91]; carbon tetrachloride, 20°C, $2.9 \cdot 10^2$ [91]; chloroform, 20°C, 2.2 [91];
benzene, 20°C, $1.35 \cdot 10$ [91]; ethanol (50%), UBM, pH 2.69, 20°C, $9.4 \cdot 10^2$
[45], pH 4.58, 20°C, $1.2 \cdot 10^2$, pH 6.53, 20°C, $1.3 \cdot 10$, pH 8.57, 20°C, $1.9 \cdot 10^{-1}$,
pH 10.17, 20°C, $5 \cdot 10^{-2}$. **69**, Ethanol, −75°C, $2 \cdot 10^5$ [44]. **71**, Ethanol,
−75°C, $2 \cdot 10^{-5}$ [44]. **72**, Ethanol, −75°C, $1 \cdot 10^{-5}$ [44]. **73**, Ethanol, −75°C,
$3 \cdot 10^{-5}$ [44]. **74**, Ethanol (50%), UBM, pH 2.40, 20°C, $3.6 \cdot 10^2$ [43], pH 12.50,
20°C, $1.07 \cdot 10$. **75**, Ethanol, −75°C, $2 \cdot 10^{-5}$ [45]; ethanol (50%), UBM, pH
2.86, 20°C, 6.1 [47], pH 4.96, 20°C, 6.4, pH 6.35, 20°C, 3.3, pH 7.66, 20°C,
$2.5 \cdot 10^{-1}$, pH 9.50, 20°C, $2.7 \cdot 10^{-2}$. **77**, Ethanol, −75°C, $1 \cdot 10^{-5}$ [45];
ethanol (50%), UBM, pH 2.68, 20°C, 5.3 [47], pH 4.88, 20°C, $9 \cdot 10^{-1}$, pH 6.38,
20°C, 1.3, pH 8.04, 20°C, $1 \cdot 10^{-1}$, pH 10.07, 20°C, $2.5 \cdot 10^{-2}$ [47].

The decoloration of 2,4- [7] and 2,6-dinitrotoluene [8] anions is dependent on general acidic catalysis, and the experimental rate constant is described by the equation

$$k_{exp} = k_{H_2O}[H_2O] + k_{HA}[HA] + k_{H^+} + [H^+]$$

where k_{H_2O}, k_{HA}, and k_{H^+} are bimolecular rate constants of the reaction with water, undissociated acids, and protons, respectively.

For 2,4-dinitrotoluene anions k_{H_2O} is $1.8 \cdot 10^{-2}$, k_{CH_3COOH} is $4.8 \cdot 10^3$, and k_{H^+} is $7.4 \cdot 10^4$ liters/mole·s at 30°C [7]. Upon the decoloration of the 2,6-dinitrotoluene anion by HA acids, whose pK_a changes from −1.75 to 16.75, Brönsted's coefficient changes from zero to unity as the pK_a of the acid increases [8]. Decoloration rate constants of the 2-nitrotoluene and 2,4-dinitrotoluene anions generated by flash photolysis in 0.1 N NaOH are equal to 1.9 [4] and 1.0 [7] s^{-1}, respectively, which accounts for the slight influence of the second nitro group on the kinetic basicity of the anions.

The isomerization of acinitro acids to nitrotoluenes in aqueous solutions should involve a proton in accordance with the scheme

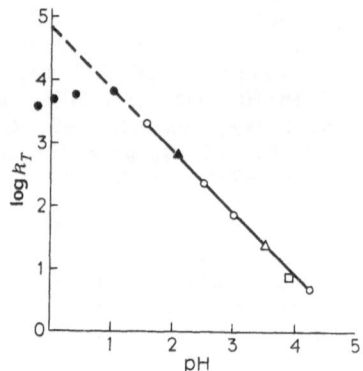

Acinitro acids of 2-nitrotoluene and 2,4- and 2,6-dinitrotoluenes decolorize with rate constants of $7.6 \cdot 10$ [4], $6.5 \cdot 10^3$ [7], and $2 \cdot 10^3$ [8] s^{-1}; it is seen that the structure of acinitro acids strongly influences their kinetic basicity. The rate of transition of acinitro acids into isomeric toluene is independent of pH (Figure 9). Contrary to the scheme cited

Fig. 9. The pH dependence of the logarithm of the rate constant of decoloration of photoinduced forms of 2,4-dinitrotoluene. O and • − for solutions with different concentrations of perchloric acid; □ − values of k_T calculated by means of extrapolation of the dependence $k_T = f([CH_3COOH])$, obtained in an acetate buffer solution with an equal concentration of acetic acid and sodium acetate; △ − the same as in the acetate buffer solution with a ratio of the concentration of acetic acid and sodium acetate of 10:1. △ − the same, $k_T = f([H_3^+NCH_2COOH])$, in solutions of a mixture of perchloric acid and glycine with a mole ratio of 1:2.

above, the authors of the papers [4,7,8] attributed it to the intramolecular course of the decoloration.

For cases involving conversions of 2,6-dinitrotoluene, isomeric aci-nitro acid, and the corresponding anion in water, the following formal kinetic scheme is proposed [8]:

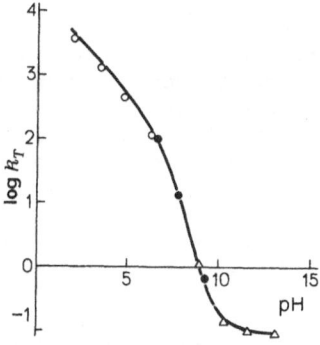

This scheme should be supplemented by a series of protolytic stages.

In (nitroaryl)(hetaryl)alkanes, because of the presence of the key heteroatom, the pH dependence of the rate constant of the decoloration of the colored form is far more complex than in the case of nitroarylalkanes. As the data on the kinetics of decoloration of benzazole 64, 75, 77 and pyridine 26 derivatives given in Figures 8 and 10 indicate, the curve consists of several highly characteristic sections [47,71]. These add up to the overall result of the interaction of the anion and azamerocyanine with water, acids, and protons. The region of existence of anion and azamerocyanine for each particular material is determined by the pK_a value of the equilibrium between them. In this pH zone the curves show an inflection at pH 7 for benzimidazole derivatives 64 (pK_a 7.4 [46]), at 6.0 for benzoxazole and benzothiazole derivatives 75 and 77 (pK_a 5.4 and 6.4, respectively) [47] (see Figure 8), at 6.8 for aqueous solutions of 2-DNBP and, according to the data reported by Mosher et al., at pH 6.3 for solutions of 2- and 4-DNBP in 2-propanol with $HClO_4$ [41] (pK_a 6.8 [22] and 7.4 [40] for 2- and 4-DNBP in 50% ethanol).

At high pH values (> 9) the decoloration rate constant of the anions of compounds 26, 64, 75, and 77 is independent of the pH of the medium (see Figures 8 and 10). This is due to the preferential participation in the decoloration of the anion of the solvent molecules RH (B⁻ + RH → BH + R⁻), which with its high concentration is, naturally, only slightly dependent on pH. The contribution made by the interaction with the proton (B⁻ + H⁺ → BH)

Fig. 10.   The pH dependence of the logarithm of the rate constant of the decoloration of photoinduced forms of 2-DNBP 26 in water at 24°C in acetate (O), phosphate (•), and borate (Δ) buffer solutions.

199

is insignificant on account of its low concentration. As the latter increases, the interaction with H^+ should be dominant and k_T should be linearly dependent on $[H^+]$, which is confirmed experimentally in the pH region 7-9 [45].

The decoloration kinetics of azamerocyanines vary widely from compound to compound in 50% aqueous ethanolic phosphate—acetate—borate buffer solutions (Figure 8). This is attributed to the fact that in this pH region the concentrations of undissociated acids are such that the contribution of the reaction with them becomes commensurate with the protic contribution. A complex pattern of dependences of k_T on pH results. In simple buffer systems containing fully dissociated acids only, e.g., HCl in water, k_T is directly proportional to $[H^+]$ for compounds **64**, **75**, and **77** (see Figure 6) [47].

In interpreting the pH dependence of log k_T for 2-DNBP in aqueous solutions (see Figure 10), Wettermark believed that the provision of an anion at pH > 9 occurs as a result of its monomolecular conversion into a "negatively charged colorless form corresponding to the starting structure but without a proton," with its subsequent addition in other than a limiting stage [71]. The increase in the decoloration rate as the solvent acidity rises (pH 7-9) is attributed by Wettermark to the increasing contribution made by a process involving the azamerocyanine $BH^\pm \rightarrow BH$, which proceeds faster than $B^- + H^+ \rightarrow BH$. However, judging by his findings, an increase in k_T occurs at higher pH values (~9) than one would expect from the pK_a value (4-6) of the protolytic equilibrium between azamerocyanine and anion cited in [71]. The precise value of the pK_a of the equilibrium $BH^\pm \rightleftharpoons B^- + H^+$ in 2-DNBP determined by the spectrophotometric technique with the pulsed generation of colored particles [22] is 6.8 ± 0.1 in 50% aqueous alcohol and 6.5 ± 0.1 in water. At these pK_a values azamerocyanine is absent from solutions with pH > 8.5 and does not contribute to the decoloration.

At pH < 5, following the acid-catalyzed decoloration of azamerocyanine from 2-DNBP, Wettermark, on the basis of the overlapping of the pH dependence of log k_T for various buffer systems, believes that k_{HA} of different acids are the same [71]. This is not in agreement with the data obtained for aqueous formate—acetate buffer solutions of meso-dinitrobenzyl derivatives of benzazoles **64**, **65**, and **67**. For instance, the k_{HA} values of acetic and formic acids for 1-methyl-2-(2,4-dinitrobenzyl)benzimidazole **67** are $2.6 \cdot 10^4$ and $5.8 \cdot 10^4$ liters/mole·s at 20°C [91]. At −78°C there is no influence of acid additions on the decoloration rate of ethanolic solutions of 2-DNBP; at −46°C this influence is weak [69].

It has been established by stationary photolysis of ethanolic solutions at −55°C that substitution of positions 4,4' in 2,2'-dinitrodiphenylmethane as well as the introduction of methyls into positions 5,5' of tetranitro-diphenylmethane increases the rate of decoloration of the colored forms of these compounds, while an increase in the number of nitro groups slows it down [10,34]. The decoloration rate of the photocoloring in 2-DNBP at room temperature is highest in isooctane and slows down in the series benzene, ether, alcohols [18].

The dependence between the nature of the heterocycle and the decoloration rate in (nitroaryl)(hetaryl)alkanes has a qualitative character (see Figure 6) [47], while the effect of para substituents in the benzene ring in 2-(2-nitro-4-R'-benzyl)pyridines is described by the Hammett equation (ρ −2.75) [16]; this influence is linked with differences in the energy of activation of decoloration. The values of the decoloration activation energies are given in the paper [18] for ethanolic solutions of 2-DNBP (24.28 kJ/mole) and 2-(2-nitro-4-cyanobenzyl)pyridine (19.67 kJ/mole) without a rigorous identification of the colored form. Generally speaking, for solutions and crystals of 2-DNBP the decoloration activation energies were determined repeatedly [17,18,20,21,41]. In different media they ranged from 10.46 to 104.65 kJ/mole. It is extremely difficult to analyze these values since in each individual case the question of the nature of the decolorizing particle was not resolved correctly.

This error was allowed for only during the measurements of the E_a of nitrotoluene derivatives, where the E_a of decoloration of the photoinduced anion in aqueous solutions of 1 M NaOH was 58.19 for 2,4-dinitrotoluene [7] and 48.14 kJ/mole for the 2,6-isomer [8]; in aqueous solutions of 1 M HCl E_a of the corresponding acids were 39.77 [77] and 42.28 kJ/mole [8]. In ethanolic solutions the E_a of decoloration were 24.44 for methylbis(2,4-dinitrophenyl) acetate, 22.60 for the ethyl analog and 2,4,4'-trinitrodiphenylmethane, and 23.02 kJ/mole for the 2,2'4,4'-tetranitro-substituted form [12].

The Quantum Yields of Photocoloration

The quantum yield of photocoloration is a major characteristic of any photochromic system. Its determination requires knowledge of the extinction coefficients ϵ of the colored forms. Although the extinction coefficient of the anion ϵ_{B^-} can be easily found from the data of the ionization of starting (nitroaryl)(hetaryl)alkanes BH, the determination of the extinction coefficients of azamerocyanine $\epsilon_{BH^\pm}$ and nitronic acid ϵ_{iso-BH} is difficult. While attempting to determine the extinction coefficients of the colored forms, some authors assumed them to be equal to the extinction coefficient of the models [17,70]. The error introduced by this approximation was compounded by the fact that in some cases the actual structure of the registered form was not correctly established and could deviate from the structure of the model. In other papers ϵ was computed using the hypothesis about 100% conversion of the starting compound into the colored form, something that was never proved definitely [20,28]. Other indirect methods were used [7]. As Table 4 indicates, the data cited in the papers [17,21,40,69, 70] are rather contradictory.

In the spectral-kinetic method developed by V.P. Bazov, the photocoloration BH $\xrightarrow{h\nu}$ C and photodissociation of the starting and photoinduced forms [21,68,93] are taken into account:

$$P_1 \xleftarrow{\ h\nu_1\ \varphi_{P_1}\ } BH \underset{\Delta}{\overset{h\nu\ \varphi_C}{\rightleftarrows}} C \xrightarrow{\ h\nu_1\ \psi_{P_2}\ } P_2$$

This method applies provided only one colored form is registered [see the note to 4-DNBP (solvent — ethanol) in Table 4]. This is why its application requires a careful selection of appropriate experimental conditions. The calculations performed for ethanolic solutions of benzoxazole 75 and benzothiazole 77 derivatives are incorrect because of the simultaneous presence of the corresponding anion and azamerocyanine; however, they may be useful for a comparative evaluation of decoloration in identical positions [47]. The action of light on alcoholic solutions at −78°C of benzimidazole derivatives 64-67 and 69-73 produces only azamerocyanines, if the absorption spectrum is any guide. Therefore, the spectral-kinetic method is not applicable in this instance, and the photochromism parameters calculated by this method are not correct (see Table 5).

In protic media at room temperature the primary products of the photoreaction, in accordance with the ratio of the constants of acid–base equilibria, quickly convert into an equilibrium mixture of azamerocyanine and anion for (nitroaryl)(hetaryl)alkanes or acinitro acids and anion for nitroarylalkanes [7,46]. The equilibrium mixture of these particles does not change composition in the course of the dark decoloration and behaves as a "pseudoparticle." In the case of heterocyclic derivatives the spectral and

Table 4. Values of ϵ for the colored forms of 2-DNBP 26 and 4-DNBP 54

Compound	Solvent	$\epsilon \cdot 10^4$ (λ, nm)	Note
		2-(2,4-dinitrobenzyl)pyridine	
26	amines of the aliphatic series ethanol + potassium hydroxide solution	3.4(472) 3.4(472)	Corresponds to the 2-DNBP anion; measurements made at 25°C [20]
	ethanol	1.2 1.5(260)	Equated with ϵ of the model 1-methyl-2-(2,4-dinitrobenzyl)-1,2-dihydropyridine; measurements made at 25 to −60°C [17,70]
	isopropanol	1.97(567)	
	acetone	4.5(565)	
	isopropanol	0.053(565)	Equated with ϵ for the deeply colored reaction product of the reaction between 2-DNBP or 4-DNBP with epichlorohydrin [70]. Abnormally small value of ϵ, in the opinion of Hiraoka and Hardwick [70], is due to the low degree of conversion into reaction product
		4-(2,4-dinitrobenzyl)pyridine	
54	isopropanol	2.2(575)	Obtained on the assumption of 100% conversion upon acidification of cooled solutions of anion of 2-DNBP or 4-DNBP into a form identical to the photoinduced one; measurements made at −78°C [69]
26	"	2.3(568)	
54	ethanol	3.6(575)	Spectral-kinetic method; corresponds, in the authors' view, to nitronic acid; judging by the electronic spectrum cited, this is azamerocyanine [21]; measurements made at −78°C
26	ethanol	4.4(567)	
26	50% ethanol	4.3(565)	Corresponds to azamerocyanine; obtained from the constancy of the total molar concentration of anion and azamerocyanine in solutions of the phototolysis product with varied pH; measurements made at 20°C [40]
54	isopropanol	2.4(580)	Equated with ϵ of the model 1-methyl-4-(2,4-dinitrobenzyl)-1,4-dihydropyridine [70,92]
		2.4(588)	Equated with ϵ of the deeply colored product of the reaction between 2-DNBP or 4-DNBP with epichlorohydrin [70]

kinetic characteristics of the "pseudoparticle" depend on the ratio of the components B⁻ and BH±, which is determined solely by the equilibrium constant between them and the acidity of the medium.

The overall concentration of the photoinduced components immediately after the flash or sometime afterwards, during which it has no time to

Table 5. The results of calculations by the spectral-kinetic method[21]

Number of compound* in Table 1	ϵ_C (at λ_{max})	$\phi_C \cdot 10^2$	$\phi_{P_1} \cdot 10^2$	$\phi_{P_2} \cdot 10^2$	n, %	$k_T \cdot 10^3$ s^{-1}
10	3.9	6.0	1.2	0.4	50	0.65
26	4.4	6.0	1.2	0.4	50	0.2
54	3.6	12	4.0	0.5	50	0.02
64	3.45	7.4	1.0	0.07	83	0.002
65	3.56	36.0	3.5	0.5	78	0.9
66	2.06	40.0	6.0	0.1	79	0.55
67	4.57	4.1	0.7	0.08	70	0.01
69	4.67	7.8	1.2	0.06	82	0.02
70	4.77	7.4	0.9	0.06	89	0.01
71	5.09	5.0	1.3	0.09	79	0.02
72	5.05	5.3	0.7	0.06	83	0.01
73	3.63	0.3	0.1	0.08	54	0.03
75	4.77	0.7	0.7	0.04	26	0.02
76		30.0			60	0.60
77	4.80	4.1	0.5	0.4	75	0.01

* Compound 10, in the opinion of Bazov [21], is acinitro acid. Judging
by the electronic absorption spectrum, however, this is an anion[11,34];
26 and 54 — see the note to 4-DNBP with ethanol as solvent in Table 4; 64-
67, 69-73 — azamerocyanine[46]; 75-77 — an azamerocyanine and anion mixture[47].

significantly diminish as a result of the dark decoloration, depends only on
the quantum yields of their formation and the photolysis conditions. When
the pH of the medium changes in an interval that does not affect the region
of the protolytic reactions of the starting compound and excited particles
preceding the photoinduced forms, the quantum yields should remain constant
[94].

The photocoloration of (nitroaryl)(hetaryl)alkanes is responsible for
the excited state of the nitroaryl fragment [44]. The pK_a^* of proton addition to the nitro group [95] and the pK_a^* of proton elimination from the
methylene group [45,96] are known to be considerably lower than the pK_a values of the azamerocyanine $\rightleftharpoons$ anion equilibrium. Therefore, under constant
photolysis conditions in the pH interval near the pK_a of the equilibrium
$BH^{\pm} \rightleftharpoons B^-$, the overall molar concentration of the photoinduced particles $BH^{\pm}$
and B^- should also be constant, regardless of whether or not an equilibrium
has been reached between them. The composition of the equilibrium photoinduced mixtures will be established in accordance with the pH of the medium. Then the series of experiments involving photocoloration with variation in the pH of the medium only during the registration of the spectra of
the photoinduced forms in the time intervals when the decoloration can be
disregarded could be used, as in conventional spectrophotometry, for determining the characteristics of conjugated components [97]. The absorption
spectra of the photogenerated forms at the limits of the pH intervals under
discussion are the spectra of the individual forms of $BH^{\pm}$ and B^- with the
molar concentration of both being the same. The ratio of the extinction
coefficients of both forms can easily be calculated from these spectra:

$$\epsilon_\lambda^{B^-}/\epsilon_\lambda^{BH^{\pm}} = D_\lambda^{B^-}/D_\lambda^{BH^{\pm}}$$

where $D_\lambda^{B^-}$ and $D_\lambda^{BH^{\pm}}$ are the optical densities of coloring of the photoinduced anion B^- and azamerocyanine $BH^{\pm}$ at any wavelength with pH values

of the medium such that only the anion and azamerocyanine forms are registered.

Knowledge of ϵ of the anion is essential for determining the ϵ of the
azamerocyanine. It can be found in many cases during the dark ionization of
the starting photochromes in basic media [20,44].

The preceding equation was used to determine the extinction coefficients of the azamerocyanines from compounds 26, 64, 67, 75, and 77 [40].
The pH of the media in which the measurements were made was chosen with due
regard for the pK_a values of the $BH^{\pm} \rightleftharpoons B^- + H^+$ equilibrium; they were determined spectrophotometrically in 50% ethanol, generating colored particles
with a light pulse [22,46,47]. For instance, for compound 26 the pK_a is 6.8
[22]; for 54, 7.4 [40]; for 64, 7.4 [46]; for 67, 6.9 [46]; for 75, 5.4
[47]; and for 77, 6.4 [47]. For these compounds the optical densities of
coloring, measured right after the light pulse in the pH region where only
the forms $BH^{\pm}$ or B^- are generated, correspond to the same molar concentration of these forms since their decoloration during the flash is so insignificant as to be disregarded. That is why the optical densities D^{B^-} and
$D^{BH^{\pm}}$ were determined directly after the flash.

The use of this equation for compounds 26, 64, and 67 is hindered
because of the complexity involved in the determination of $D^{BH^{\pm}}$ in the
region where only the azamerocyanine form is generated, due to the high
basicity of the BH and its protonation into BH_2^+ [45,71]. The pK_a value of
of the $BH_2^+ \rightleftharpoons BH + H^+$ equilibrium for compound 67, determined potentiometri-
cally in ethanol, is 3.95 [46], and for compound 26 in water it is 4.1
[71]. Therefore, for instance, the N-methyl derivative 67 at a pH of about
5, corresponding to 99% formation of azamerocyanine, is protonated by about
10%.

The study of this compound shows that BH_2^+ has a lower photocoloration
efficiency [45], although the absorption spectra of the BH and BH_2^+ forms are
practically the same. The quantum yield of the BH_2^+ coloration was found to
be one fifth that of the BH form [46]. Therefore, the values of $D^{BH^{\pm}}$ for
compounds 26, 64, and 67 were calculated with the help of a system of equa-
tions: $D_i^{exp} = \alpha(1 - p_i) + \alpha\beta p_i$, where D_i^{exp} is the optical density of coloration of the azamerocyanine generated from the BH and BH_2^+ mixture at the i-th
pH; p_i is the fraction of the BH form in the starting solution; α is the
parameter of the efficiency of the BH_2^+ transition into the azamerocyanine;
β is the coefficient of the relative efficiency of coloration of the BH and
BH_2^+ forms. α and β for the various D_i^{exp} values were found from a system of
equations; while $D^{BH^{\pm}} = \alpha\beta$, it corresponds to the participation of the BH
form (p_i 1) only in the photocoloration.

In the general case, the dependence of the optical density of photoinduced coloration on the concentration of free hydrogen ions, allowing for
the varying photocoloration efficiency of the BH and BH_2^+ forms, present the
following picture which differs somewhat from that cited in the paper [46],
where $k_\lambda = \epsilon_\lambda^{B^-}/\epsilon_\lambda^{BH^{\pm}}$:

$$D_\lambda = \alpha\epsilon_\lambda^{BH^{\pm}} l c_0 \frac{\left([H^+] + \beta K_a^{(1)}\right)\left([H^+] + k_\lambda K_a^{(3)}\right)}{\left([H^+] + K_a^{(1)}\right)\left([H^+] + K_a^{(3)}\right)}$$

As Table 4 shows, the values of the extinction coefficients obtained
for azamerocyanines of compounds 25, 64, and 67 in 50% ethanol and the
extinction coefficients of a colored form determined by the spectral-kinetic
method for ethanolic solutions practically coincide.

For benzoxazole and thiazole derivatives 75 and 77 the calculation of
$\epsilon^{BH^{\pm}}$ is made easier when compared with the more basic compounds 26, 64, and

67 because direct experimental determination of $D^{BH^{\pm}}$ [40] is possible. The pK$_a$ of the BH$_2^+$ ⇌ BH + H$^+$ equilibrium for them is lower, −0.5 [98] and 1.2 [99], respectively, and in the pH regions 1-4 and 2-5 they should exist in the form of BH bases; the independence of $D^{BH^{\pm}}$ on the pH in these regions has been shown experimentally [47]. The values of $\varepsilon^{BH^{\pm}}$ calculated by the above equation differ from those found by the spectral-kinetic method for the reasons discussed above.

Irreversible Photodissociation

In all (nitroaryl)(hetaryl)alkanes, as a result of photolysis of shorter or longer duration, the optical density of coloration decreases as compared with the original one. In most cases this is attributed to the photodegradation of the photochrome [4,7,11,18,20,24,28,39,44,45,56,69]; only for 3,3'-dinitro-4,4'-di(2-pyridylmethyl)azoxybenzene 38 has it been noted that under the conditions of low-temperature stationary photolysis and pulsed excitation at room temperature in alcohol and cyclohexane no photodissociation occurs [32]. Using the example of 2-(2-nitro-4-cyanobenzyl)pyridine 28 it was shown that upon light photolysis with λ 254 nm no photodissociation at all takes place and at λ 365 nm it is weak; thus the influence of the wavelength of the exciting light on the efficiency of photodissociation is demonstrated [28]. The adsorption of 2-DNBP on NaCl, SiO$_2$, and LiF does not prevent its irreversible photodissociation [24].

Only one attempt has been reported in the literature to estimate the quantum yield of photodissociation and to compare it with the quantum yield of photocoloration; this has been achieved within the framework of the spectral-kinetic method [21,68]. It has been shown that in ethanolic solutions at −78°C in 2-DNBP, 4-DNBP, and tetranitrodiphenylmethane 10 both the starting BH and the photoinduced C-forms with quantum yields ϕ_{P1} and ϕ_{P2}, respectively, undergo photodissociation (see Table 5). In an alcoholic matrix at −196°C only BH is dissociated, and the value of ϕ_{P1} is constant within the temperature range from 20 to −78°C but drops at −196°C, apparently because of the increased rigidity of the matrix; in heptane at −78°C ϕ_{P1} and ϕ_{P2} are considerably higher than in alcohol at the same temperature [100]. Proceeding from the values of ϕ_C and ϕ_{P1}, two paths of the dissipation of excitation energies of these substances are comparable in terms of efficiency; the values of ϕ_{P2} are far lower on the average.

The spectral-kinetic method is used for a qualitative description of the efficiency of coloration and dissociation of various benzazole derivatives 64-67, 69-73, and 75-77 [44,47] which were found, judging by the ϕ_{P1} values (see Table 6), also to be nonphotostable (Figure 11), especially the C-methyl-substituted forms 65 and 66. At room temperature the decoloration rate increases after coloration—decoloration cycles. An analogous effect was observed upon the photolysis in absolute alcohol solutions of tetranitrophenylmethane 10, 2-(2-nitro-4-cyanobenzyl)pyridine 28, and 2-DNBP 26; in o-nitrotoluene 1 a slow down of decoloration was observed [3]. This is attributable to an increase in the acidity of the solution due to the build-up of irreversible reaction products. In buffer solutions the decoloration acceleration effect is poorly pronounced, for understandable reasons (see Figure 12) [45]. At low temperature, dissociation products have no influence on the decoloration rate [69].

Attempts have been made to identify the products of the dissociation of (nitroaryl)(hetaryl)alkanes. Upon irradiation, 2-nitrotriphenylmethane is known to give 2-nitrotriphenylcarbinol [101]. This intramolecular photoredox process was regarded as the source of dissociation products of nitroarylalkanes upon photolysis [11]. Other photoreactions involving aromatic nitro compounds described in surveys [102,103] can result in numerous dissociation products.

Table 6. Kinetic parameters of the decoloration of photoinduced forms of compounds 47 and 48 in solid solutions [26]

Number of compound in Table 1	Polymer	M, g/mole	t, °C	k_T, liter/mole·s	$\Delta H^{\neq}$, kJ/mole	$S^{\neq}$, J/mole·K	E_a, kJ/mole	log A (A in liters/mole·s)
47	polyethyl acrylate	35 000	10.1;−49.0	$4.3 \cdot 10^{-2}*;1.0 \cdot 10^{-2}$	85.4;44.4	31.4;−83.7	88.0;46.1	14.8*;9.0
48	polymethyl methacrylate	3 900	3.5;−30.7	$2.8;2.0 \cdot 10^{-2}$	57.4;57.4	−29.3;−39.0	59.5	11.7;11.7
		23 600	3.5;−30.7	$3.7 \cdot 10^{-1};8.0 \cdot 10^{-3}$	52.3;52.3	−64.1;−67.0	54.4	9.8;9.6
	polystyrene	47 000	3.5;−30.7	$2.2;3.0 \cdot 10^{-2}$	54.0;54.0	−43.1;−49.4	56.1	10.9;10.6

* In s^{-1}.

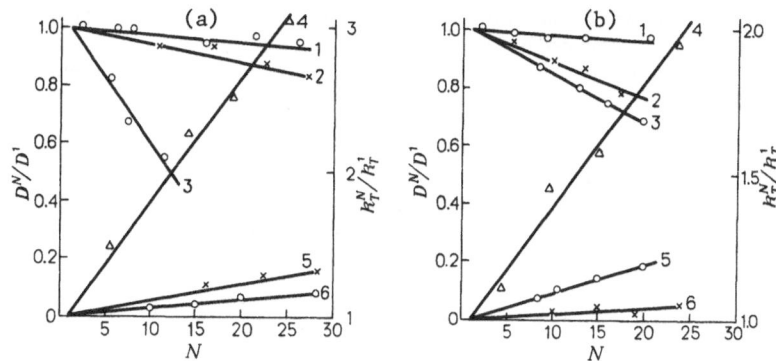

Fig. 11. The dependence of the ratio of optical densities of coloration D^N/D^1 at the band maxima of the photoinduced forms (1-3) and the ratio of the rate constants of decoloration k_T^N/k_T^1 (4-6) on the number of flashes N at 20°C (D^N and k_T^N – after the N-th flash; D^1 and k_T^1 – after the first flash). a) For compound 64: 1, 6) in 50% aqueous alcoholic buffer solution, pH 2.75; 2, 5) in 50% aqueous alcoholic buffer solution, pH 9.06; 3, 4) in 96% alcohol. b) For compound 67: 1, 5) in 50% aqueous alcoholic buffer solution, pH 7.29; 2) in 50% ethanol; 3, 4) in 96% ethanol; 6) in 50% aqueous alcoholic buffer solution, pH 2.75.

Upon prolonged stationary irradiation with ultraviolet light in 2-propanol at a low temperature 2-DNBP gave, among many other substances, a product with λ_{max} of about 350 nm [17]. At the same time, a band appeared in the infrared spectrum at 1020 cm^{-1} that did not disappear upon decoloration and which was apparently also caused by dissociation products. On another occasion, after 10-h irradiation with ultraviolet light of 2-DNBP in absolute ethanol at temperatures of from −30 to 0°C as many as 13 new compounds were found in the photolysis product by thin-layer chromatography. Dissociation products also result from the prolonged irradiation of crystals at room temperature, irrespective of whether the photolysis is conducted with the unfiltered light of a mercury lamp or with the 365 nm line. Three substances out of the dissociation products formed in each case were isolated and identified: 2-(2-amino-4-nitrobenzyl)pyridine 145, 2-(2,4-dinitrobenzoyl)pyridine 146, and 2-{α-(5-nitro-2-(2-pyridoyl)phenylimino)-2,4-dinitrobenzyl}pyridine 147.

In blue 2-DNBP crystals irradiated by ultraviolet light no change in the crystal lattice was detected [23,53]. Apparently, as in many photoreactions, the dissociation products form a solid solution in the starting crystal [104].

207

In the course of flash photolysis of 1-methyl-2-(2,4-dinitrobenzyl)-benzimidazole **67** a product builds up with a band at 320 nm. In this process the short-wavelength bands which correspond to transitions in the benzimidazole chromophore [45] do not change. Apparently, the dissociation is localized in the dinitrobenzyl fragment.

Available data indicate that a decrease in the optical density can be caused not only by the dissociation of the photochrome, its photoisomers, and anion, but also by the filtering action of the dissociation products that lower the effectiveness of photochrome excitation.

The Photochromism of (Nitroaryl)(hetaryl)alkanes in Crystals, Melts, and Polymer Films

The 2-DNBP is photochromic in crystals [17,23,53], in powder form, and in a supercooled melt [25]; according to the data cited by Chichibabin [53], the only identified form is the azamerocyanine. Crystals of 2-DNBP analogs substituted in the nuclei **39**, **40**, **43**, and **47-49** are photochromic at room temperature. 4-DNBP is photochromic in powder form and in a supercooled melt [25,92]; in this compound only the azamerocyanine was detected. It is assumed that in solid (nitroaryl)(hetaryl)alkanes, which exist in the form of associates with an intermolecular hydrogen bond, the hydrogen moves to the nitrogen of the heterocycle intramolecularly (2-DNBP) or intermolecularly (4-DNBP) [25], and the intramolecular movement is hindered by the crystal's structure; hence crystalline 4-DNBP is not photochromic [53,60]. The failure of attempts to observe photochromism in tetranitrodiphenylmethane **10** in crystals or in the supercooled melt [50] is attributed to the absence of the basic nitrogen and the hydrogen acceptor from the molecule.

These ideas have been borne out by a study of the special features of the crystal structure of 2-DNBP. It was found that the distance between the hydrogen of the CH_2 group and the oxygen of the NO_2 group of the same molecule is 0.24 nm, while in neighboring molecules within the crystal lattice it is 0.26 nm; the nitro group is rotated through 32° with respect to the plane of the ring and lies at least 0.33 nm from the pyridine nitrogen of the same molecule [105].

Clark and Lothian [23] discovered dichroism of the absorption of photocolored crystals of 2-DNBP: there was no absorption along the crystallographic α axis, along the β axis of the band maximum of the induced absorption at 545 nm, or along the γ axis at 620 nm [23]. The decoloration rate constants are the same for the forms arising from the action of polarized and nonpolarized light. This indicates the identical nature of both forms having an azamerocyanine structure.

Reflection spectra of powder groups of crystalline samples of photochromic compounds ground together with standard compounds are convenient for the study of the photochromism of crystals. They have helped to observe, for the first time, the photochemical reaction involved in the decoloration (Figure 12) [37] of crystals of substituted 2-DNBP **39**, **40**, **43**, **44**, and **47-49** upon irradiation with visible light in the absorption band of a photoinduced merocyanine, the possibility of which for solutions and crystals was discussed earlier [17,23,41,69]. Thus photochromic layers based on 2-DNBP and its substituted forms can be quickly and reversibly colored and decolorized photochemically. A high cyclicity of such systems has been observed [37].

It has been shown by the spectral-kinetic method that fluid and frozen solutions of 2-DNBP in alcohol and in polymethyl methacrylate and polystyrene films, with 2-DNBP within the temperature range of between −78 and −196°C do not display the photoinduced decoloration of the light-generated color [21,68,100].

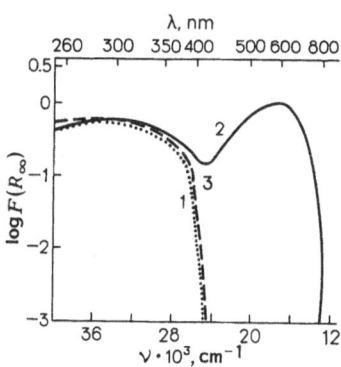

Fig. 12. The reflection spectra of powders of compounds 48 before irradiation (1), after ultraviolet irradiation (2) by light with λ 365 nm
for 30 s, and after the photochemical decoloration (3) by light
with $\lambda \geq 556$ nm, 3 min.

Polymer films with 2-DNBP and its analogs usually contain up to 5% of
the photochromic material, and their thickness is 20 μm (Table 6); the polymer matrix exerts a distinct influence on photochromism [26]. Thus, the
λ_{max} of the absorption of 2-(2,4-dinitrobenzyl)-4-methylpyridine 48 in polymer films, depending on their average molar mass M (g/mole) and glass point
T_g, is: in polymethyl methacrylate 571 nm (M 3900, T_g 85°C) and 583 nm (M
23,600, T_g −129°C); in polyethyl acrylate (M 33,000, T_g −26°C) − 571 nm; in
polystyrene (M 47,000, T_g 100°C) − 561 nm [26].

The thermal decoloration of photocolored substituted forms of dinitro-
benzylpyridines at a temperature lower than T_g runs a complex course and can
be described well by a second-order equation; at a temperature above T_g the
decoloration kinetics obey a first-order equation (Figure 13, Table 6). The
lifetime of the azamerocyanine colored form depends on the polymer structure
and increases with rising polarity and molar mass of the polymer [26]. The
values of the Arrhenius activation energies and activation enthalpies measured at temperatures above T_g are higher than those measured at temperatures below T_g; above T_g $\Delta S^{\pm}$ is positive, and below T_g it is negative.

The discussion of these experimental data is based on the concept of
reaction volume [104]. According to this concept, the decoloration of a
photoinduced form in an amorphous solid polymer occurs in a space of a
particular volume, form, and stability which is limited by the contact surface between the molecules within and without. In contrast to crystals and
fluid solutions, the environment of dissolved photochromic molecules in
glassy polymers below T_g is nonuniform; therefore, the characteristics of
the reaction volume vary from molecule to molecule. As a result, a whole
spectrum of rates is observed for decoloration and, in general, a complex
reaction course which is approximately described by a second-order equation. Above T_g local differences level out, the mobility of solid solutions
goes up sharply, and the decoloration is described by a first-order equation. The kinetic stabilization of a colored form in polymers compared with
fluid solutions can be explained within the framework of the same concepts
[106,107].

In polymer layers based on poly(N-vinylcarbazole) a complex is formed
with charge transfer from the polymer to the photochromic 2-DNBP; as a
result, the photoconductivity of the polymer increases 100-fold and the
spectral sensitivity expands up to 17,000 cm^{-1} [108].

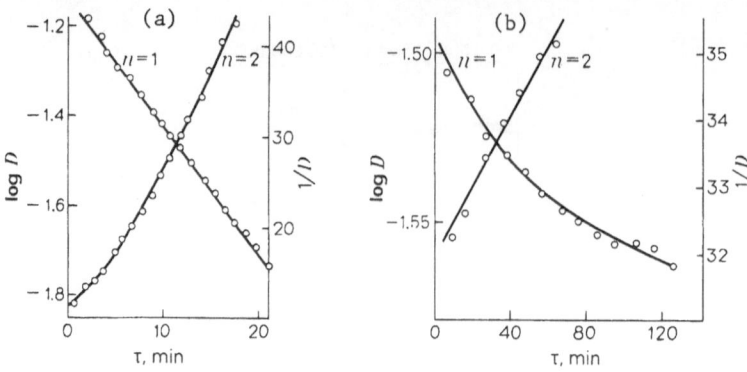

Fig. 13. The decoloration kinetics of the colored form of compound 47 in polyethyl acrylate at a temperature above (−17.3°C) (a) and below (−45.8°C) (b) the glass point T_g (−26°C) (n is the order of the reaction).

The synthesis of polymers in which the dinitrobenzyl fragment is chemically bonded to the polymer has been described. Upon the reaction of 2- and 4-polyvinylpyridine with benzyl chloride followed by nitration, deeply colored polymers were produced which were not photochromic [109]. Photochromic polymer ethers and amides, on the other hand, were synthesized through the copolymerization of 2,4-dinitrobenzyl acrylate 85, 2,4-dinitrobenzyl methacrylate 86, and 2-(4-acrylamido-2-nitrobenzyl)pyridine [31] with methacrylic acid or the methyl ester of acrylic acid, which could be used in the production of transparent polymer films and three-dimensional samples [51,52].

PHOTOCHROMIC PERI-ARYLOXY-p-QUINONES

The worldwide range of available dyes includes a great many anthraquinone derivatives used as dyes for cotton, synthetic fibers, and wool, including fast vat (indigo-blue) dyes. Interestingly, derivatives with the aryloxy group in the peri position relative to the quinoid carbonyl are completely ignored: dyes containing such substituents have been neither described nor used.

One obvious explanation of this neglect appears to be the low fastness of the aryloxy group-containing dyes. Indeed, upon investigation of the reaction to light of 1-phenoxyanthraquinone 148 it was found [110] that it changes color upon light irradiation both in the crystalline state and in solution. However, the irradiation does not lead to an irreversible change in molecular structure. The photoinduced orange-colored form disappears in the dark or upon irradiation with orange light (Figure 14), and the color of the initial 1-phenoxyanthraquinone is fully restored.

Thus, 1-phenoxyanthraquinone behaves as a typical photochromic compound, but its photochromism cannot be related to any of the well-known cases of photochromic conversion [111].

The Structure of the Photoinduced Form

The photoinduced form of 1-phenoxyanthraquinone was not stable enough for it to be isolated in pure form, and to be identified and studied. The photoinduced form 149 of 6-phenoxy-5,12-naphthacenequinone 150, which was isolated as orange-colored crystals from an ultraviolet light-irradiated benzene solution of the phenoxy derivative [112], is far more stable.

210

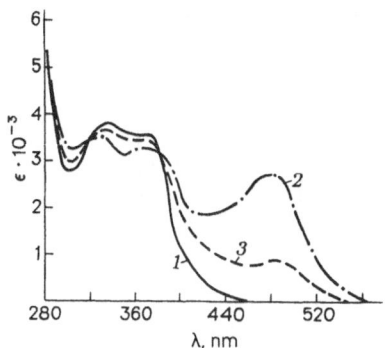

Fig. 14. The absorption spectra of a benzene solution of 1-phenoxyanthra-
quinone **148**: (1) before irradiation; (2) after 20-min irradiation
through PS-1 and SZS-9 light filters; (3) after 17-min irradiation
through SZS-9 and ZhS-18 light filters. Irradiation with a DRSh-
1000 lamp; the distance from the lamp to the photolysis product is
85 cm.

The isolated compound, with the same elementary composition as the
initial quinone, has an electronic spectrum similar to the electronic spectra of 6,12-naphthacenequinone (ana-quinone) and its 5-bromo-substituted
form. Under the action of ammonia and aniline the new compound readily
changes into the 5-amino and 5-anilino derivatives of 6,12-naphthacene-
quinone, respectively [112]:

$$
\underset{\textbf{150}}{\text{(quinone, C}_6\text{H}_5\text{O)}} \underset{h\nu', kT}{\overset{h\nu}{\rightleftarrows}} \underset{\textbf{149}}{\text{(naphthacenequinone, OC}_6\text{H}_6\text{)}} \xrightarrow{\text{RNH}_2} \underset{\textbf{151, R = H, C}_6\text{H}_5}{\text{(NHR)}}
$$

Thus, the photoinduced form is a 5-phenoxy derivative of 6,12-naphth-
acenequinone, while the photochromic process corresponds to the photoarylo-
tropic transition in the pentad tautomeric system with a reversible structural rearrangement p-quinone ⇌ ana-quinone.

Spectral measurements confirmed the formation of the ana-quinoid structure. Figure 15 presents a comparison between the infrared spectra of
solutions in CCl_4 of 1-phenoxyanthraquinone and 6-phenoxy-5,12-naphthacene-
quinone **150** before and after irradiation [113]. In both cases the intensity
of the C—O stretching vibrations band at 1685 and 1682 cm^{-1}, respectively,
drops and a new band appears in a lower-frequency region, 1635 and 1660
cm^{-1}. As in the case of the electronic absorption spectra, changes in the
infrared spectra are reversible. Analogous frequency shifts have been
detected for 2-amino-substituted forms of 1-phenoxyanthraquinone [114].

Shifts of the H^3, H^5, and H^8 proton signals and the NH group in the PMR
spectra of these same compounds and their photoinduced forms indicate an
ana-quinoid structure of the resultant compounds. This structure is confirmed by the ^{13}C NMR spectrum of 2-ethylamino-1-(p-tert-butylphenoxy)-
anthraquinone **158** (Table 7), whose photoinduced form has two carbon signals
for the C—O groups at 173.5 and 177.3 ppm relative to tetramethylsilane,
while for the starting compound analogous signals appear at 180.9 and 181.5
ppm [114].

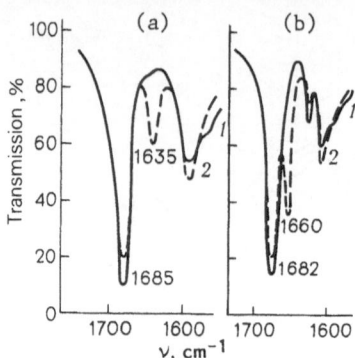

Fig. 15. The infrared spectra of 1-phenoxyanthraquinone **148** (a) and 6-phenoxy-5,12-naphthacenequinone **150** (b) in CCl_4: (1) before irradiation; (2) after irradiation through PS-1 and SZS-10 light filters. Irradiation by a PRK-4 lamp.

The 9-phenoxy-1,10-anthraquinone that forms from 1-phenoxyanthraquinone is similar in the nature of its electronic absorption spectrum to the unstable pentamethyl derivatives of 1,10-anthraquinone [115] and to the far more stable halogen-substituted forms of 1,10-anthraquinone [116].

The Relation between the Structure and Photochromism of Aryloxyquinones

Table 7 presents the spectral characteristics and data on the photochromism of aryloxy derivatives of p-quinones, which have been studied as potential photochromic compounds. These data show that the ability of phenoxyanthraquinones to photoisomerize into the corresponding ana-quinones depends both on the position and the electronic nature of the substituent. 2-Amino derivatives of 1-phenoxyanthraquinone **152-154** photoisomerize irreversibly into the corresponding derivatives of 1,10-anthraquinone. 2-Alkylamino derivatives **155-159** proved to be sufficiently stable compounds and were isolated in crystal form. 3-Amino derivatives **164-166** photoisomerize irreversibly, and their photoinduced forms quickly disintegrate. 4- and 5-amino derivatives of 1-phenoxyanthraquinone (**168, 169, 178, 179**) are photochromic but have a low sensitivity, while 4-methylamino derivatives **170** and **171** isomerize irreversibly into 4-methylamino-9-phenoxy-1,10-anthraquinone, and 5-methylamino- and 4- and 5-dimethylamino derivatives (**180, 181, 176, 182**) do not undergo photoisomerization. 8-Amino- and 8-methylamino-1-phenoxyanthraquinone (**187, 188**) do not photoisomerize either. All acylamino derivatives (**163, 167, 172-174, 177, 183-185, 189, 190**) are photochromic compounds, with 4- and 5-acylamino-ana-quinone being the most stable.

Aryloxy-substituted forms of 5,12-naphthacenequinone and 6,13-pentacenequinone readily photoisomerize into ana-quinones and are relatively more stable than the derivatives of 1,10-anthraquinone and are isolated in the crystalline state. Reverse isomerization occurs upon visible light irradiation or thermally – during melting, while for pentacenequinone it occurs also while a benzene solution of the photoinduced form is held in the dark. Ana-naphthacenequinone derivatives under the same conditions practically do not isomerize.

Like the corresponding 4-derivatives of 1-phenoxyanthraquinone, 6-dimethylamino and 6-phenylamino derivatives were found to be nonphotochromic, as were indeed 6-hydroxy- and 6-acetoxy-11-phenoxy-5,12-naphthacenequinone (**201, 202, 207, 210**).

212

The photoisomerization of 6-phenoxy derivatives of naphtho[2,1,8-qra]-naphthacene-7,12-quinone occurs upon ultraviolet or diffused daylight irradiation (Figure 16); in the dark the green color of the solution very slowly disappears. The absorption spectra of both the starting and the photoinduced forms change insignificantly under the influence of the substituent in the phenyl nucleus. 11-Phenoxy-substituted forms were found to be nonphotochromic.

Absorption and Luminescence

The special features of the spectral and luminescent characteristics of the photochromic system p-quinone ⇌ ana-quinone were examined using p- and ana-naphthacenequinones and their phenoxy-substituted forms as an example. The absorption spectrum of 5,12-naphthacenequinone in hexane is characterized by a long-wavelength band of the $\pi \to \pi^*$-transition with a vibrational structure and a relatively weak shoulder — apparently, a manifestation of the $n \to \pi^*$-transition band. 5,12-Naphthacenequinone does not fluoresce either at room temperature or at 77°K. At 77°K a prolonged green fluorescence was observed in methylpentane [122]. In hexane the fluorescence spectrum has a well-resolved structure with an intensity peak at 1940 cm^{-1} [123]. A considerable interval between the absorption and fluorescence bands (6000 cm^{-1}) and the duration of phosphorescence (τ_{ph} about 0.3 s [122]) made it possible to relate the fluorescence to the π,π^*-state.

The introduction of a phenoxy group into position 6 results in a slight bathochromic shift (4000 cm^{-1}) of the long-wavelength π,π^*-absorption band and an even smaller shift in the long-wavelength phosphorescence band from the π,π^*-triplet. 6-Phenoxy-5,12-naphthacenequinone 150 lacks fluorescence of its own. The introduction into the para position of the phenoxy of a nitro or acetyl group changes neither the position of the long-wavelength absorption band nor the phosphorescence band.

The long-wavelength absorption band of 5,11-naphthacenequinone is the $\pi \to \pi^*$-transition. The study of luminescent properties carried out by I.L. Belaits showed that quinone solutions do not display luminescence at room temperature or at 77°K. The weak fluorescence, which becomes more intense over time, was attributed to the appearance of 6-hydroxy-5,12-naphthacenequinone — the product of the conversion of 5,11-naphthacenequinone. In acetic acid, which quenches the fluorescence of the hydroxy derivative, a bright green fluorescence of 5,11-naphthacenequinone was recorded at 77°K. The "activation" of fluorescence by a polar solvent indicates that the

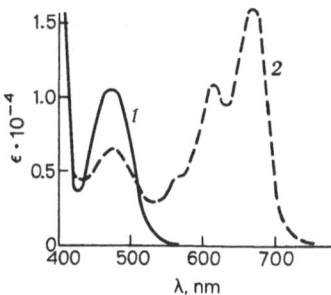

Fig. 16. The absorption spectra of a solution of 6-phenoxy-1,3-dichloronaphtho[2,1,8-qra]naphthacene-7,12-quinone 213 in chlorobenzene: (1) before irradiation (solution prepared in the dark); (2) solution after 10-min irradiation with diffused daylight.

Table 7. Spectral characteristics of photochromic aryloxyquinones

Compound number	R; R'	λ_{max}, nm (log ϵ) for solutions in benzene Before irradiation	After irradiation	Reference

1-Phenoxyanthraquinones

Compound number	R; R'	Before irradiation	After irradiation	Reference
148	H; H	364(3.70)	480	110, 117
152	2-NH$_2$; H	415(3.70)*	616	114, 118
153	2-NH$_2$; m-CH$_3$	402(3.69)**	587(3.81)	
154	2-NH$_2$; p-C(CH$_3$)$_3$	404(3.73)**	587(3.87)	114
155	2-NHCH$_3$; H	424(3.78)**	615(3.94)^{3*}	
		438(3.81)	648	118
156	2-NHCH$_3$; m-CH$_3$	424(3.72)**	615(3.93)^{3*}	
157	2-NHCH$_3$; p-C(CH$_3$)$_3$	423(3.86)**	616(3.94)^{3*}	
158	2-NHC$_2$H$_5$; p-C(CH$_3$)$_3$	426(3.90)**	628(4.03)^{3*}	114
159	2-NHCH$_2$C$_6$H$_5$; p-C(CH$_3$)$_3$	427(3.89)**	621(4.02)^{3*}	
160	2-N(CH$_3$)$_2$; p-C(CH$_3$)$_3$	430(3.75)**	-	
		448(4.70)	665	118
161	2-N(C$_2$H$_5$)$_2$; p-C(CH$_3$)$_3$	468(3.77)**	-	114
162	2-N(CH$_2$C$_6$H$_5$)$_2$; p-C(CH$_3$)$_3$	445(3.83)*	-	
163	2-NHCOC$_6$H$_5$; H	376(3.85)	552$^{3\bar{*}}$	118
164	3-NH$_2$; p-C(CH$_3$)$_3$	415(3.64)	434(3.88)	
165	3-NHCH$_3$; p-C(CH$_3$)$_3$	438(3.64)	442(3.81)	
166	3-N(CH$_3$)$_2$; p-C(CH$_3$)$_3$	459(3.74)	443(3.89) 466(3.89)	
167	3-NHSO$_2$C$_6$H$_4$CH$_3$-p; p-C(CH$_3$)$_3$	380(3.66)	466(3.81)	
168	4-NH$_2$; H	478(3.88)	581	119
169	4-NH$_2$; p-C(CH$_3$)$_3$	478(3.86)	581	
170	4-NHCH$_3$; H	518(3.91)	621	
171	4-NHCH$_3$; p-C(CH$_3$)$_3$	518(3.91)	621	
172	4-NHCOCH$_3$; p-C(CH$_3$)$_3$	436(3.87)	532(4.11)	
173	4-NHCOCH$_3$; H	436(3.79)	532(4.06)	
174	4-NHCOC$_6$H$_5$; p-C(CH$_3$)$_3$	445(3.86)	541(4.11)	
175	4-NHCOC$_6$H$_5$; H	441(3.76)	538	118
176	4-N(CH$_3$)$_2$; H	511(3.71)	-	
177	4-OC$_6$H$_5$; H	384	434	117
178	5-NH$_2$; H	466(3.81)	582	118
179	5-NH$_2$; p-C(CH$_3$)$_3$	466(3.76)	575	119
180	5-NHCH$_3$; H	505(3.88)	-	118
181	5-NHCH$_3$; p-C(CH$_3$)$_3$	504(3.89)	-	119
182	5-N(CH$_3$)$_2$; H	502(3.72)	-	118
183	5-NHCOCH$_3$; p-C(CH$_3$)$_3$	415(3.86)	502(4.04) 523(4.02)	119
184	5-NHCOC$_6$H$_5$; H	419(3.87)	508, 532	118
185	5-NHCOC$_6$H$_5$; p-C(CH$_3$)$_3$	418(3.84)	506(3.97) 532(3.95)	119
186	5-OC$_6$H$_5$; H	370	480	117
187	8-NH$_2$; p-C(CH$_3$)$_3$	472(3.75)	-	
188	8-NHCH$_3$; p-C(CH$_3$)$_3$	510(3.79)	-	
189	8-NHCOCH$_3$; p-C(CH$_3$)$_3$	416(3.84)	556(3.56)	119
190	8-NHCOC$_6$H$_5$; p-C(CH$_3$)$_3$	421(3.88)	566(3.53)	
191	8-OC$_6$H$_5$; H	374	494	117

Table 7. (continued)

<table>
<tr><td rowspan="4">Compound
number

R; R'</td><td colspan="2">λ_{max}, nm (log ϵ)</td><td rowspan="4">Reference</td></tr>
<tr><td colspan="2">for solutions in benzene</td></tr>
<tr><td>Before
irradiation</td><td>After
irradiation</td></tr>
</table>

6-Phenoxy-5,12-naphthacenequinones[4*]

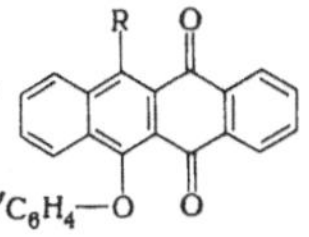

p-R'C$_6$H$_4$—O

Compound number	R; R'	Before irradiation	After irradiation	Reference
150	H; H	400(3.78)	480(4.23)[3*]	
192	p-NHC$_6$H$_5$; H	396(3.81)	476(4.26)	
193	p-OH; H	400(3.72)	480(4.23)	
194	p-OCH$_3$; H	400(3.76)	480(4.20)	120
195	2,4,6-(CH$_3$)$_3$; H	410(3.83)	498(4.11)	
196	p-COCH$_3$; H	400(3.78)	480(4.26)	
197	p-NO$_2$; H	395(3.78)	480(4.20)	
198	H; Cl	398(3.74)	444(4.20)[3*]	
199	H; NH$_2$	468(3.99)	538(4.17)[3*]	
200	H; NHCH$_3$	502(4.09)	559(4.07)[3*]	
201	H; N(CH$_3$)$_2$	465(3.60)	–	
202	H; NHC$_6$H$_5$	496(3.95)	558(3.86)[3*]	
203	H; NHCOCH$_3$	429(3.87)	504(4.20)[3*]	124
204	H; N(CH$_3$)COCH$_3$	396(3.85)	476(4.20)[3*]	
205	H; NHSO$_2$C$_7$H$_7$	422(3.77)	509	
206	H; NO$_2$	394(3.70)	479(4.13)	
207	H; OH	443(4.00)	–	
208	H; OCH$_3$	400(3.82)	469(4.16)	
209	H; OC$_6$H$_5$	404	472	117
210	H; OCOCH$_3$	396(3.89)	–	124
211	5-phenoxy-6,13- pentacenequinone	404(3.68)	461, 490	121

6-Phenoxynaphtho[2,1,8-gra]naphthacene-7,12-quinones[5*]

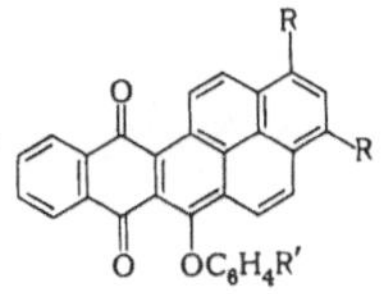

OC$_6$H$_4$R'

Compound number	R; R'	Before irradiation	After irradiation	Reference
212	H; H	476	604, 652	
213	Cl; H	482	606, 662	
214	Cl; 2-Cl	480	608, 658	
215	Cl; 2-CH$_3$	480	610, 662	
216	Cl; 2-OCH$_3$	484	610, 664	
217	Cl; 4-Br	480	612, 662	117
218	Cl; 4-OC$_2$H$_5$	483	612, 662	
219	Cl; 4-OC$_6$H$_5$	480	610, 660	
220	Cl; 4-NO$_2$	480	610, 666	
221	Cl; 2,4,6-(CH$_3$)$_3$	494	615, 664	
222	11-phenoxynaphtho[2,1,8-gra]- naphthacene-7,12-quinone	464	–	

* In chloroform. ** In heptane. [3*] Isolated in crystal form. [4*] For solutions of compounds 150 and 192-210 in toluene. [5*]For solutions of compounds 212-222 in chlorobenzene.

215

$^3n,\pi*$-level, which lies below the $^1\pi,\pi*$-level in a nonpolar solvent, in polar acetic acid is found to be above the $^1\pi,\pi$-level and no longer participates in the radiationless deactivation of the latter [123].

The absorption spectrum of photochromic 6-phenoxy-5,11-naphthacene-quinone 150 is close to the spectrum of the unsubstituted 5,11-naphthacene-quinone. However, no fluorescence was observed in acetic acid. The introduction of substituents into the phenoxy group does not change appreciably the spectral-luminescent properties of the molecule. The substituted forms are also characterized by a lack of luminescence. This difference in the behavior of 5,11-naphthacenequinone and its 6-phenoxy-substituted forms may be due to a competing process of photoisomerization. Figure 17 shows a plot of the states and the more likely paths of the degradation of excitation energies of 5,12- and 5,11-naphthacenequinones.

The Mechanism of Photoisomerization

It is a fair assumption that reversible photoisomerization proceeds as an intramolecular substitution initiated by light (in both directions) and thermally (in the reverse direction), in all probability, by an associative mechanism (although the possibility of a dissociative mechanism involving the intermediate formation of a contact ionic or radical pair cannot be entirely ruled out).

An attempt to detect radicals by EPR spectroscopic methods upon irradiation of the p-quinone ⇌ ana-quinone system [117] was unsuccessful. This could be treated only as indirect evidence of the heterolytic cleavage of the C–O bond. Bond breaking must be preceded by the formation from the S_1 and (or) T_1 excited state of the $(\pi,\pi*)\sigma$-complex in the ground state 223 whose stabilization could lead either to the p- or ana-quinoid structure:

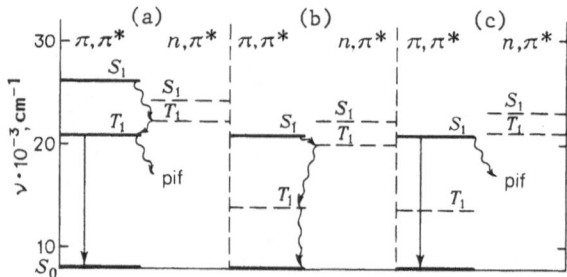

Fig. 17. The energy of the electronic levels of p- and ana-naphthacene-quinones: (a) 5,12-naphthacenequinone and its 6-phenoxy-substituted form 150 in hexane and toluene; (b) 5,11-naphthacenequinone and its 6-phenoxy-substituted form in hexane and toluene; (c) 5,11-naphthacenequinone and its 6-phenoxy-substituted form in acetic acid. pif) photoinduced form.

Some results of the measurements of the quantum yields of the photo-conversion 6-aryloxy-5,12-naphthacenequinone $\underset{\phi_A}{\overset{\phi_B}{\rightleftharpoons}}$ 5-aryloxy-6,12-naphth-acenequinone in toluene carried out by A.A. Parshutkin are as follows (compound No., ϕ_B, ϕ_A): **150**, 0.30, 0.05; **202**, 0.02, 0.015; **208**, 0.08, 0.05; **210**, 0.27, 0.10; **206**, 0.20, 0.05.

It is interesting to note that the isomerization p-quinone ⇌ ana-quinone can occur not only photochemically but also thermally in the presence of protic and Lewis acids [124]. Upon isomerization of 11-phenoxy-5,12-naphthacenequinone and its 6-chloro-, 6-methoxy-, and 6-phenoxy-substituted forms, the equilibrium mixture is dominated by the para isomer, while upon the isomerization of its 6-amino-, 6-methylamino-, 6-phenylamino-, and 6-acetylamino-substituted forms, the equilibrium mixture is dominated by the ana isomer. 1-Phenoxy-4-amino-9,10-anthraquinone isomerizes under the action of aluminum chloride to 9-phenoxy-4-amino-1,10-anthraquinone.

Potential Practical Applications

Aryloxy-p-quinones, as photochromic compounds, possess a good enough potential for practical applications since their properties make it possible to develop photochromic systems based on them. These properties include a high quantum yield, a good color contrast, a high "cyclicity" (the conversion of p-quinone into ana-quinone and back can be repeated any number of times without a noticeable degradation of the properties of the material), and a negligible rate of the back dark reaction at room temperature.

The activity of ana-quinones reacting with nucleophiles is far higher than that of p-quinones. This makes it possible to irreversibly fix the resulting aryloxy-ana-quinone in the form of amino-ana-quinone. 6-Phenoxy-5,11-naphthacenequinone readily substitutes the phenoxy group with various nucleophiles; e.g., it reacts with amines at room temperature. This property has been used to determine the molar absorption coefficient of 6-phenoxy-5,11-naphthacenequinone without its isolation. The 6-phenoxy-5,12-naphth-acene solution was irradiated until the photoinduced form achieved an optical density of 0.4-0.8. Then decylamine was added, and the solution was again irradiated until a permanent optical density of the resulting 6-decyl-amino-5,11-naphthacenequinone [120] was achieved; the latter can easily be isolated.

The photochromic materials obtained upon the introduction of 6-phenoxy-5,12-naphthacenequinone into polymethyl methacrylate and other polymers can be used for recording holograms by means of an He—Cd laser (λ 441.6 nm) and nondestructive image restoration from the hologram by the irradiation of an He—Ne laser (λ 632.8 nm) [125].

Approaching in light sensitivity to the best organic photochromic materials (spiropyrans), aryloxynaphthacenequinone-based photochromic materials are more stable: after 500 photocoloration—photobleaching cycles the material fully retains its properties; dark decoloration at 25°C is characterized by the value $k_T = 10^{-8}$ s^{-1}, i.e., the lifetime of the photoinduced form in the dark at room temperature will span many years [126].

The light sensitivity of photochromic materials to radiation (emission) with λ 366 nm is not less than 10 cm^2/J, and to visible light — 1 cm^2/J (for a photoinduced optical density gradient at λ 480 nm of ΔD 1.0). Interestingly, the quantum yields upon the transition from the solution in toluene to the solid polymer matrix practically do not change [126].

Reversible photoinduced acyl migrations in triad and pentad systems have a recent history, although irreversible photorearrangements of acyl groups in similar systems have been studied fairly well. Best known among the latter are O,C-photo-Fries rearrangements of O-acylated phenols [127], whose mechanism is generally associated with the formation of close radical pairs and the subsequent intramolecular recombination leading to photoproducts, primarily o-acylphenols. The radical mechanism is not the only possible channel of O,C-phototransfers of acyl in the triad systems, as Veirov et al. [128] demonstrated, using the example of 1,3-photoregroupings of acylated enols which model the photo-Fries rearrangements:

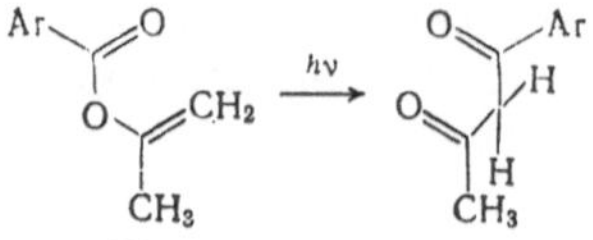

224   225

The reaction proceeds through the singlet excited state of the molecule, with a high quantum yield of the rearrangements being reached only in those compounds where excitation energy is not localized in the aryl group of the migrant.

Studies carried out by Hoffmann and Eicken showed the possibility of N,C-photoacylotropic rearrangements. During their careful study of the photochemical properties of enamides, they [129] discovered that the long-wavelength irradiation ($\lambda > 300$ nm) of enamide solutions leads to E/Z-isomerization through the T_1-state (E_T about 219 kJ/mole), with the reaction being sensitized by triplet donors:

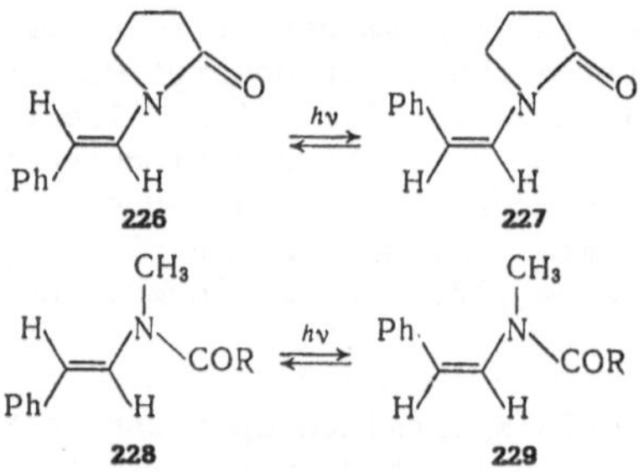

226   227

228   229

Irradiation with the unfiltered light of a high-pressure mercury lamp results in an irreversible 1,3-transfer of the acyl group:

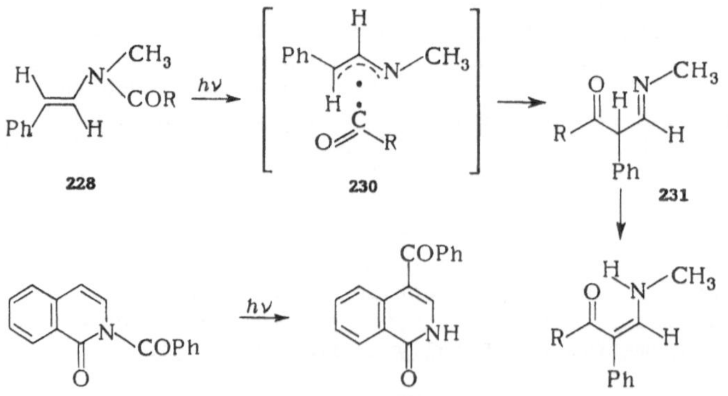

228   230   231

233   234   232

The 1,3-acyl rearrangement that has been studied does not depend on the geometry of the starting molecule because the transfer product forms at almost the same rate both from Z- and from E-enamides. They assumed that the 1,3-acyl shift reaction occurs through the higher excited states (S_2 or T_2) with the formation of a radical pair. The use of isotopomeric substrates showed that the reaction proceeds intermolecularly only by 8 ± 2%.

Reversible photoregroupings have so far been found for cases of the migration of acyl groups between two heteroatomic centers. Wachsen and Harthe [130] established the following sequence of rearrangement stages in 1-(acyloxymethylene)-2-indanones:

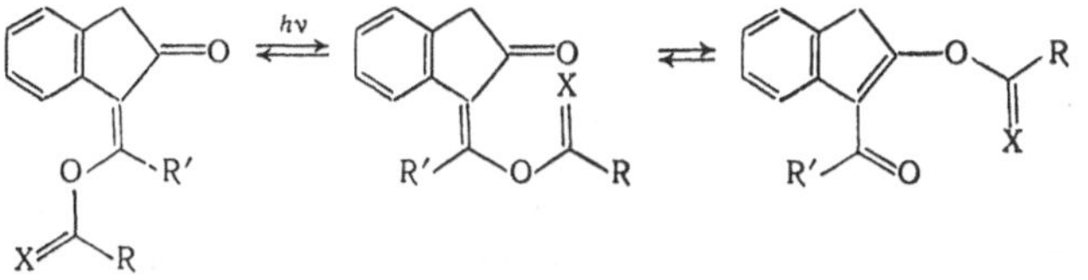

235   236   237

R = Alk, Ar, OAlk;   R' = CH$_3$, Ar;   X = O, S

The role of photoexcitation here is limited to the realization of the E → Z-isomerization which secures a favorable orientation of nucleophilic centers [131] for the acyl transfer with the subsequent transfer of the acyl group in the ground electronic state, i.e., the acylotropic rearrangement proper proceeds non-photochemically.

Such photoinitiated acylotropic regroupings, including the E/Z-photoisomerization stage, were also found in the series of O-acylated isoamides [132].

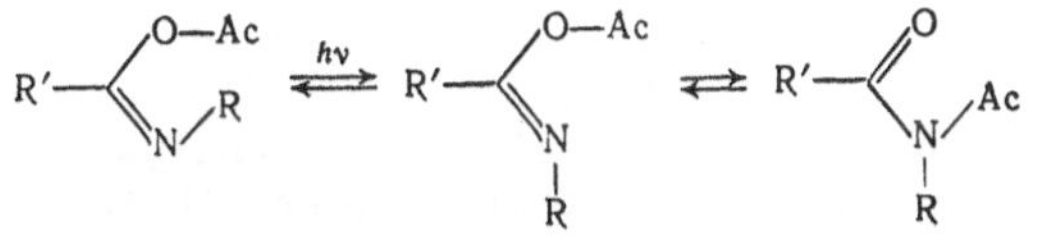

238   239   240

Recently, photoacylotropic rearrangement of 2-(N-acyl-N-arylaminomethylene)-3(2H)benzo[b]furan(thiophen)ones 241 and their aza analogs 244 and 2-(acyloxybenzal)anilines 248 [133,134] were discovered. In these reactions the photoinitiated acyl transfer does not occur homolytically but rather as an intramolecular nucleophilic substitution:

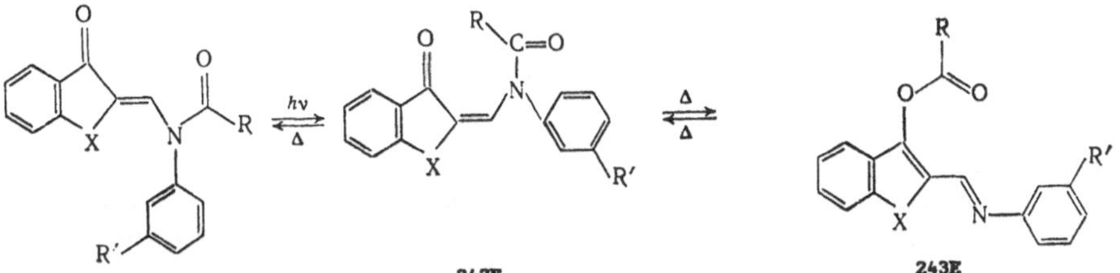

241Z   242E   243E

OCOR

N—Ar

246E

COR

Δ

hv₁ / Δ

O

COR

$=N-N-Ar$

245E

hv₂ Δ

OCOR Aₗ

$-N=N$

X

247Z

O

$=N-N-COR$
Ar

244Z

O

$=N-COR$

X

hv₁ / Δ

OCOR

$=N-C_6H_4R'$

248

hv / Δ

O

COR

$=N-C_6H_4R'$

249

Compounds 241 and 244, whose steric structures do not favor intramolecular 1,5-acyl transfer, undergo Z → E-photoisomerization before the formation of 242E(S_0) and 245E(S_1), which are capable of rearrangement into the
end products 243E and 246, 247.

If compounds exist in a geometric configuration favorable to acyl
transfer, photoexcitation leads directly to acyl group transfer, which is
what we discovered in the case of the O-acylated imines of salicylaldehyde
248.

The rearrangements of compounds 241, 244, and 248 that have been studied are reversible. Reverse reactions proceed thermally in all cases. The
irradiation of solutions of compound 241 (X — S) at room temperature in the
region of the long-wavelength absorption band with a mercury lamp or with
sunlight results in the formation of structure 243 with a long-wavelength
band in the 340-360 nm region (Figure 18). The maxima of the long-wavelength absorption bands of the compounds with X — S (in nm) are: 241Z, 425;
242E, 450; 243E, 340-360; 244Z, 420; 246E, 370; 247Z, 340.

Compounds 243E are the only photoreaction products and are isolated
preparatively with a yield of up to 95%. Their structures have been confirmed by spectral data and X-ray crystallographic analysis. Moreover, they
can be obtained by back synthesis from 3-acetoxybenzo[b]thiophen-2-aldehyde
and the corresponding amines, e.g., 243 (X — S, R — CH₃, R' — m-NO₂, m-Cl).

The acylation of the sodium salts of 2-(N-arylaminomethylene)-3(2H)-
benzo[b]thiophenones (R' — m-NO₂, p-COCH₃) by acetyl chloride in anhydrous
benzene produces a mixture of compounds 241 and 243 which can be isolated by
separation techniques.

The data on PMR spectra with the use of the shift-reagent tris(dipiva-
loylmethanate)europium (for 241, X — S, R — CH₃, R' — H) and X-ray diffrac-
tion analysis (for 241 and 243, X — S, R — CH₃, R' — m-NO₂) [135] make it
possible to determine the most stable conformations 241Z and 243E shown in
the scheme above (X — S).

The kinetics of the photoreaction 241 → 243 is described with a curve
close to the exponential type. Reaction rate constants for compounds 241
under experimental conditions have values on the order of 10^{-1} s^{-1}, and the
quantum yields amount to 0.60 ± 1. The photoreaction rate is practically
independent of the polarity of the solvent (benzene, acetone, dimethylformamide) and variation of the substituent R'.

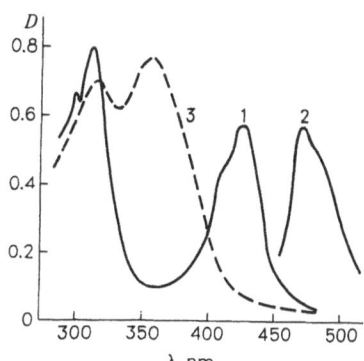

Fig. 18. The absorption and luminescence spectra of compounds 241 and 243 in benzene (c_0 5 $\cdot$ 10^{-5} mole/liter): (1) absorption of compound 241 (X = S, R = CH_3, R' = H), 293°K; (2) luminescence of compound 241 (X = S, R = CH_3, R' = H), 77°K; (3) absorption of compound 243 obtained by means of 2-min irradiation of the solution of compound 241, λ_{irr} 436 nm, 293°K.

At 77°K the yield of the photoreaction drops down to zero and an intense fluorescence band which is mirror-symmetrical to the long-wavelength absorption band of the form 241 is observed in the process. The photoreaction observed on this occasion was not sensitive to either sensitization or desensitization via triplet—triplet energy transfer. No concentration dependence on the reaction rate was observed either, and this suggests a unimolecular character of the reaction.

Upon the irradiation of solutions of compounds 244 (X = S) with light of 405 or 436 nm wavelength, the formation of a mixture of Z- and E-geometric isomers 247 and 246, capable of mutual conversions characteristic of azobenzenes was observed. Irradiation with light of 365 nm wavelength leads to a build-up of the Z-isomer, which at room temperature changes completely into the form 246E within 15-20 min. The rate constants of the photoreaction 244 → (246, 247) are on the order of 10^{-2} sec^{-1}.

On the basis of the results set out above the photoconversion mechanism in compounds 241 is described by the following scheme (Figure 19).

The excitation of the Z-isomer of the N-acyl form 241Z into the S_1-state results in the formation of the intermediate product 242E registered spectrally using flash photolysis (λ_{max} 450 nm, τ_{25} 5 $\cdot$ 10^{-2} s). The mechanism of the Z → E-isomerization has not been studied in detail, but the results described above (the observed absorption of the short-lived form 242 and the absence of triplet sensitization) suggest a nonadiabatic reaction (241Z)* → 242E that proceeds from a singlet excited state. The reaction proceeds with a low energy barrier, which is indicated by a notable decrease in the yield of the photoreaction and an increase in the yield of the fluorescence of the starting form 241Z with decreasing temperature.

This assumption regarding the nature of the photoreaction is in good agreement with available data on the nonadiabatic mechanism of the E/Z-isomerization around the C=C double bonds and the possible formation of intermediates common to both isomers in the singlet excited state [136-139].

Subsequently, the O-acyl form 243E is formed from the ground state of the intermediate product 242E. The high quantum yield of the photoproducts

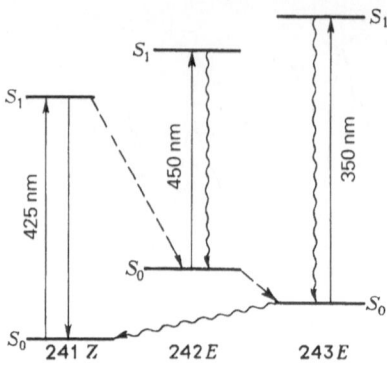

Fig. 19. A plot of the photoreaction and the reverse thermal reaction 241 ⇌ 243.

243 indicates the low activation barrier of this reaction and at the same time the considerable barrier of the thermal E → Z-isomerization of the intermediate product to the starting form. The low barrier of the reaction 242E → 243E is attributable to the structure of the E-configuration being close to the transition state of the acyl transfer [141].

The photoproduct 243 displays no luminescence due to the high rate of intersystem crossing typical of this structure [135]. The reverse photo-reaction 243 → 241 was not observed. It is assumed that the reverse thermal reaction should proceed with a high activation barrier. Indeed, upon heat-ing solutions of 243 in high-boiling inert solvents up to 140-180°C a re-verse dark O → N-transfer of the acyl group is observed which is registered by means of electronic absorption spectra. The reaction product (X = S, R = CH_3, R' = H) was isolated preparatively. Its melting point and infrared and ultraviolet spectra are identical to the characteristics of the starting ketoamine form 241, which shows the full reversibility of the 241 ⇌ 243 cycle, which involves direct photo and reverse thermal reactions. The cycle can be repeated many times.

The reverse dark reaction is subject to acid catalysis. At a 1:1 ratio of 243 and CCl_3COOH the back reaction occurs within 15-20 min at room tem-perature. An analysis of the relevant kinetic data shows that $\Delta G^{\neq}$ = 105-126 kJ/mole, depending on the variation of R and R'. Compounds 246 undergo a reverse dark reaction starting at 70-90°C.

The study of the spectral-luminescent characteristics of compounds 248 and of the photochemical conversion 248 → 249 gives the results described below. Solutions of 0-acylsalicylalarylimines 248 at room temperature do not luminesce. An increase in the solvent polarity or a drop in the temper-ature to 77°K does not change the absorption spectra of 248; i.e., no tran-sition to the quinoid form 249 in the ground state occurs, unlike the case of salicylalaniline. This is attributable to an increased stability of the benzenoid form 248 as compared with the form 249 ($\Delta\Delta H_{S_0}$ 33.5 kJ/mole) caused by the replacement of a proton with an acyl group [140].

Irradiation of frozen solutions of compound 248 (R = CH_3, R' = H) at 77°K in the 365 nm region causes two competing processes: 1) fluorescence (Figure 20), which because of the abnormally large Stokes' shift cannot be related to the starting form 248; 2) the formation of a colored photoform. Other compounds of the 248 type are indifferent to irradiation.

The calculation of the absorption spectra of the quinoid form of com-pound 248 (λ_{calc} 465 nm, λ_{exp} 470 nm) and the similarity with the spectral-luminescent characteristics of salicylalaniline are the grounds for attri-

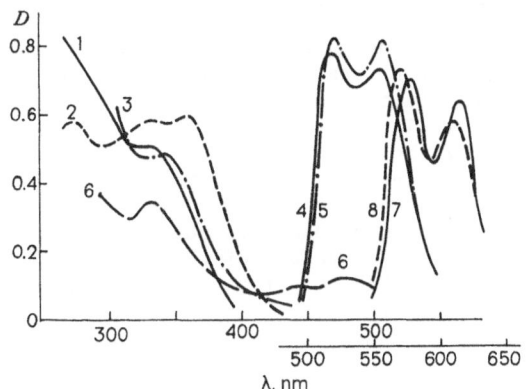

Fig. 20. The absorption and luminescence spectra of O-acetylsalicylalaniline in isopentane (c_0 $5 \cdot 10^{-5}$ mole/liter). Before irradiation:
(1) and (2) absorption of O-acetylsalicylalaniline 248 (R – R' =
H) and salicylalaniline 250, respectively, 293°K; (3) absorption
of 248 (R – R' – H), 77°K; (4) and (5) luminescence of 248 (R –
R' – H), 250, 77°K; λ_{exc} 340 nm. After irradiation: (6) and (7)
the absorption and luminescence of the photoproduct from 248 (R =
R' – H), 77°K; λ_{exc} 470 nm; (8) luminescence of 250, 77°K, λ_{exc}
470 nm.

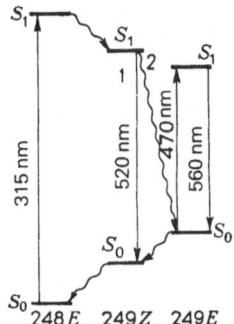

Fig. 21. A plot of the photoreaction 248 → 249 according to the data of the
quantum-chemical calculations and electronic absorption spectra:
(1) fluorescence of the 249Z form; (2) nonadiabatic transition to
the colored form.

buting the quinoid structure 249 to the photoproduct. The color of the
frozen solution is retained when irradiation stops, but upon thawing, the
original spectral properties are restored.

According to the foregoing data, the photoinduced processes in
O-acetylsalicylalaniline follow the scheme shown in Figure 21. From the
excited state S_1 of the starting form 248, the transfer of acetyl to the
nitrogen atom results in the formation of the excited state of an intermediate form which seems to have the structure 249Z (λ_{em} 520 nm).

The possibility of acyl transfer in the excited state is in good agreement with the data of quantum-chemical calculations, which indicate an
inversion of the relative position of the energy levels of the benzenoid and
quinoid forms in the excited state as compared with the ground state ($\Delta\Delta Hs_1^*$

−42 kJ/mole). Subsequently, deactivation occurs in two ways: 1) fluorescence of the excited state of the intermediate form 249Z, analogous to the
fluorescence of the salicylalaniline anion [141]; 2) nonadiabatic transition

to the colored form. The conclusion regarding the nonadiabatic nature of the transition follows from the lack of fluorescence of the colored form upon the excitation of the starting benzenoid form 248. The photoinduced form of the acylated salicylalaniline is interpreted by us as a noncoplanar quinoid with the nucleus of the aldehyde residue withdrawn from the molecule's plane, just as in the case of the salicylalaniline itself [142].

Structure 249, as already mentioned, is unstable at room temperature, unlike compounds 243 obtained as a result of photoacylotropic regrouping, which in the crystalline state can keep for years and in solution for weeks. This makes it possible to "store" light energy, e.g., solar energy, and release it, as needed, by means of the dark-catalyzed O → N reversion reaction of the acylic migrant.

REFERENCES

1. G. N. Brown, in: Photochromism, Wiley, New York (1971), Vol. 6, pp. 557-632.
2. E. Klemm and D. Klemm, Wiss. Friedrich-Schiller-Univ. Jena, Math-natur-wiss. R., 22:887-901 (1973).
3. G. Wettermark, Nature, 194(4829):677-680 (1962).
4. G. Wettermark, J. Phys. Chem., 66:2560-2562 (1962).
5. G. Wettermark, E. Black, and L. Dogliotti, Photochem. Photobiol., 4:229-239 (1965).
6. J. D. Margerum, J. Am. Chem. Soc., 87:3772 (1965).
7. G. Wettermark and R. Ricci, J. Chem. Phys., 39:1212 (1963).
8. M. E. Langmuir, L. Dogliotti, E. D. Black, et al., J. Am. Chem. Soc., 91:2204-2207 (1969).
9. K. Suryanarayanan and C. Capellos, Int. J. Chem. Kinet., 6:85-102 (1974).
10. D. Klemm, E. Klemm, and H. Jerusel, et al., J. Prakt. Chem., 320:767 (1978).
11. J. Margerum, L. Miller, E. Saito, et al., J. Phys. Chem., 66:2434-2438 (1962).
12. A. L. Bluhm, J. A. Sousa, and J. Weinstein, J. Org. Chem., 29:636-640 (1964).
13. D. Klemm, E. Klemm, A. Graness, et al., Chem. Phys. Lett., 55:503-505 (1978).
14. D. Klemm, E. Klemm, A. Graness, et al., Z. Phys. Chem., 260:555-560 (1979).
15. J. D. Margerum and R. C. Brault, 157th Meeting, Minneapolis, Paper Phys. (1969), p. 142.
16. J. Weinstein, A. Bluhm, and J. Sousa, J. Org. Chem., 31:1983-1985 (1966).
17. R. Hardwick, H. S. Mosher, and P. Passailaigue, Trans. Faraday Soc., 56:44-50 (1960).
18. J. Sousa and J. Weinstein, J. Org. Chem., 27:3155-3159 (1962).
19. D. Klemm, E. Klemm, S. Winkelman, et al., J. Prakt. Chem., 317:761-770 (1975).
20. E. Inoue, H. Kokado, and K. Yoshida, Kogyo Kagaku Zasshi., 71:1792-1797 (1968).
21. V. P. Bazov, Author's Abstract of Candidate's Dissertation, L. Ya. Karpov Scientific-Research Physicochemical Institute, Moscow (1974).
22. É. R. Zakhs and A. I. Ponyaev, in: Problems of the Chemistry and Technology of Organic Dyes and Intermediate Products, Issue 3, Dep. VINITI (1979), p. 119.
23. W. C. Clark and G. F. Lothian, Trans. Faraday Soc., 54:1790 (1958).
24. G. Kortum, M. Kortum-Seiler, and S. D. Bailey, J. Phys. Chem., 66:2439-2442 (1962).
25. D. Klemm and E. Klemm, J. Prakt. Chem., 321:404-406 (1979).

26. D. Klemm, E. Klemm, K. Schuler, et al., J. Prakt. Chem., 319:647-654 (1977).
27. D. Klemm, E. Klemm, A. Graness, et al., Chem. Phys. Lett., 55:113-115 (1978).
28. G. Wettermark and J. Sousa, J. Phys. Chem., 67:874 (1963).
29. A. Bluhm, J. Weinstein, and J. A. Sousa, J. Org. Chem., 28:1989 (1960).
30. E. Klemm, D. Klemm, and H.-H. Horhold, Wiss. Z. Friedrich-Schiller-Univ. Jena, Math-naturwiss. R., 22:903-906 (1973).
31. D. Klemm, E. Klemm, and H.-H. Horhold, Z. Chem., 15:61-62 (1975).
32. J. Weinstein, J. A. Sousa, and A. Bluhm, J. Org. Chem., 29:1586-1588 (1964).
33. J. Weinstein, A. L. Bluhm, E. Langmuir, et al., US dep. Comm. AD 634651, Avail CFST (1966), p. 14; C. A., 66:115126a (1967).
34. E. Klemm and D. Klemm, J. Prakt. Chem., 321:407-414 (1979).
35. E. Klemm, D. Klemm, J. Reichardt, et al., Z. Chem., 13:375-376 (1973).
36. D. Klemm and E. Klemm, Z. Chem., 17:416-417 (1977).
37. D. Klemm and E. Klemm, Z. Phys. Chem., 258:1179-1182 (1977).
38. D. Klemm, E. Klemm, A. Graness, et al., J. Prakt. Chem., 321:415-419 (1979).
39. H. Mosher, C. Souers, and R. Hardwick, J. Chem. Phys., 32:1888 (1960).
40. A. V. El'tsov, É. R. Zakhs, and A. I. Ponyaev, Zh. Org. Khim., 16: 2176-2185 (1980).
41. H. Mosher, R. Hardwick, and D. Ben-Hur, J. Chem. Phys., 37:904 (1962).
42. É. R. Zakhs, A. V. El'tsov, and E. V. Lyashenko, Zh. Org. Khim., 14: 1992-1998 (1978).
43. M. A. Subbotina, É. R. Zakhs, A. I. Ponyaev, and A. V. El'tsov, Zh. Org. Khim., 17:203 (1981).
44. A. M. Sergeev, T. I. Frolova, V. P. Bazov, et al., Zh. Org. Khim., 12: 2436-2440 (1976).
45. A. I. Ponyaev, T. I. Frolova, É. R. Zakhs, et al., Zh. Org. Khim., 13: 1548-1556 (1977).
46. A. V. El'tsov, É. R. Zakhs, A. I. Ponyaev, et al., Zh. Org. Khim., 14: 1760-1769 (1978).
47. É. R. Zakhs, A. I. Ponyaev, A. M. Sergeev, et al., Zh. Org. Khim., 15: 2129-2135 (1979).
48. A. V. El'tsov, V. P. Martynova, and É. R. Zakhs, Zh. Org. Khim., 14: 203-204 (1978).
49. E. Klemm, J. Reichardt, and D. Klemm, Z. Chem., 17:259-260 (1977).
50. G. N. Utenkova and M. A. Gal'bershtam, Khim. Geterotsikl. Soedin., 523-530 (1969).
51. Patent 161403, GDR (1972).
52. Patent 161405, GDR (1972).
53. A. E. Tschitschibabin, B. M. Kuindschi, and S. W. Benewolenskaja, Ber., 58:1580-1586 (1925).
54. É. R. Zakhs, M. A. Subbotina, and A. V. El'tsov, Zh. Org. Khim., 15:200 (1979).
55. J. D. Margerum and C. T. Petrusis, J. Am. Chem. Soc., 91:2467 (1969).
56. J. D. Margerum and R. G. Brault, J. Am. Chem. Soc., 88:4733-4735 (1966).
57. Patent 3649549, USA (1972).
58. C. F. Bernasconi, J. Org. Chem., 36:1671-1678 (1971).
59. K. Schofield, J. Chem. Soc., 2408 (1949).
60. A. I. Nunn and K. Schofield, J. Chem. Soc., 583 (1952).
61. A. G. Turovets and V. I. Danilova, Izv. Vyssh. Uchebn. Zaved., Fiz., No. 4, 68-72 (1972).
62. B. M. Kuindzhi, L. A. Igonin, Z. P. Gribova, et al., Opt. Spektrosk., 12:220-223 (1962).
63. R. Hurley and A. C. Testa, J. Am. Chem. Soc., 88:4330-4332 (1966).
64. E. Lippert and J. Kelm, Helv. Chim. Acta, 61:278-285 (1978).
65. A. A. Pashayan and A. L. Prokhoda, Khim. Vys. Energ., 12:349-353 (1978).

66. A. A. Pashayan, Author's Abstract of Candidate's Dissertation, L. Ya. Karpov Scientific-Research Physicochemical Institute, Moscow (1976).
67. A. A. Pashayan, A. L. Prokhoda, and S. A. Sarkisyan, Khim. Vys. Energ., 11:55 (1977).
68. A. A. Parshutkin, V. P. Bazov, and V. A. Krongauz, Khim. Vys. Energ., 4:131-136 (1970).
69. D. Ben-Hur and R. Hardwick, J. Chem. Phys., 57:2240-2246 (1972).
70. H. Hiraoka and R. E. Hardwick, Bull. Chem. Soc. Jpn., 39:380-386 (1966).
71. G. Wettermark, J. Am. Chem. Soc., 84:3658 (1962).
72. P. P. Chegodaev, E. G. Strukov, and V. I. Tupikov, Khim. Vys. Energ., 8:406-409 (1974).
73. E. F. Caldin, Fast Reactions in Solution, Wiley (1966).
74. L. P. Hammett, Physical Organic Chemistry, 2nd ed., McGraw-Hill (1970).
75. V. I. Danilova and A. G. Turovets, Izv. Vyssh. Uchebn. Zaved., Fiz., No. 3, 141-145 (1970).
76. B. Tinland and C. Decoret, Tetrahedron Lett., No. 27, 2467-2470 (1971).
77. A. G. Turovets and V. I. Danilova, Izv. Vyssh. Uchebn. Zaved., Fiz., No. 1, 154-160 (1971).
78. V. M. Ryaboi and V. P. Bazov, Teor. Eksp. Khim., 11:585-591 (1975).
79. V. M. Ryaboi and V. P. Bazov, Teor. Eksp. Khim., 12:178-187 (1976).
80. L. G. S. Brooker, G. A. Keyes, R. H. Spraque, et al., J. Am. Chem. Soc., 73:5332-5339 (1951).
81. R. A. Jeffreys, J. Chem. Soc., 2394-2402 (1955).
82. R. A. Jeffreys and E. B. Knott, J. Chem. Soc., 4632-4641 (1952).
83. A. I. Kiprianov, Usp. Khim., 29:1336-1352 (1960).
84. A. V. El'tsov, É. R. Sakhs, and T. I. Frolova, Zh. Org. Khim., 12: 1088-1093 (1976).
85. A. Streitwieser, C. J. Chang, and A. T. Young, J. Am. Chem. Soc., 95: 4888 (1972).
86. A. I. Kiprianov, G. G. Dyadyusha, and F. A. Mikhailenko, Usp. Khim., 35:823 (1966).
87. L. G. S. Brooker, A. L. Sklar, H. W. J. Gressman, et al., J. Am. Chem. Soc., 67:1875-1889 (1945).
88. A. I. Kiprianov and I. K. Ushenko, Izv. Akad. Nauk SSSR, Otd. Khim. Nauk, No. 5, 492 (1950).
89. E. E. Coldin, M. Kasparian, and G. Tomalin, Trans. Faraday Soc., 64: 2802 (1968).
90. M. Eigen, G. G. Hummes, and K. Kustin, J. Am. Chem. Soc., 82:3482-3483 (1960).
91. A. I. Ponyaev, Author's Abstract of Candidate's Dissertation, Lensovet Leningrad Institute of Technology (1980).
92. H. Hiraoka and R. Hardwick, J. Chem. Phys., 41:2568-2569 (1964).
93. H. S. Bagdasaryan and V. A. Krongauz, Zh. Vses. Khim. Ova., 18:6-15 (1973).
94. C. A. Parker, Photoluminescence of Solutions, Am. Elsevier (1968).
95. R. H. Hurley and A. C. Testa, J. Am. Chem. Soc., 89:6917 (1967).
96. M. G. Kuz'min and V. L. Ivanov, in: Contemporary Problems in Physical Chemistry, Izd. Mosk. Gos. Univ., Moscow (1970), p. 193.
97. A. Ya. Bershtein and Yu. L. Kaminskii, Spectrophotometric Analysis in Organic Chemistry, Khimiya, Leningrad (1975).
98. L. R. Snyder and B. E. Buell, J. Chem. Eng. Data, 11:545-552 (1966).
99. W. H. Poesche, J. Chem. Soc., B, 469-477 (1966).
100. V. P. Bazov, A. A. Parshutkin, and V. A. Krongauz, Khim. Vys. Energ., 4:174-180 (1970).
101. I. Tanasescu, Bull. Soc. Chim. France, 39(4):1449-1453 (1926).
102. A. N. Frolov, N. A. Kuznetsova, and A. V. El'tsov, Usp. Khim., 45: 2000-2019 (1976).
103. J. A. Barltrop and N. J. Bunce, J. Chem. Soc., C, 1467-1477 (1968).
104. M. D. Cohen, Angew. Chem., 87:439-444 (1975).

105. K. Seff and K. N. Trueblood, Acta Cryst. Copenhagen, 24:1406-1412 (1968).
106. D. Klemm, Diss. B. Friedrich-Schiller-Univ., Jena (1976).
107. E. Klemm, Diss. B. Friedrich-Schiller-Univ., Jena (1976).
108. J. Opfermann, D. Klemm, E. Klemm, et al., Z. Phys. Chem., 258:1112-1116 (1977).
109. C. Mercier and P. J. Dubosc, Bull. Soc. Chim. France, 268-271 (1969).
110. Yu. E. Gerasimenko and N. T. Poteleshchenko, Zh. Vses. Khim. Ova., 16: 105 (1971).
111. V. A. Barachevskii, G. I. Lashkov, and V. A. Tsekhomskii, Photochromism and Its Applications, Khimiya, Moscow (1977).
112. Yu. E. Gerasimenko and N. T. Poteleshchenko, Zh. Org. Khim., 7:2413-2415 (1971).
113. V. N. Kostylev, B. E. Zaitsev, V. A. Barachevskii, et al., Opt. Spektrosk., 30:86-89 (1971).
114. E. P. Fokin, S. A. Russkikh, L. S. Klimenko, et al., Izv. Sib. Otd. Akad. Nauk SSSR, Khim., 110 (1978).
115. P. Boldt and A. Topp, Angew. Chem., 82:174-178 (1970).
116. M. V. Gorelik, S. P. Titova, and V. A. Trdtyan, Zh. Org. Khim., 15: 157-166 (1979).
117. N. T. Poteleshchenko, Author's Abstract of Candidate's Dissertation, Scientific-Research Institute of Organic Intermediates and Dyes, Moscow.
118. Yu. E. Gerasimenko, N. T. Poteleshchenko, and V. V. Romanov, Zh. Org. Khim., 14:2387-2391 (1978).
119. E. P. Fokin, S. A. Russkikh, and L. S. Klimenko, Izv. Sib. Otd. Akad. Nauk SSSR, Khim., 117-123 (1979).
120. Yu. E. Gerasimenko, A. A. Parshutkin, N. T. Poteleshchenko, et al., Zh. Prikl. Spektrosk., 30:954-956 (1979).
121. Yu. E. Gerasimenko and N. T. Poteleshchenko, Zh. Org. Khim., 15:393-396 (1979).
122. M. Nepras and A. Novak, Coll. Czech. Chem. Commun., 42:2343-2351 (1977).
123. R. N. Nurmukhametov, The Absorption and Luminescence of Aromatic Compounds, Khimiya, Moscow (1971).
124. Yu. E. Gerasimenko, N. T. Poteleshchenko, and V. V. Romanov, Zh. Org. Khim., 16:1938-1945; 2022-2023 (1980).
125. V. V. Belov, V. A. Barachevskii, N. I. Bolondaeva, et al., Summaries of the Papers presented at the II All-Union Conference on Holography, Vol. 1, Kiev (1975), p. 50.
126. A. A. Parshutkin, P. P. Kisilitsa, Yu. E. Gerasimenko, et al., Summaries of the Papers presented at the III All-Union Conference on Silverless and Unconventional Photographic Processes, Vilnius (1980), pp. 163-165 (1980).
127. D. Bellus, Adv. Photochem., 8:109 (1971).
128. D. Veierov, T. Bercovici, E. Fischer, et al., Helv. Chim. Acta, 58:1240 (1975).
129. R. W. Hoffmann and K. R. Eicken, Chem. Ber., 102:2987 (1969).
130. E. Wachsen and K. Harthe, Chem. Ber., 108:138, 683 (1975).
131. V. I. Minkin, L. P. Olekhnovich, and Yu. A. Zhdanov, Zh. Vses. Khim. Ova., 22:273 (1977).
132. D. G. McCarthy and A. F. Hegarty, J. Chem. Soc. Perkin Trans. II, 1085 (1977).
133. A. É. Lyubarskaya, G. D. Palui, V. A. Bren', et al., Zh. Org. Khim., 12:918 (1976).
134. G. D. Palui, A. É. Lyubarskaya, B. Ya. Simkin, et al., Zh. Org. Khim., 15:1348 (1979).
135. S. M. Aldoshin, O. A. Dyachenko, L. O. Atovmyan, et al., Tetrahedron, 38:2214-2217 (1982).
136. H. Görner and D. Schulte, Frohlinde, Chem. Phys. Lett., 66:383 (1979).
137. T. Karstens et al., Chem. Phys. Lett., 48:540 (1977).

138. R. M. Weiss and A. Warshell, J. Am. Chem. Soc., **101**:6131 (1979).
139. M. Sumitani, N. Nakashima, and K. Yoshihara, Chem. Phys. Lett., **68**:255 (1979).
140. V. I. Minkin, V. A. Kosobutski, B. Ya. Simkin, et al., J. Mol. Struct., **24**:237 (1975).
141. M. I. Knyazhanskii, M. B. Stryukov, V. I. Minkin, et al., Izv. Akad. Nauk SSSR, Ser. Fiz., **36**:1102 (1972).
142. V. I. Minkin, B. Ya. Simkin, L. P. Olekhnovich, et al., Teor. Eksp. Khim., **10**:668 (1974).

Chapter 5

ORGANIC PHOTOCHROMES FOR SOLAR ENERGY STORAGE

V. I. Minkin, V. A. Bren', and A. É. Lyubarskaya

Scientific-Research Institute of Physical Organic
Chemistry, Rostov-on-Don University, USSR

Current research efforts in the field of photochemical conversion of solar energy are focusing on the following options:

1) photochemical dissociation of water under the influence of metal-complex catalysts [1-5];

2) photoelectrochemistry based on photoelectronic transfers, the photogalvanic effect, stabilization of the products of photoelectronic transfers in the electric double layer at the phase boundary, etc. [4-10];

3) reversible thermal dissociation of inorganic compounds (e.g., NH_4HSO_4) upon exposure to light with the use of the latent heat of melting [9-11];

4) photosynthesis [9,10,12].

A detailed comparative analysis of the potential use and outlook for these methods of solar energy utilization is to be found in a series of recently published monographs [13-17].

The accumulation of solar energy in the form of the strain energy of metastable but kinetically stable photoisomers of organic compounds is a less studied avenue which has recently been actively pursued by researchers in the United States, Australia, Britain, and some other countries. One advantage of this approach is the possibility of long-term storage and, therefore, the accumulation of solar energy for subsequent release in the form of thermal energy.

This chapter examines the physical principles of this method of storing light energy, the criteria of its potential use and economics, and the basic chemical mechanisms of the principal photo and thermal reactions involved. A comparative analysis of systems of this type is also given.

THE PRINCIPLE OF ACCUMULATING LIGHT ENERGY AS THE ENERGY OF STRAINED METASTABLE STRUCTURES

The idea of accumulating solar energy as a result of the formation of sterically hindered strained metastable structures in the course of a photochemical reaction was first broached by Weigart as early as 1909 [18].

Weigart observed the photoconversions of anthracene crystals exposed to solar radiation. At the time, the physical or chemical mechanisms of the conversions induced by light quanta were beyond his understanding. Today, the basic scheme of photoconversions resulting in the formation of metastable photoisomers follows the pattern described by the data given in Figure 1.

A substance A capable of absorbing radiation within the range of solar radiation undergoes isomerization into form B, normally nonadiabatically, after transition into the electronically excited state A*. Form B is unfavorable in energy terms as compared to form A; however, the spontaneous thermal transition B → A requires overcoming the considerable activation barrier $\Delta H^{\neq}$. This ensures the very slow course of the process B → A and, consequently, the accumulation of part of the solar energy $h\nu$ in the form of the strain energy of structure B with its subsequent liberation in the form of heat ΔH. The reaction

$$A \underset{\Delta}{\overset{h\nu}{\rightleftharpoons}} B \tag{1}$$

describes the photochromic process [19]. Therefore, substance A must be a photochrome with a very high activation barrier for the reverse dark reaction.

It should be noted that scheme (1) and Figure 1 reflect the simplest variant of photochromic back reactions, which in reality also include sensitized and catalyzed stages and can also be complicated by the addition of side and equilibrium processes. These cases are examined in more detail in the survey [20].

A photochromic material capable of accumulating solar energy effectively must meet a series of mandatory requirements. Considerable attention is given to an analysis of these requirements [21-24]. The results of this analysis could be summed up as follows:

1. The starting material A must be capable of absorbing in the ultraviolet and visible regions of the spectrum since more than 50% of the solar energy reaching the earth is in the 300-700 nm range (380-170 kJ per einstein). The presence of long-wavelength absorption in the 400-650 nm region in a photochromic material is considered optimal in view of the nature of the energy distribution and intensity of solar radiation (see [13,25]; Figure 2).

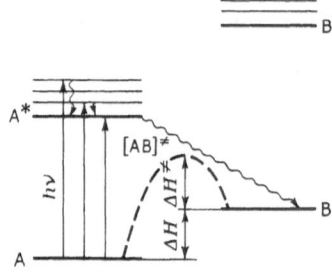

Fig. 1. A plot of the photoisomerization A $\overset{h\nu}{\rightleftharpoons}$ B: A*, B* — electronically excited states of molecules A and B; [AB]$^{\neq}$ — intermediate state of the thermal reaction; $\Delta H^{\neq}$— thermal activation barrier; ΔH— thermal effect of the back reaction; wavy arrows indicate radiationless transitions.

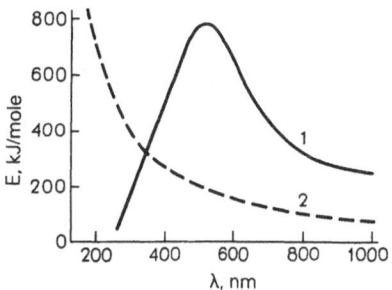

Fig. 2. The distribution of solar radiation (1) and the energy content of quanta (2) [25].

    2. The energy level of substance B in the ground state must lie sufficiently high enough above the level of the ground state of the starting material A for heat transfer ΔH to be considerable. Values of the heat effect starting from 50-60 kJ/mole are acceptable. It is obviously desirable to have a material with the least molecular mass M. Thus, for a compound with a molecular mass of 100 at ΔH 60-100 kJ/mole the accumulated heat per gram will be 600-1000, while for a compound with a molecular mass of 200-300 it will be as little as 500 J/g. These figures exceed the accumulated heat upon the solar heating of water (ΔT 50°C) (209), sand (42), and salt hydrates (250 J/g). In the series of organic photochromes benzene—prismane values on the order of 380 kJ/mole, i.e., 5280 J/g, are attainable.

    3. The activation barrier of the thermal transition $B \rightarrow A$ ($\Delta H\ddot{)}$ must be sufficiently high — on the order of 85-105 kJ/mole. Then the reaction rate at room temperature will be between 10^{-3} and 10^{-5} s^{-1}, which will make it possible to store photoisomer-accumulated energy for prolonged periods (the required period ranges from a few days to several weeks).

    4. The transition $B \rightarrow A$ that releases accumulated thermal energy must be subjected to catalytic acceleration or thermal initiation (preferably at temperatures not higher than 50-100°C).

    5. Photoisomer B must not absorb in a region close to the maximum of the intensity distribution of the solar spectrum, and the back reaction $B \rightarrow A$ must not be subjected to photochemical initiation by sunlight. Thus, substance A must be colored when substance B is colorless.

    6. The photochemical reaction $A \rightarrow B$ must have a high quantum yield. When sensitizers are used, this requirement also applies to sensitized photoreactions. This requirement implies that intramolecular reactions are preferable to intermolecular reactions, whose quantum yields are normally small. Another requirement is the absence of luminescence in substances A and B.

    7. High chemical yields without the formation of by-products are required for both the foward and reverse reactions of the $A \rightleftharpoons B$ cycle. Thus, 1% of a side reaction in each cycle gives, in 10 cycles, 18% of the by-product, which does not participate in the energy accumulation process.

    8. Substances A and B must be sufficiently low-cost and readily available. Neither should be toxic or highly inflammable. The substances should be chemically resistant to atmospheric water and oxygen.

231

The qualitative measure of the efficiency of the conversion of solar energy into thermal energy (the so-called Q factor) which has been proposed by Calvert [26] is defined by the expression $Q = (\Delta H \phi \cdot 100)/E_{h\nu}$, where ϕ is the quantum yield of the photoreaction; $E_{h\nu}$ is the energy of photons (per einstein) of the absorbed light (usually for the maximum of the long-wavelength absorption band of substance A).

Clearly, Q is an implied equivalent of efficiency and may only equal 100% provided that $\Delta H = E_{h\nu}$ with ϕ equalling unity. Needless to say, this condition is practically unfulfillable since it would require the lower excited level A* to coincide in energy terms with the level of the ground state B (see Figure 1). For this reason, the value of Q for real systems will always be far less than 100%.

Factor Q does not reflect many of the peculiarities of the photochromic processes affecting the efficiency of solar energy use. Porter and Archer [27] introduced an additional coefficient η_A to take into account the extent to which solar radiation is used for the ideal photochromic material which absorbs sunlight completely in the region $\lambda \leq \lambda_b$ (here λ_b is the long-wavelength absorption boundary). The coefficient η_A equals the ratio of the energy of one einstein (i.e., 1 mole of photons) with a wavelength of λ_b multiplied by the number of einsteins in the solar spectrum† with $\lambda \leq \lambda_b$ to the total energy of the solar radiation:

$$\eta_A = \frac{\int_0^{\lambda b} I_\lambda \, (\lambda/\lambda_b) \, d\lambda}{\int_0^\infty I_\lambda \, d\lambda}$$

where I_λ is the wavelength distribution of solar radiation intensity.

The coefficient η_A depends only on λ_b (Figure 3); its value is much less than unity for two reasons: 1) photons with $\lambda \geq \lambda_b$ are not absorbed by the materials; 2) the energy of photons with $\lambda < \lambda_b$ is partially dissipated during transitions from the higher excited states of the form A to the states $A^*_{S_1,T_1}$ (see Figure 1).

To evaluate the absorption spectra of real photochromes, Scharf et al. [20] have suggested that one estimate their total energy efficiency by the value η_e based on the expression $\eta_e = Q\eta_A\eta_{abs}$, where the absorption coefficient η_{abs} equals the ratio of the number of photons absorbed by substance A to the total number of photons in the solar spectrum with $\lambda \leq \lambda_b$. The coefficient η_{abs} depends on the absorption spectrum of material A and on such characteristics of the system as the concentration of the working solution, the thickness of the solution layer in the manifold, and the absorption of form B and other products.

A more complete evaluation of the efficiency of photochromes as accumulators of solar energy also requires that one take into account the energy capacity of the material W, i.e., the amount of accumulated energy per gram of starting material with a given intensity of incident light [20], the number of operating cycles (O_{cycle}) without an appreciable degradation of the material [20], and the thermal and photostability.

---

†We refer to the spectrum of the solar radiation reaching the surface of the Earth (for details see [20]).

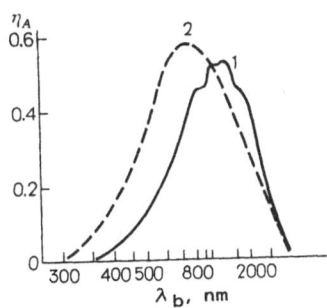

Fig. 3. The dependence of coefficient η_A on the long-wavelength absorption
boundary λ_b for solar radiation at the surface of the Earth (1) and
for diffused daylight (2) [20].

A critical study of the economics and engineering requirements of
photochromic solar-energy storage systems and an analysis of their potential
advantages and disadvantages, as compared with conventional thermal radiation receivers, have been carried out by a group of experts within the
framework of a research program funded as part of the Battelle project. The
results of these studies were published [28] and later supplemented with new
data, some of which are cited in the survey [23]. The following basic advantages of photochromic systems as compared to water heating—cooling systems already operational in the United States are listed: 1) a more light-
weight and low-cost solar manifold; 2) the possibility of storing solar
energy and using it on overcast and cloudy days; 3) substantially smaller
size of the tank containing the active material; 4) the possibility of
storing accumulated energy at normal temperatures; 5) a uniform energy level
of the solar energy storage system.

The main drawbacks of the system are the high cost of the active material and its possible destruction after a large number of operating cycles.

The authors came to the general conclusion that under the weather conditions of the southern United States photochemical systems would easily
outperform water heating systems as long as the values of Q were in excess
of 15%, and with a heat transfer of not less than 315 J/g (summer) or 670
J/g (winter).

VARIOUS PHOTOCHROMIC SYSTEMS FOR SOLAR ENERGY ACCUMULATION
AND THEIR MECHANISMS

To date, a fairly large number of photoactive compounds have been studied. Detailed surveys of their characteristics are contained in a series of
articles and monographs [9,10,16,20,22-24]. We are going to examine some of
the more important examples, classifying them by the type of mechanism
underlying the A ⇌ B reaction.

We may as well make a prior assumption as to the kind of conversions
which must fulfill the two major requirements generally imposed on this type
of reaction (1): the rate of the exothermic B → A reaction and its high activation barrier. Since isomer B is less favorable in energy terms than
isomer A and the high activation barrier implies thermally forbidden reactions [30], we may single out (on the theoretical plane) just two known
types of intramolecular reactions: geometric and valence isomerizations. To

233

these may be added migration reactions of polyatomic groups and intermolecular reversible conversions when A and (or) B in equation (1) are not one
molecule.

Geometrical Isomerizations

These reactions are based on the principle that the stable E-isomer (A)
upon irradiation changes into an unstable Z-isomer (B). The thermal Z → E-
isomerization usually requires a high activation barrier.

A large series of photoactive compounds, capable of converting solar
energy into thermal energy on the basis of reversible photochromic reactions
of geometrical isomerization, have been patented in the United States [31]:
indigo and thioindigo derivatives; derivatives of perinaphthoindigo and
perinaphthothioindigo and stilbene; and cyanine dyes.

Isomerization with Respect to the C=C Bonds. The geometrical isomerization A → B for indigoids and stilbenes is light-induced in the 350-800 nm
region. The absorption of the photoproduct B is slightly shifted towards
shorter wavelengths as compared with the initial form A. The quantum yield
ϕ of the photoreaction for different compounds varies from 0.01 to 0.2, and
the thermal effect ΔH of the back reaction from 8 to 34 kJ/mole. The activation barrier of the reverse thermal reaction for certain compounds
exceeds 85 kJ/mole, reaching 170-190 kJ/mole. An increase in the barrier
lengthens the energy storage time, but it also hinders the B → A back reaction upon thermal initiation. Organic and inorganic acids, the oxides of
certain metals, and so on have been suggested as catalysts of the back reaction for indigo and thioindigo derivatives. The use of triplet sensitizers
(benzophenone, fluorenone) has been recommended for stilbene derivatives.

According to estimates [32], the Q factor for systems based on geometrical isomerization cannot exceed 17% primarily because of the low thermal effect and the reversibility of the reactions upon exposure to sunlight.

The most promising system of this type is represented by indigo and
thioindigo derivatives, more specifically E-N,N'-diacetylindigo 1 $\underset{\Delta}{\overset{h\nu}{\rightleftharpoons}}$
Z-N,N'-diacetylindigo 2.

The advantage of this system lies in the fact that its absorption
spectrum nearly coincides with the photochemically active part of the solar
spectrum, which indicates high values of η_A. The catalyzed back reaction
proceeds almost quantitatively [33]. The system's disadvantages include the
low quantum yield of the direct photoreaction (see Table 2), the poor solubility of the material in most solvents, and the instability of the Z-form 2
in solution, which quickly isomerizes into the starting E-form 1. However,
Japanese chemists [34] have succeeded in identifying suitable conditions for
isolating the Z-isomer 2 of 2-diacetylindigo in the crystalline state. This
makes it possible to store accumulated energy liberated upon the dissolution
of Z-diacetylindigo crystals for prolonged periods.

Steiner et al. [35] have put forward an interesting new model based on
the concept of geometrical isomerization:

3   4

$h\nu$ φ0.43 $h\nu$ φ0.33 $h\nu_1$ φ0.57

(structures 3, 4, 5, 6 with $-H^+$ interconversion)

6   5

Betaine **3** exists in the E-form. In a neutral medium (pH 7) it is thermally and photochemically stable, while in acid media it is protonated into **4** and upon irradiation changes into a Z-isomer **5** with ϕ 0.33. Deprotonation of **5** produces the Z-isomer **6** of betaine, which is more advantageous in energy terms than the E-isomer **3** and which is kinetically stable (k $6.4 \cdot 10^{-5}$ s^{-1}, $E^{\neq}$ 120 kJ/mole 25°C).

The sequence of processes $3 \rightleftharpoons 4 \rightleftharpoons 5 \rightleftharpoons 6 \rightarrow 3$ represents a closed cycle and may serve as a convenient data-recording and storage system. From the standpoint of the use of these types of compounds for storing solar energy, the outlook is uncertain in the absence of data on the thermal effect of the cycle involved.

Isomerization about N=N Bonds. A prototype of these reactions is provided by the isomerization E-azobenzene **7** $\underset{\Delta}{\overset{h\nu}{\rightleftharpoons}}$ Z-azobenzene **8**.

The thermal effect of this reaction was estimated at 49 ± 5.4 kJ/mole [36], which at ϕ - 0.45 corresponds to a Q efficiency of 10%; the solar irradiation use coefficient η_A - 0.25, and η_e - 2.5%. The lifetime of the photoinduced form **8** — the Z-isomer — is considerable (k $1.6 \cdot 10^{-6}$ s^{-1}, and the half-life at room temperature is 34 h), but the activation of the reverse transition is a difficult problem.

Recently [37], an ingenious solution has been proposed, viz., to use a nickel complex **9** as an azo component:

$\frac{1}{2}M^{2+}$ ... $-CO-$ ... $-N=N-C_6H_5$ $\overset{h\nu}{\underset{\Delta}{\rightleftharpoons}}$

9

$\rightleftharpoons \frac{1}{2}M^{2+}$ ... $-CO-$ ... (Z-azobenzene with C_6H_5)

10

A cobalt chelate complex with a tetraaza ligand serves as the catalyst of the back reaction. This complex combines only with the Z-azobenzene derivative **10** and, consequently, catalyzes only the reverse reaction. Fischer et al. [37] do not cite data on the quantum yields and the thermal effect.

We could sum up the merits of systems based on geometrical isomerization as follows: 1) the possibility of achieving high activation barriers of the heat-releasing back reaction; 2) the presence of a conjugated bond system, ensuring absorption in the right spectral region by the photoactive form, i.e., a high value of coefficient η_A; 3) the resistance of compounds to side reactions, and thus a large number of cycles may be performed.

The disadvantages of these systems are as follows: 1) relatively small thermal effects either found or expected upon Z → E-transitions; 2) photochemical reversibility of the photoisomer (the Z-form), leading to light-induced forward and reverse reactions (see system 3 ⇌ 6); 3) complicated engineering approaches and solutions required for the initiation of reverse thermal reactions in those cases when the thermal Z → E-isomerization proceeds with a high activation barrier; 4) a high molecular mass and, hence, the relatively low calorific value of the incombustible fuel.

Valence Isomerizations

Upon the valence isomerization of conjugated molecules, as a result of the transition into a valence isomer with saturated bonds, a double effect results, which makes it possible to achieve a particularly high thermal effect in the back reaction: 1) energy resonance loss and 2) severe steric strain in the photoisomer. For this reason, valence isomerizations have been attracting particularly close attention in the context of the task in question.

The photoisomerization of nitrones was first examined by Calvin [38], who as early as 1958 broached the idea of storing light energy in the form of the strain energy of the cycle:

11   12   13

The photoreaction is light-induced at a wavelength of about 500 nm and ensures a high thermal effect (ΔH of about 85 kJ/mole). However, the reaction has a very low ϕ of about 0.05. Besides, its chief effect is the side conversion of the oxazirane 12 into amide 13, which is irreversible and results in a rapid unproductive consumption of nitrone 11. It is possible to eliminate this reaction and raise the quantum yield of the initial stage in cyclic nitrones. Work on this is well in hand [21].

The Photoisomerization Benzene → Prismane and Their Derivatives. The thermal transition prismane 14 → benzene 15

15   14

gives a value of the heat effect of 380 kJ/mole or 5280 J/g). The half-life at 90°C is about 11 h [39] for the thermal back reaction. These characteristics of the system would be highly attractive, except for some of its defects: 1) the initial benzoid form 15 does not absorb beyond the 270 nm region (see Figure 2) and the reaction 15 → 14 is poorly sensitized; 2) ϕ is extremely small — on the order of 0.001.

Because of this, activation of the 15 ⇌ 14 system is only achievable in its substituted derivatives. The most effective of the latter are perfluoroalkyl derivatives with pentafluoroethyl groups, but such products are too expensive and have a very high molecular mass. For instance, the molecular mass of hexapentafluoroethylprismane 16 is 786, and its calorific value drops down to 50 J/mole, which does not make this compound cost effective.

The isomerization norbornadiene 17 ⇌ quadricyclene 18 was first investigated by Hammond et al. [40], and at the moment it is among the more important isomerizations considered in the context of the problems discussed. The following advantages of this photochromic system are apparent: 1) both forms are fluids, which offers a number of engineering advantages (see below); 2) although norbornadiene 17 (R = R' = H) is not commercially produced, the starting products for its synthesis are (for instance, cyclopentadiene dimer); 3) the thermal effect of the back reaction is rather high (110 kJ/mole) [41], which corresponds to a particularly high calorific value of the "fuel" (1200 J/g), considering its low molecular mass; 4) ϕ of the photosensitization reaction reaches 0.5-0.7, with a high value of ϕ being retained in the presence of air; 5) the back reaction features a high activation barrier: the half-life of 18 is 14 h at 140°C [42]; 6) the back reaction is catalyzed by transition-metal complexes [43].

The main drawback of the reaction, which is not easy to overcome, is the fact that norbornadienes 17, with their nonconjugated π-bonds, absorb at short wavelengths (below 300 nm). The introduction of the substituents R and R' shifts the spectrum slightly in the desired direction, but at the same time sharply increases the molecular mass and decreases the possible energy content per gram. Besides, the chemical modification of the structure of norbornadiene 17 results in undesirable irreversible photoreactions [21]. It should also be noted that norbornadiene is combustible [20].

The main thrust of current efforts is towards the sensitization of the photoreaction 17 → 18 towards long-wavelength irradiation by means of triplet sensitizers. This sensitization has been carefully studied by Hautala et al. [29], and the results are presented in Table 1.

Although sensitizers help shift the absorption region of the absorbed light towards longer wavelengths, their use also activates the back photochemical reaction 18 → 17, and thus results not in a complete photoconversion of norbornadiene 17 but in the establishment of a photostationary state, which is an undesirable effect. Moreover, the introduction of appreciable quantities of the photosensitizer causes a chemical reaction between them and the photochromic components [44].

Another interesting direction of the sensitization of the 17 → 18 reaction is suggested in a series of papers by Schwendiman and Kutal [45,46]. They demonstrated that the reaction can be sensitized towards radiation with a wavelength of up to 350 nm in the presence of univalent copper salts, which, unlike the salts of other metals, do not cause a side dimerization or any other undesirable processes. Univalent copper does not catalyze the back reaction, and ϕ of the photoreaction is 0.3-0.4.

The chief disadvantage of the process under consideration is the extreme sensitivity of the intermediate copper complexes to moisture in the air. A recent report [47] points out that the triarylphosphine cuprous borohydride complex $[Cu(PPh_3)_2]BH_4$ is free of this defect while retaining a high quantum yield for the reaction.

Table 1. The quantum yields of the 17 → 18 reaction (R = R' = H) with different sensitizers [29]

Sensitizer	ϕ	λ_b, nm
$C_6H_5COCH_3$	0.9	370
$[-CH_2-CH(-C_6H_4COCH_3-p)-]_n$	0.26	380
$(CH_3)_2NC_6H_4COC_6H_5-p$	0.5	420
$[-CH_2-CH(-C_6H_4COC_6H_4NMe_2-p)-]_n$	0.55	420

A heterogeneous sensitizer – o-benzoylbenzoic acid adsorbed on aluminum oxide, which increases the reaction's sensitivity to light with a wavelength of 360 nm – is of interest [5]. The following are the most recent available data on the photoisomerizations of norbornadiene derivatives [23,48]: 17, R = R' = $COOCH_3$, λ_{max} 320 nm, ϕ 0.5, ΔH 77 kJ/mole; 17, R = $COOCH_3$, R' = C_6H_5, λ_{max} 360 nm, ϕ 0.6, ΔH 96 kJ/mole.

Thus, with norbornadiene derivatives particularly encouraging results may be achieved. Examining the system 19 → 21 → 18 and assuming for this system λ_b 380 nm, ϕ 0.4, and ΔH 110 kJ/mole, it is possible to calculate approximately: Q 14%, η_A 0.015, and η_e 0.21%. Thus, at a high level of energy efficiency this system has a low solar radiation use coefficient η_A and, consequently, a low total efficiency η_e. The main task in this area is still the development of efficient methods for sensitizing direct reactions to visible light. Another task, the search for an effective catalyst of the back reaction, seems to be closer to a successful completion [49].

<u>Other Valence Isomerizations</u>. A fairly large number of examples are known of other intramolecular valence isomerizations suitable for light energy storage. Guided by the principles set out in this chapter, we can select these isomerizations from among the photoisomerizations of heterocyclic and carbocyclic compounds [30,50*,51]. Here are just a few systems considered [5,21-23,52] to be potentially workable:

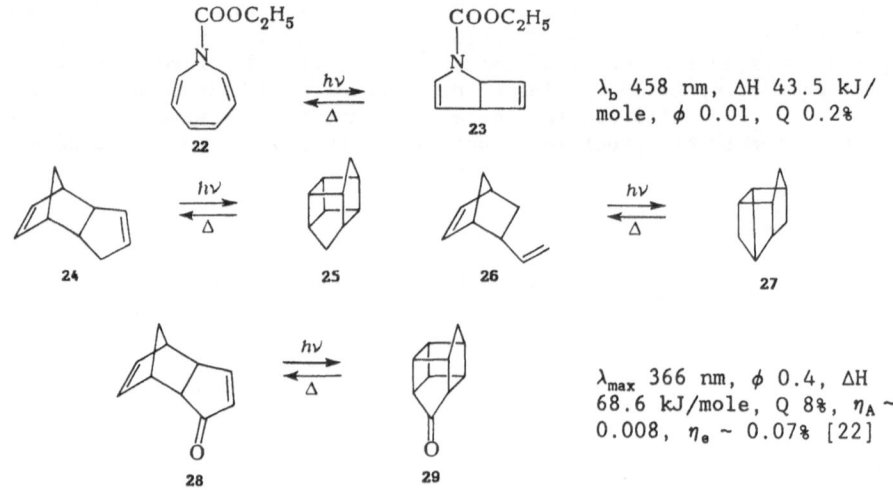

λ_b 458 nm, ΔH 43.5 kJ/mole, ϕ 0.01, Q 0.2%

λ_{max} 366 nm, ϕ 0.4, ΔH 68.6 kJ/mole, Q 8%, η_A ~ 0.008, η_e ~ 0.07% [22]

The reactions 24 ⇌ 25 and 26 ⇌ 27 have not been studied in any detail, even though they are known to be induced by ultraviolet radiation [53], and for the conversion 24 ⇌ 25 a rough estimate of the thermal effect is avail-

*The book [50] contains a comprehensive survey of work on the photoisomerizations of heterocyclic compounds carried out until 1975.

able (67 kJ/mole). Nonetheless, they deserve special attention since 24 (cyclopentadiene dimer) and 26 (the adduct of cyclopentadiene and butadiene) are both low-cost, commercially produced, sufficiently stable when exposed to the air, incombustible, nontoxic, and noninflammable.

Other photoisomerizations of related substances are known [21]:

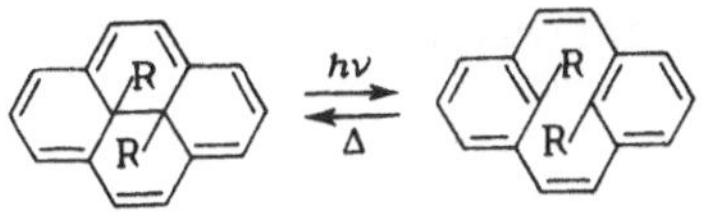

The 30 ⇌ 31 system has not been studied in detail, but its unquestionable merit is that isomer 30 absorbs in the long-wavelength region of 630 nm, while isomer 31 is colorless (λ_{max} of about 280 nm). One advantage of the system 32 ⇌ 33 is the exceptional thermal stability of photoisomer 33. Other characteristics are virtually unknown.

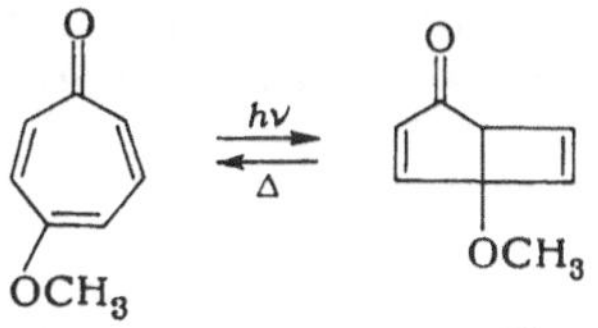

Dissociation Reactions and Dimerization

Let us begin by examining the photodissociation–thermal formation of nitrosyl chloride:

$$\text{NOCl} \underset{\Delta}{\overset{h\nu}{\rightleftharpoons}} \text{NO} + \tfrac{1}{2}\text{Cl}_2$$

34   35

The disadvantages of this reaction are the too rapid back reaction, which proceeds in parallel with the forward reaction and which, in turn, makes separation of the photoproducts NO and Cl_2 necessary, and the high toxicity of the starting compound and products. The most important representative of the dimerization reactions is the already mentioned photodimerization of anthracene 36 upon its exposure to sunlight:

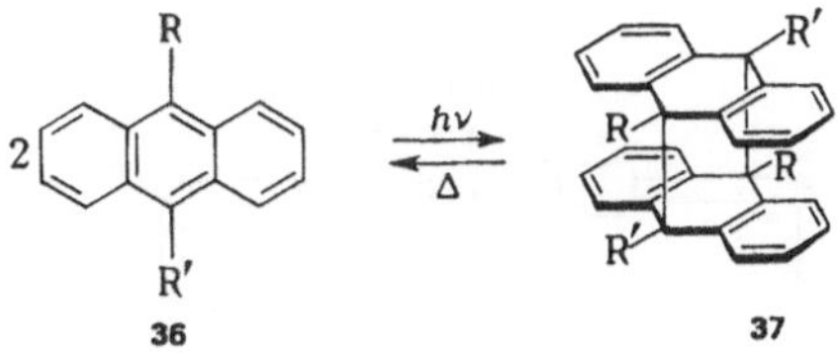

For the thermal reaction 37 → 36 the thermal effect is 65.2 kJ/mole (R = R' = H) and 85 kJ/mole (R = H, R' = CN) [20]. However, anthracene is extremely sensitive to atmospheric oxygen admixtures, while the back reaction 37 → 36 is not readily amenable to control. Anthracenes linked together by one or two bridges were found to be the most promising:

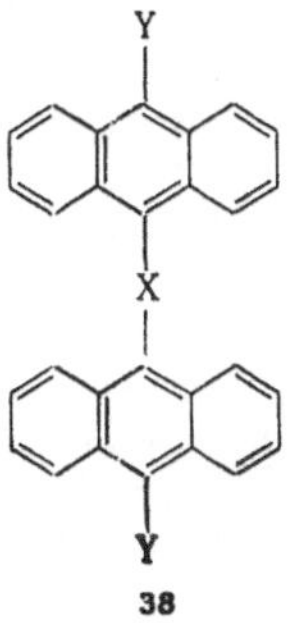

Undesirable photochemical processes in the back reaction and a competing fluorescence are observed for the systems described. This, as well as
the high molecular masses of the components, which leads to a low energy
capacity per gram of material, along with the relatively low values of Q and
η_A reduce the value of reactions of the type **36** ⇌ **37** for the stated purposes.

A BASIC SCHEME OF THE ENGINEERING SOLUTION TO THE PROBLEM OF SOLAR
ENERGY STORAGE AND RELEASE IN THE FORM OF THERMAL ENERGY

The following scheme has been proposed [21,22] for the storage and
liberation of energy in the photothermal-chemical cycle (Figure 4). Suitable concentrations of photoactive solutions and the thickness of the manifold are chosen, depending on the spectral and photochemical characteristics
of the working material. The rate of flow of the solution through the
manifold is regulated, subject to the intensity of solar radiation. The
proposed scheme is simple enough. It should be emphasized that the scheme
incorporates the idea of immobilized sensitizers and catalysts, i.e., fixed
on a stationary carrier such as a polymer base. The papers [49,54] examine
the principles of catalyst immobilization.

A United States registered patent [55] describes another path of the
dark reaction. A certain amount of heat essential for initiating the back
reaction in the local volume of a reaction is supplied to the reaction,
following which the reaction is sustained by the liberated heat.

ACYLOTROPIC AZOMETHINE SYSTEMS − A NEW TYPE OF SOLAR ENERGY
STORAGE BATTERY

An analysis of the available literature and patent data brings us to
the following conclusions:

1. The use of valence isomerizations seems to be the focus of current
efforts in this field. The primary objective is to identify suitable sensitizers. Another central task is a detailed study of the photophysical
mechanisms involved, the establishment of the role of exciplexes, triplet
sensitizing agents, and so forth [5,23].

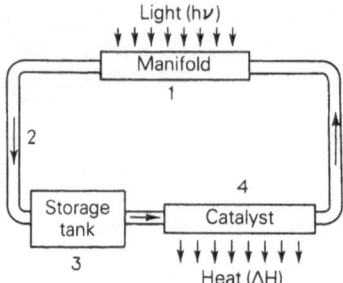

Fig. 4. A basic diagram of a device for the absorption of solar energy with
        subsequent release in the form of thermal energy (heat) [22]: 1 −
        manifold (a version involving an immobilized sensitizer is also
        possible); 2 − photochemical fuel circulating within a closed sys-
        tem; 3 − tank in which the stored-up solar energy may be kept in the
        form of structure B (see reaction (1)); 4 − catalyst immobilized on
        a polymer. The forcing through of the solution of substance B from
        tank 3 via the catalyst initiates the B → A reaction and releases
        heat ΔH. Upon reverting to form A the active material enters the
        manifold.

2. The mechanism of geometrical isomerization suffers from a basic drawback, which is that the photoisomer B (see reaction (1)) is usually colored. Reactions of this type can become far more attractive objects of research if isomerizations are found which would produce uncolored photo-isomers B that would not compete with the starting form A for the absorption of solar radiation in the 400-700 nm region.

3. No new mechanistic ideas other than those examined in this chapter have been put forward.

We assumed that it should be possible to complement the primary E → Z-isomerization with a secondary reaction which would lead to an un-colored product. The general idea, based on the principles of acylotropic tautomerism [56], was that upon the formation of the Z-isomer a structure should arise whose geometry would be favorable for a particular intramolec-ular rearrangement. Such a rearrangement was found during the study of the photoacylotropic systems of the 40–41–42 type [57]. It is possible to es-timate how suitable such a system is for solar energy accumulation, using as an example 2-(N-phenyl-N-acetylaminomethylene)-3(2H)-benzo[b]thiophenone (40, X – S, Y – 0, Ar – Ph, Ac – COCH₃) — one of the photoacylotropic azo-methine compounds:

The starting substance, whose molecules are in the Z-form 40, absorbs light energy and changes into the E-isomer 41, which has a conformation suitable for the transfer of the acyl group; it then undergoes a conversion by means of an intramolecular regrouping into the stable 0-acyl isomer 42. The energy of this molecule, according to quantum-chemical calculations, is approximately 34 kJ/mole higher than the energy of the starting ketoamine structure 40.

To overcome the high activation barrier between forms 41 and 42, the reverse reaction proceeds by heating or by means of acid catalysis. At this stage the stored-up energy is released as heat. After that the substance reverts to the starting form 40 and is ready for another cycle.

The system 40–41–42 meets the basic requirements set out in this chap-ter:

1. The starting material 40 has an intense absorption band with a maximum wavelength of absorption of 425 nm, which ensures a sufficiently efficient utilization of solar radiation (η_A 0.11).

2. The activation barrier of the thermal transition 42 → 41 is between 105 and 125 kJ/mole, which is sufficient to retain the energy accumulated by photoisomer B for several weeks at room temperature.

3. The reverse reaction 42 → 41, which releases the energy accumulated in the system, undergoes catalytic acceleration and thermal initiation, which makes the process controllable.

4. While the starting material 40 is colored, isomer 42 is colorless (λ_{max} 350 nm). For this reason, the back reaction 42 → 41 requires no sen-sitization; the luminescence of solutions 40, 42 → 41 is not subject to

Table 2. Parameters of some of the more effective photochromic systems capable of storing solar energy

System	λ_b, nm	ϕ	ΔH, kJ/mole (J/g)	Q,%	η_A	η_e, ($\eta_{abs} \cdot \tau$)	W, $\frac{mW}{g}$ ($I \cdot 100 \frac{W}{m^2}$)	Thermo-stability	Photo-stability
1 = 2 [20,58]	610	0.2	34(109)	3.5	0.27	0.95	300	200 s (25°C)	0.1
7 = 8 [36]	545	0.45	49	10	0.25	2.5		34 h (25°C)	0.24
17 = 18 (R = R = CO$_2$CH$_3$) [20]	334	0.5	77(390)	11	0.0015	0.02	0.035⎫	14 h (140°C)	about 0
19 = 18 (R = R = H) [20]	388	0.4	110(1200)	14	0.015	0.21	1.75⎬		
34 = 35 + ½Cl$_2$ [20]	600	1.0	37.6(580)	19	0.26	4.95	3	necessary separation of photoproducts	
38** [20]									
X = CH$_2$, Y = H	500	0.15	72.3	3.6	0.04	0.2			0.76
X = CH$_2$CH$_2$, Y = H		0.26	61.1	5.2	0.15	0.8		2 h (25°C)	0.55
39** [22]	460	0.36	35.5(100)	5.4					0.6
40 = 42		0.6	34(113)³*	8	0.11	0.9	300	20-30 days (25°C)	about 0
(X = S, Y = O, Ar = C$_6$H$_5$, Ac = COCH$_3$) [57]									
Perinaphthothioindigo [58,59]	650	0.1	21	1.0	0.3	0.3		2 h (70°C)	0.1
	700	0.1	467	26	0.36	8⁴*			
Photosynthesis: CO$_2$ + H$_2$O $\xrightarrow{h\nu}$ $^1/_6$C$_6$H$_{12}$O$_6$ + O$_2$ [20]*									

* Thermostability – the half life for the reaction B $\xrightarrow{\Delta}$ A at the indicated temperature ($\tau_{1/2}$); photostability – the quantum yield of the reverse reaction B $\xrightarrow{h\nu_1}$ A (ϕ_{rev}).

** ϕ_f 38(X = CH$_2$, Y = H) – 0.06; 38(X = CH$_2$CH$_2$, Y = H) – 0.16; 39 – 0.001. $E_{abs(rev)}$, kJ/mole: 38-137; 39-93.

³* Estimate based on the results of quantum-chemical calculations.

⁴* A theoretical value of the efficiency is given; in real systems η_e is about 0.1-0.3% [20].

photochemical initiation and no undesirable competition for the absorption of the energy of incident light is observed.

5. The forward intramolecular photoreaction 40 → 42 requires no sensitization; no luminescence of solution 40 at room temperature is observed, which ensures a high ϕ of the photoreaction − 0.6.

6. The system 40–41–42 withstands several scores of cycles without showing noticeable signs of degradation (irradiation by means of a DRSh-250 mercury lamp, thermal initiation of the back reaction by means of heating to 140°C).

7. Substances 40–41–42 are nontoxic. Products of heavy organic synthesis provide the starting material for synthesizing compounds 40 of the benzo[b]thiophene series.

Compared with the structurally related molecule of diacetylindigo 1, in which only E → Z-isomerization occurs, compound 40 has a highly resistant isomer form 42 and a higher thermal effect ΔH; in comparison with norbornadiene the 40–41–42 system features a higher coefficient η_A and a considerable energy capacity W while being inferior to norbornadiene in thermal effect.

Some basic characteristics of the systems recommended for use in solar energy storage are summarized in Table 2.

REFERENCES

1. V. Balzani, L. Moggi, M. F. Manfrin, et al., Science, 189:852 (1975).
2. D. M. Watkins, Plat. Met. Rev., 22:118 (1978).
3. J. R. Bolton, J. Solid State Chem., 22:3 (1977).
4. N. Sutin, J. Photochem., 10:19 (1979).
5. J. R. Bolton, ed., Solar Power and Fuels, Academic Press, New York—London (1977).
6. W. D. K. Clark and J. A. Eckert, Solar Energy, 17:147 (1975).
7. R. Gomer, Electrochim. Acta, 20:13 (1975).
8. A. Marzin, Prax. Naturwiss. Chem., 27:68 (1978).
9. M. D. Archer, in: Photochemistry Spec. Per. Report, Vol. 7, Burlington, London (1976), pp. 559-584.
10. M. D. Archer, in: Photochemistry Spec. Per. Report, Vol. 8, Burlington, London (1977), pp. 569-590.
11. W. E. Wentworth, A. F. Hildebrandt, C. F. Batten, et al., Rev. Int. Hautes Temp. Refract., 15:231 (1978).
12. O. V. Heath, Physiological Aspects of Photosynthesis, Stanford Univ. Press (1969).
13. B. J. Brinkworth, Solar Energy for Man, Halsted Press (1973).
14. K. L. Coulson, Solar and Terrestrial Radiation, Academic Press, New York (1975).
15. J. O'M. Bockris, Solar Energy. The Hydrogen Alternative, Wiley, New York (1975).
16. H. Messel and S. Butler, eds., Solar Energy, Pergamon Press, Oxford (1975).
17. S. Claessen and W. Engstrom, Solar Energy — Photochemical Conversion and Storage, Lyeber Tryck, Stockholm (1977).
18. F. Weigart, Eder's Jahrbuch (1909), p. 111; Chem. Abstr., 4:3170 (1910).
19. V. A. Barachevskii, G. I. Lashkov, and V. A. Tsekhomskii, Photochromism, Khimiya, Moscow (1977).
20. H.-D. Scharf, J. Fleischhauer, H. Leissman, et al., Ann. Chem., 91:696 (1979).

21. T. Laird, Chem. Ind., 186 (1978).
22. G. Jones, T. E. Reinhardt, and W. R. Bergmark, Solar Energy, 20:241 (1978).
23. G. Jones, Sheau-Hwa Chiang, and Phan Thanh Xuan, J. Photochem., 10:1 (1979).
24. G. Stein, Israel J. Chem., 14:213 (1975).
25. C. H. Paleocrassas, Solar Energy, 16:45 (1974).
26. J. G. Calvert, in: Introduction to the Utilization of Solar Energy, ed. by A. M. Zarem, McGraw-Hill, New York (1963).
27. M. D. Archer, Solar Energy, 20:167 (1978).
28. S. G. Talbert, D. H. Frieling, J. A. Eibling, et al., Solar Energy, 17:365 (1975).
29. R. R. Hautala, J. Little, and E. Sweet, Solar Energy, 19:503 (1977).
30. R. B. Woodward and R. Hoffmann, Conservation of Orbital Symmetry, Verlag Chemie (1970).
31. Patent 4004572, USA (1977).
32. M. Almgren, Photochem. Photobiol., 28:603 (1978).
33. R. E. Schwerzel, R. J. Barlett, J. R. Kelly, et al., Poster-Contribution, 2nd Conference on Solar Energy Conversion, Cambridge (England), Aug. 1978.
34. Y. Omote, S. Imada, and H. Ayoama, Chem. Ind., 415 (1979).
35. U. Steiner, M. H. Abdel-Kader, and P. Fischer, J. Am. Chem. Soc., 100:3190 (1978).
36. A. W. Adamson, A. Vogler, H. Kunkely, et al., J. Am. Chem. Soc., 100:1298 (1978).
37. D. P. Fischer, Y. Piermattie, and J. C. Dabrowiak, J. Am. Chem. Soc., 99:2811 (1977).
38. J. S. Splitter and M. Calvin, J. Org. Chem., 23:651 (1958).
39. J. F. Liebman and A. Greenberg, Chem. Rev., 76:311 (1976).
40. G. S. Hammond, P. Wyatt, C. D. De Boer, et al., J. Am. Chem. Soc., 86:2532 (1964).
41. K. B. Wiberg and H. A. Connon, J. Am. Chem. Soc., 98:5411 (1976).
42. G. S. Hammond, N. J. Turro, and A. Fischer, J. Am. Chem. Soc., 83:4674 (1961).
43. K. C. Bishop, Chem. Rev., 76:461 (1976).
44. A. A. Gorman, R. L. Leyland, M. A. Rodgers, et al., Tetrahedron Lett., 5085 (1973).
45. D. P. Schwendiman and C. Kutal, Inorg. Chem., 16:719 (1977).
46. C. Kutal, D. P. Schwendiman, and P. Grutsch, Solar Energy, 19:651 (1977).
47. P. Grutsch and C. Kutal, J. Am. Chem. Soc., 99:646 (1977).
48. G. Kaupp and H. Prinzbach, Helv. Chim. Acta, 52:956 (1969).
49. R. B. King and E. M. Sweet, J. Org. Chem., 44:385 (1979).
50. O. Burchard, ed., The Photochemistry of Heterocyclic Compounds, Wiley, New York (1976).
51. J. A. Barltrop and J. D. Coyle, Excited States in Organic Chemistry, Mir, Moscow (1978).
52. M. V. Alfimov and O. B. Yakusheva, Adv. Chem., 48:585 (1979).
53. W. L. Dilling, Chem. Rev., 55:373 (1966).
54. I. V. Berezin, A. M. Klibanov, and K. Martinek, Adv. Chem., 44:17 (1975).
55. Patent 4004573, USA.
56. V. I. Minkin, L. P. Olekhnovich, and Yu. A. Zhdanov, Molecular Design of Tautomeric Systems, Izd. Rostovsk. Univ., Rostov-on-Don (1977).
57. A. É. Lyubarskaya, G. D. Palui, V. A. Bren', et al., Zh. Org. Khim., 12:918 (1976).
58. D. L. Ross, in: Photochromism, ed. by G. H. Brown, Wiley, New York (1971), p. 486.
59. M. A. Mostoslavskii and M. M. Shapkina, J. Phys. Chem., 44:2708 (1970).

Chapter 6

LUMINESCENCE OF PHOTOCHROMIC COMPOUNDS

M. G. Kuz'min and M. V. Koz'menko

Moscow State University
Moscow, USSR

The photosensitivity of photochromic materials is limited by the values of the quantum yield of the photoreaction and the absorption coefficients of the starting compound and the product. These limitations are of a fundamental nature. It is possible to increase the sensitivity of a material if the image obtained is detected not by the light absorption of the photoproduct formed but rather by its photoluminescence. It is well known that luminescent methods of analysis are several orders of magnitude more sensitive than spectrophotometric methods. This is why the formation of a far smaller amount of the photoproduct is enough to obtain a luminescence image.

Back in 1965, Bukatin [1] noted the possibility of increasing the photosensitivity of photographic materials significantly by means of luminescent toning, i.e., the conversion of the developed silver image into a luminescent compound. In the early 1970s work began in a number of countries, including the USSR, on the development and careful study of photosensitive systems with luminescent image reading [2-11]. Zweig [2,3] examined many instances of organic reactions suitable, in principle, for the photochemical production of luminescent images. Later, different authors proposed a large number of reversible and irreversible photosensitive systems for luminescent image recording [6-11]. Since the achievement of a substantial increase in the photosensitivity of photochromic systems is quite feasible, this chapter will examine the luminescent properties of various types of photochromic compounds. Although there has been a massive spate of publications devoted to the study of photochromic systems, their luminescent properties remain little studied.

The photosensitivity of luminescent image recording systems is predicated not only on the properties of the material but also on the capabilities of the equipment used to observe the resulting luminescing image. Current estimates of the maximum photosensitivity of luminescent materials used in contemporary photoelectric equipment and for visual observation under conditions of dark adaptation are extremely optimistic (up to 10^{10} cm^2/J). However, in practice the photosensitivity of such materials is likely to be limited (as well as that of luminescent analysis methods) by the presence of luminescent impurities in the materials, and by the luminescence of the film base and the optical components of the system. In estimating practically achievable parameters, one should proceed from the fact that the concentration of the luminescent material possessing a high quantum yield of luminescence ϕ_1 should be about 10^{-12} mole/cm^2, which corresponds to the sensitivity of the existing luminescent methods of analysis. It is possible to achieve a high photosensitivity of luminescent materials, provided a luminescent substance forms in the process of data recording apart from its changing concentration.* Therefore, in the simple photochromic system

$$A \xrightleftharpoons[h\nu',\ kT]{h\nu} B$$

the thermal equilibrium should be strongly shifted towards the starting form A ($K < 10^{-5}$ mole/liter) or the thermal conversion of A into B must be extremely slow ($k < 10^{-10}$ s^{-1}). It is desirable that only form B be luminescent or that the absorption spectra and the spectra of B luminescence not be overlapped by the absorption spectrum of A. This would make it possible to excite luminescence in B selectively. In principle, the wavelength of the light used for image reading (excitation of fluorescence in form B) can be either longer or shorter than the wavelength of the light necessary for recording that image (photoconversions of A into B). Image reading by means of a longer-wavelength light is easier to achieve in practice.

The conditions described above limit the choice of photochromic systems suitable for luminescent image recording, but at the same time they also simplify the examination of the kinetics of the process involved and lower the level of requirements that side photoprocesses are expected to meet. The need for multiple reading of recorded images imposes certain limitations on the quantum yield of the back reaction. Generally, it should not exceed $1/n$, where n is the number of reading cycles. Erasure of a fluorescent image is possible if the back reaction proceeds thermally or photochemically.

The photosensitivity of a luminescent photochromic system is higher the greater the quantum yields ϕ_B of the formation of B and the quantum yield ϕ_l' of luminescence of B and its absorption coefficient ϵ at the wavelength λ' of the excitation of luminescence. The following expression can be used for an approximate estimate of the energy photosensitivity S (cm^2/J): $S \approx \alpha\lambda\phi_B\phi_l'\epsilon'$, where α is the coefficient determined by the equipment used for the image reproduction ($\alpha = 1$ for the above-cited value of the minimal concentration of luminophore 10^{-12} mole/cm^2; λ is the wavelength of recording, nm).

POLYACENES AND THEIR ANALOGS

The photochromism of polyacenes is due to their photoisomerization; for instance, for anthracene 1 (X = C):

1   2

The absorption spectra of photodimers have large hypsochromic shifts relative to polyacenes. The absorption, fluorescence, and phosphorescence bands of polyacenes [12] refer to the π,π^*-type. For the long-wavelength absorption band, we have approximately ϵ^{max} 10^3, ϕ_f 0.3, τ_S 10 ns,

---

*For luminescent image recording systems based on the disappearance of the luminescence of form A [10,11], the maximum achievable photosensitivity is a mere 10^{-2} cm^2/J, which is typical for photochromic systems based on light absorption.

k_f 10^7 s^{-1}. Internal conversion is practically absent, and $\phi_f + \phi_{isc} \approx 1$. Photodimers apparently do not fluoresce — no data on their fluorescence have been reported in the literature.

The quantum yield of the forward reaction is dependent on the polyacene concentration. It is generally believed that photodimerization proceeds from state S_n and is the cause of the concentration quenching of the fluorescence of polyacenes [14-16]; 9,10-substituents in the anthracene nucleus prevent dimerization. Photodimerization occurs not only in solutions but also in polymer films, an important consideration for practical applications. The reverse reaction proceeds photochemically and thermally [17-23]. Below 100°C photodimers are practically stable. The quantum yields of photodissociation of dimers range from 0.3 to 1.0 [23]. Upon the photolysis of a crystalline photodimer, the emission of an anthracene excimer was observed [24,25].

Tomlinson et al. were the first to propose the use of photodecomposition of polyacene dimers for luminescent image recording [23]. The high quantum yields of the photodecomposition of dimers and fluorescence of polyacenes favor higher photosensitivity. One drawback is the need to use short-wave ultraviolet light for recording purposes since polyacene dimers absorb only under 300 nm. No data have been reported in the literature on the availability of such materials.

A luminescent material based on the photodimerization of anthracene is described in the paper [11]; the best results were obtained in a polyethylene film. Dimerization occurs upon irradiation at 365 nm, and anthracene luminescence disappears to be replaced by a negative luminescent image. The sensitivity of this material is a mere 0.025-0.2 cm^2/J.

Reversible photodimerization of the $1 \rightleftharpoons 2$ type is also inherent in azapolyacenes, for instance, for the acridizinium ions 1 (X — $\overset{+}{N}$) and benzacridizinium 3 [23,26-29]. Acridizinium 1_+(X — $\overset{+}{N}$) and benzacridizinium 3 salts fluoresce well, while dimers 2 (X — $\overset{+}{N}$) and 4 do not. These salts and their dimers have spectral characteristics analogous to those of related hydrocarbons but shifted towards a longer-wavelength region. Because of such spectral properties, acridizinium and benzacridizinium salts are of considerable practical importance. The following are the spectral-kinetic characteristics of certain polyacenes and their analogs and also of photodimers [data are given in this order: compound number; name; solvent; conditions; λ_{abs}^{max}, nm (log ϵ^{max}); λ_f^{max}, nm (ϕ_f); τ_f, ns; ϕ_B (concentrations in mole/liter at which ϕ is measured) (the last two values are italicized)]:

1 (X — C), anthracene; ethanol; 293°K; 379 (3.8) ($S_0 \rightarrow S_1$), 252 (5.1) ($S_1 \rightarrow S_2$) [30]; 400 [31] (0.2 [27], 0.39 [20]); 5.2 [27]; 0.13 (0.02) [16].

2 (X — C), anthracene dimer [30]; ethanol; 293°K; 260-280 (3.1) 218 (4.3); 0.3-1.0 [23].

5, tetracene; ethanol; 293°K [32]; 471 (4.1) ($S_0 \rightarrow S_1$), 277 (5.1) ($S_0 \rightarrow S_2$); 485 (at 77°K) [33].

6, tetracene dimer; ethanol; 293°K; 280 [32].

7, pentacene; benzene; 293°K [32]; 576 (4.1) ($S_0 \rightarrow S_1$), 310 (5.6) ($S_0 \rightarrow S_2$); 577 (in liquid paraffin) [340].

8, pentacene dimer, benzene; 293°K, 280 [32].

1 (X — $\overset{+}{N}$), acridizinium perchlorate; methanol; 293°K, 365 (3.99) [23]; 430 (0.50 [25]); 7.0, 0.13 (0.02) [29].

1 (X — $\overset{+}{N}$), acridizinium toluenesulfonate; 0.7 [17] (crystals), 0.5 [17] (polymer matrix).

2 (X — $\overset{+}{N}$), dimer of acridizinium perchlorate; methanol, 293°K, 313 [23], 0.49 (0.002) [29].

2 (X — $\overset{+}{N}$), dimer of acridizinium toluenesulfonate (crystals, polymer matrix); 0.3 [17].

3, benzacridizinium toluenesulfonate; polymer matrix; 436 [23]; *0.5* [23].

4, dimer of benzacridizinium toluenesulfonate, polymer matrix; 365 [23]; *0.3* [23].

Another type of photochromic system based on polyacenes is the photo-addition of oxygen to polyacenes [14-16]. Photooxides, like photodimers, absorb in a much shorter wavelength region and do not fluoresce. Upon irradiation, photooxides decomposed to form O_2 and the starting polyacene. Photooxidation and photodimerization usually proceed in parallel. Many 9,10-disubstituted anthracenes (9,10-dimethyl-, 9-methyl-10-methoxy-, 9,10-dipropyl-, and 9,10-diphenylanthracene [14,16] and others) give only photooxides without forming photodimers. Unlike photodimers, photooxides arise from the triplet state of the polyacenes in accordance with the law of conservation of a system's total spin. The formation of photooxides was observed upon the photolysis of anthracene films exposed to the air and is one of the reasons for the disappearance of fluorescence upon the irradiation of such films [10,11]. Prolonged irradiation brings about irreversible oxidation until the formation of anthraquinone [11].

DIARYLETHYLENES

Diarylethylenes are typical representatives of compounds capable of Z/E-photoisomerization. Although the absorption spectra of the E- and Z-isomers differ but slightly, their luminescent properties are totally different [35-45]. As a rule, only the E-form fluoresces with a high quantum yield, while the Z-form does not fluoresce at all [46-58]. Only in the solid state and in frozen solutions is it sometimes possible to observe the fluorescence of Z-isomers [69]. The absorption and luminescence spectra of E-stilbene **9** (Table 1) and its derivatives in crystal matrices (octane, dibenzyl, tolane) at 4-77°K exhibit a well-resolved vibrational structure [36,83-86] with progressions characteristic of the C=C bond, and vibrations of the benzene ring. The longer-wavelength band of the absorption spectrum is polarized along the molecule's long axis, the second band — perpendicular to it. The fluorescence spectrum is polarized much less, which may indicate a substantial change in the geometry in the excited state, although this is contradicted by the slight Stokes' shift.

The luminescence of E-stilbene and its analogs has been studied in detail in connection with the mechanism of Z/E-photoisomerization [35,37, 38,45,69,87]. For E-stilbene at 293°K in nonpolar solvents, ϕ_f is about 0.08 [35,54]. It rises sharply with a drop in temperature (ϕ_f of about 1.0 at 77°K) or an increase in solvent polarity. With falling temperature the quantum yield of the photoisomerization, which competes with emission, declines. The activation energy of energy degradation determined from the dependence of ϕ_f and the fluorescence decay time on the temperature is 15 ± 2 kJ/mole (in a 3:2 mixture of methylcyclohexane with isohexane [54]), which coincides with the value of the activation energy of photoisomerization, which equals 17 kJ/mole [36]. An increase in ϕ_f of the E-stilbene was observed also upon an increase in the pressure up to 1 GPa. [56]. Brey et al. attributed it to an increase in the solvent's viscosity.

Difficulties in interpreting the luminescence data stemmed from the short lifetime (τ less than 0.1 ns) of the excited state of E-stilbene. Initial measurements by single-photon counting gave a value from 0.5 to several nanoseconds which was little dependent on temperature and solvent. It followed from such a value that the radiation rate constants determined from ϕ_f and from the absorption spectra differ by one order of magnitude or more. This led to the conclusion about the considerable differences existing between the Franck—Condon and fluorescent states. Another explanation assumed the existence of an equilibrium in the excited state between the planar and orthogonal isomers, provided the latter had a longer lifetime.

The available data on the kinetics of fluorescence pointed to a non-exponential character of decay and the presence of a contribution from a faster process. However, the study of the fluorescence kinetics upon laser picosecond excitation showed that in the case of pure samples of E-stilbene a monoexponential decay with a lifetime of between 75 and 108 ps occurred and a possible contribution from slower processes does not exceed 1% [54, 57,71]. Apparently, the large values of decay time which were found by the single-photon counting technique with a resolution of 1 ns are an artefact; possibly what was actually measured in these kinetic experiments was the decay of the fluorescence of impurities. At any rate, these data should be treated with great caution.

The luminescence of polarized p- and p-substituted E-stilbenes [73, 74,88] has a number of peculiarities due to the fact that the dipole moment in such molecules increases dramatically upon excitation. Thus, 4-diethyl-amino-4'-nitrostilbene 27 has a dipole moment of 7.6 D in the ground state and 32 D in the excited state. This accounts for the considerable Stokes' shift and the strong dependence of the quantum yields of fluorescence and photoisomerization on the polarity and polarizability of the solvent. The excited state in such molecules has a CT (charge-transfer) character.

The simplest model of the ethylene π-bond indicates that in the S_1- and T-states the minimum on the potential surface should correspond to the orthogonal configuration [58]. In this case, neither the E- nor Z-isomers should fluoresce and the quantum yield of photoisomerization in both directions should approach 0.5. For diarylethylenes the potential surface is more complex. According to the latest theoretical works [60,61], the potential surface of the lower excited state (^1B) has minima in the planar configuration, but it intersects the potential surface of the next (^1A) excited state with a minimum in the orthogonal configuration. In this model the difference between the fluorescent properties of the E- and Z-isomers is due to the presence of an appreciable potential barrier for the $^1B \rightarrow {}^1A$ transition in the first instance and a low potential barrier, in the second. According to the earlier model proposed by Fischer [35], the reason for the difference between the luminescent properties of the Z- and E-isomers lies in the differing relative arrangements of the S_1- and T_2-states. The lower energy of the T_2-state as compared with the S_1-state in the case of Z-stilbene may contribute to a rapid intersystem crossing to a triplet state with a potential minimum in the orthogonal configuration. It has been demonstrated [89] that no other intermediate states, except the excited singlet state of E-stilbene itself with a lifetime of about 0.1 ns, are observed upon the photoisomerization of E-stilbene.

The luminescent properties and thermal stability of diarylethylenes favor their use for luminescent image recording. The introduction of various substituents helps to shift the absorption and luminescence spectra into the visible region. Unfortunately, the comparatively slight differences between the absorption spectra of the Z- and E-isomers make it possible to achieve photochemically a sufficiently deep conversion of the E-isomer into the Z-isomer and thus erase the recorded image, which is necessary for repeated use of the material. Another difficulty is the photocyclization of the Z-diarylethylenes into the corresponding phenanthrenes, which competes with the Z/E-photoisomerization.

Crystallization occurs upon the irradiation of the supercooled drops of Z-diarylethylenes with a small radiation dose corresponding to less than 1% of the photoisomerization. Apparently, the irradiation produces microcrys-

Table 1. Spectral-luminescent and kinetic characteristics of E-isomers of diarylethylenes and indigoid dyes

No. of compound* Name	Solvent**	λ_f^{max}, nm	ϕ_f	τ_f, ns	$\phi_{E \to Z}$	$\phi_{Z \to E}$ (k, s^{-1})
9 Stilbene	glycerin isobutyl alcohol [71] methylcyclohexane– methylpentane [54,57]	335	0.26[70] 0.032 0.037	0.65[71] 0.058 0.075	0.52[41] 0.5	0.35[37]
10 4,4'-Distyrylstilbene [72]	dioxane	449	0.60	1.66		
11 p-Diphenylphosphenyl-p'-methoxystilbene [44]	propanol	412	0.014	0.2	0.45	0.4
12 1-Phenyl-2-(1-naphthyl)ethylene [62]	hexane		0.16		0.3	0.2
13 1-Phenyl-2-(2-naphthyl)ethylene [62]	"	356	0.60		0.05	0.03
14 1,2-Di(1-naphthyl)ethylene [38,40,62,73]	"	400	0.55		0.07	0.03
15 1,2-Di(2-naphthyl)ethylene [38,62,73]	"	365	0.60		0.32	0.20
16 Thioindigo [74–78]	benzene (benzene,toluene)	687;587 (benzene,toluene)	0.56	13.0	0.042	0.45 (5.4·10^{-4})

			λ_{abs}^{max}				
17	6,6'-Diethoxythioindigo [76,79-81]	"		0.03-0.04	0.9	0.25	0.45
18	Selenoindigo [76,80]	"		0.03	1.7	0.032	0.81
19	N,N'-Dimethylindigo [82]	"		0.01	0.4		
20	Perinaphthothioindigo [74,77]	"	692;655 (toluene)	0.07	1.3;4.0 (toluene)	0.013	0.058 (5.6·10⁻⁷)
21	Acenaphthenethionaphthenindigo [74,77]	"	604;582 (toluene)	0.30	8.1;8.5 (toluene)	0.20	0.40 (1.9·10⁻⁷)
22	Biacenaphthylidenedione [74]	"	585(toluene)	0.04	2.6		
23	Thionaphtheneperinaphthothiopyran-indigo [74]	"	635	0.35	6.0	0.027	0.18 (1.9·10⁻⁷)
24	Indirubin [74]	"	603	2.7·10⁻³			
25	4-Methyl-6-chloro-6'-methoxythio-indigo [74]	"	574	0.15	4.0	0.18	0.45 (1.6·10⁻⁶)
26	5,5'-Dibromothioindigo[74]	"	600	0.43	10.2	0.032	

* For the following compounds λ_{abs}^{max}, nm (log ε^{max}): 9E-320(4.4) [38]; 9Z-285(4.1); 10E-385(4.9); 13E-317; 15E-355; 15Z-320 [38]; 22E-476(4800); 478 and 513; 24E-551; 25E-527(14 200); 25Z-473(12 000).
**In all media except isobutyl alcohol, methylcyclohexane– methylpentane, and dioxane, the results were obtained at 298°K.

talline nuclei which induce crystallization of the sample. The crystalline Z-isomer, as distinct from a solution, luminesces, so even weak irradiation causes luminescence. This photoinduction of phase transitions [42] can be used for image recording purposes. Its merit is the relatively small radiation dose required for such a transition to take place.

THIOINDIGOIDS

In the case of indigoid dyes (see Table 1) only the E-isomers fluoresce both in fluid and in frozen solutions and in polymer matrices [72,79,80,90-92]; no luminescence is observed in the crystalline and colloid states. In rigid matrices it is possible to obtain the structured fluorescence spectra of thioindigo 16, which are dominated by a vibrational frequency of 1540 cm^{-1} assigned to the central double bond [72,90-92]; no carbonyl vibrations are observed in the spectra. This is evidence of the fluorescence being brought about by the $\pi \rightarrow \pi^*$-transition. The presence of intense absorption bands in the visible region for thioindigoid dyes, coupled with the high values of ϕ_f and $\phi_{Z \rightarrow E}$, make these compounds a promising basis for use in luminescent data recording systems. A major obstacle to this is the thermal instability of the Z-isomers, which makes it impossible to isolate and keep them in pure form. The rate constants of the thermal Z $\rightarrow$ E-isomerization of thioindigo and its analogs is between 10^{-3} and 10^{-6} s^{-1}, and the activation energy is between 70 and 100 kJ/mole [74]. To synthesize indigoid compounds whose rate constant of the thermal Z $\rightarrow$ E-isomerization does not exceed 10^{-10} s^{-1} is a tempting idea.

ARYLMETHANE DERIVATIVES

Derivatives of mono-, di-, and triarylmethanes having a polarized bond between the central carbon atom and the functional group X - CN, OH, OCO, etc., undergo heterolytic cleavage upon irradiation to produce salts of colored conjugated carbonium ions including triarylmethane, xanthene, and acridine dyes [93-97] (Table 2), for instance:

Such reactions can be used to create reversible or irreversible photochromic systems.

Aromatic groups in the molecules of leuco derivatives are separated by the saturated carbon atom. Their absorption spectra are close to the sum of the absorption spectra of the corresponding aromatic groups [111,112]. Unsubstituted arylmethane derivatives have the lowest excited states of the π,π^*-type. In the presence of electron-donor substituents (e.g., amino groups) having lone electron pairs, the latter may participate in the electronic transition (the $1,\pi^*$-excited state). Many leuco derivatives fluoresce well in nonpolar media, where no photodissociation occurs [111]. Their fluorescence spectra are similar to the spectra of related aromatic compounds $ArCH_3$. If the aromatic groups are different, an intramolecular energy transfer takes place and the fluorescence is emitted by the group having the least excitation energy [111,112]. Like their more simple analogs, leuco derivatives containing either a nitro or acetyl group in their aromatic system do not fluoresce but undergo phosphorescence due only to the high value of k_{isc} of the process $^1(n, \pi^*) \xrightarrow{kisc} \, ^3(\pi, \pi^*)$.

The fluorescence of the Rhodamine B leucolactone 42 is characterized by a very large Stokes' shift, which is strongly dependent on solvent polarity [113-115]. From this dependence one may determine that the dipole moment upon excitation increases by 25 D. Therefore, it is assumed that the fluorescent state of the leucolactone 42, unlike the ground state, has a bipolar structure with an orthogonal arrangement of the lactone and xanthene rings (analogous to some of the benzene derivatives having donor and acceptor groups in the para position), which form the orthogonal excited state with intramolecular charge-transfer [113-115]:

49

Many arylmethyl cations fluoresce with a sufficiently high ϕ_f. The cations of triarylmethane dyes containing strong electron-donor substituents generally do not fluoresce or the values of ϕ_f are very small because of the effective energy degradation caused apparently by the flexibility of the aryl groups. Rigid cations of xanthene and acridine dyes fluoresce with high ϕ_f (see Table 2).

The photolysis of the leuco derivatives of arylmethanes has been investigated in some detail both in connection with their photochromic properties and as a model reaction of heterolytic photodissociation [95]. The heterolytic dissociation may start from the lower state S_1 and from the T_1-state, depending on the nature of the X group, the substituents in the aryl groups, and the conditions [95]. Electron-donor substituents facilitate dissociation irrespective of their position with regard to the central carbon atom. No formation of excited reaction product molecules was detected on photolysis of triarylmethane compounds. However, in the case of leucolactone 42 and Rhodamine B leuconitrile 43 and N-methyl-9-cyano-9-phenylacridan 37 [96,97,99,107-109], it was possible to observe an adiabatic heterolytic photodissociation, yielding excited singlet states of the Rhodamine B zwitterion 44 and of N-methyl-9-phenylacridinium cation 39, respectively. In the case of the Rhodamine B leucolactone 42, heterolysis occurs from the nonrelaxed state rather than from the fluorescent state [111,113].

The quantum yields of the photodissociation of the leuco derivatives of triarylmethane dyes are strongly dependent on solvent polarity [98,116]. The dipole moments of the transition state of the heterolytic dissociation calculated from this dependence reach 10-23 D. This indicates a considerable charge separation in the transition state, which can be represented as a state with an intramolecular charge transfer from the triarylmethyl group to the X group which is eliminated in the form of an anion.

Electrophilic catalysis of the photodissociation of leuco derivatives by acids and alcohols was also observed [98,116,117]. The activation energies of photodissociation did not exceed 20 kJ/mole [98,107-109,112-115].

The carbonium ions formed by the photodissociation can interact with the solvent or the nucleophiles present in the system [94,95].

Table 2. Spectral-luminescent and kinetic characteristics of triarylmethanes in different solvents

No. of compound	Name	λ_{abs}^{max}, nm (log ε^{max})	λ_f^{max}, nm	ϕ_f	τ_f, ns	ϕ_B	k, s^{-1}
28	Triphenylacetonitrile [98]	270,ethanol			10,ethanol	$<10^{-3}$,ethanol	
29	Triphenylcarbonium perchlorate [98,99]	410,methylene chloride	560,methylene chloride				
30	p-Methoxyphenyldiphenyl-acetonitrile [98]	275,ethanol	300,ethanol	0.06,ethanol 0.02,ethanol-water	4.1,ethanol	0.3,ethanol 0.5,ethanol-water	
31	p-Methoxyphenyldiphenyl-carbonium chloride [98,100]	475,perchloric acid 480,sulfuric acid	550,perchloric acid 570,sulfuric acid				$2.6 \cdot 10^{-3}$, ethanol-water
32	o-Methoxyphenyldiphenyl-acetonitrile [98]	280,ethanol	∿300,ethanol	0.26,ethanol	5.6,ethanol	0.26,ethanol	
33	o-Methoxyphenyldiphenyl-carbonium chloride [98]						$3.7 \cdot 10^{-4}$, ethanol-water
34	m-Methoxyphenyldiphenyl-acetonitrile [98]	280,ethanol	∿300,ethanol	4.8,ethanol	0.2,ethanol		
35	m-Methoxyphenyldiphenyl-carbonium chloride [98]	440,perchloric acid					$5.8 \cdot 10^{-4}$, ethanol-water
36	Di(p-dimethylaminophenyl)-phenylacetonitrile [98, 101-103]			0.04,ethanol	0.1,ethanol	0.9-1.0,ethanol	

No.	Compound					
37	N-Methyl-9-cyano-9-phenylacridan [104,105]	290,ethanol 310(3.8),heptane	360,methanol, ethanol,heptane	0.001,methanol 0.01,acetonitrile	0.2,acetonitrile	0.1,methanol 0.35,acetonitrile
38	N-Methyl-9-hydroxy-9-phenylacridan [104,105]	315,heptane		0.002,methanol 0.03,acetonitrile	0.7,acetonitrile	0.45,methanol 0.03,acetonitrile
39	N-Methyl-9-phenylacridinium chloride [104,105]	366(4.3),water	520,water			
40	9-Cyano-3,6-bis(dimethyl-amino)xanthene [105,106]		360,ethanol	0.005,ethanol 0.01,toluene	2.0,toluene	0.4,ethanol 1.0,acetonitrile
41	3,6-Bis(dimethylamino)-xanthylium chloride [105, 106]	549(4.96), ethanol	560,ethanol		2.1,ethanol	
42	Rhodamine B, leucolactone [107,108] [109]	317,chloroform 315(4.25), benzene	490,chloroform 460,benzene		16.0,chloroform	0.13,chloroform 0.08,dioxane-water
43	Rhodamine B, leuconitrile [96]	$\sim$310(3.9), ethanol		<10^{-2},ethanol		0.2,ethanol,acetonitrile
44	Rhodamine B, zwitterion [107-109] [110]	542(5.05), ethanol 548,chloroform	560,ethanol 580,chloroform		4.2,chloroform	10^6,chloroform

Mono-, di-, and triarylcarbonium ions, containing comparatively weak electron-donor substituents, can only be observed by flash photolysis because of the rapid solvolysis. The presence of electrolytes slows down the ionic recombination because of the increased ionic strength of the solvent. The more stable cations of triphenylmethane dyes are slow to react with such nucleophiles as cyanide or hydroxide ion. By modifying the concentration of the nucleophile, the polarity, and other properties of the environment, it is possible to vary the rate of the reverse reaction over a wide range.

The leuco derivatives of triarylmethane dyes possess many properties that make them suitable for luminescent image recording. Their main drawback though is the too short wavelength absorption boundary (under 300 nm) of leuco derivatives.

SPIROPYRANS

The photochromism of spiropyrans absorbing in the ultraviolet region of the spectrum is based on the photocleavage of the C—O bond, which results in the formation of colored merocyanines (see Chapter 1), for instance:

The luminescence of a series of indoline spiropyrans has been investigated [118-126]. This luminescence is generally believed to be attributable to the chromene part of the molecule [118], which has a lower excitation energy than the indoline. The values of ϕ_f are low due to photoisomerization. The nitro derivatives do not fluoresce due to the rapid intersystem crossing typical for all nitro compounds. The structure of the phosphorescence spectra of these compounds indicates the $n,\pi*$-nature of the lower triplet state [119-122].

Merocyanines can exist in several stereoisomeric forms which undergo a fairly rapid mutual conversion and possess several different absorption bands [127]. That is why the absorption spectra obtained upon pulse laser-induced excitation differ from the spectra obtained under steady-state conditions. Fluorescence is known to exist only for the indoline merocyanines [127-131]; the ϕ_f values are low because of the rapid deactivation of the excited state due to stereoisomerization. Such energy degradation processes are characteristic of many cyanine dyes. The low ϕ_f values of the merocyanines constitute the main drawback of the spiropyrans from the standpoint of their potential use for luminescent image recording purposes.

The following are the spectral-kinetic characteristics of a number of spiropyrans and merocyanines [128-130] given in the order: compound number; substituents; λ_{abs}^{max}, nm; λ_f^{max}, nm (italicized); $\underline{\phi_f}$; τ_f, ns; ϕ_B (italicized):

50; R, R', R" = H; 296 (trimethylpentane, 77°K); 336 (trimethylpentane, 77°K); ~0.001 (propanol—isopropanol, 193°K); 0.06 (trimethylpentane, 77°K).
51; R = CH₃, R' = H, R" = NO₂; 550 (ethanol, 293°K), 580 (toluene); 623 (polymethyl methacrylate), 645, 673 (polystyrene); 0.016 (ethanol, 293°K); 0.4 (ethanol, 293°K), 4.1 (ethanol, 77°K); 1.03 (polystyrene).
51; R = CH₃, R' = Br, R" = NO₂; 623 (polymethyl methacrylate), 650, 668, 691 (polystyrene); 0.014 (ethanol, 293°K); 0.84 (ethanol, 293°K); 0.52 (polystyrene), 0.55 (polymethyl methacrylate).

51; R = CH$_3$, R' = CH$_2$-CH-CH$_2$-, R" = NO$_2$; 630 (polymethyl methacrylate);
<u>0.012</u> (ethanol, 293°K), 0.37 (ethanol, 293°K); 1.07 (toluene, 293°K).

AZOMETHINEIMINES

Upon irradiation, azomethineimines undergo cyclization into diaziridines [133-137] (Table 3):

The reverse reaction proceeds thermally (k is equal to approximately
$5 \cdot 10^{-6}$ s^{-1}) and photochemically.

Azomethineimines are intensely colored and fluoresce with a slight
Stokes' shift, and the ϕ_f values are very low (about 10^{-3}) due not only to
photocyclization since its quantum yields ϕ_{cycl} are considerably less than
unity. Apparently, it is accompanied by an effective energy degradation
process (see Table 3).

In diaziridines 53 the aromatic substituent is isolated from the rest
of the system by a quaternary carbon atom. The absorption and fluorescence
spectra of the diaziridines are similar to those of the related aryl substituents. However, the ϕ_f and the lifetime of the excited states are substantially less than those of the related aromatic hydrocarbons both as a
consequence of photochemical ring opening and because of the attendant
excitation energy degradation. The quantum yields of the cyclization of
azomethineimines into diaziridines are about 0.1; they are almost independent of temperature and the nature of the solvent. Even in frozen solutions
and in polymer matrices they have the same value. Oxygen and other fluorescence quenchers have no influence on ϕ_{cycl}. These data show that the photoisomerization occurs not from the fluorescent state but apparently from the
unrelaxed Franck—Condon state [137]. The values of ϕ_{cycl} and ϕ_f are slightly dependent on the wavelength of excitation, with these dependences being
antibatic. In frozen solutions ϕ_{cycl} falls with the depth of conversion.
Upon thawing and repeated freezing, the initial value of ϕ_{cycl} is restored.
It is a fair assumption that only a particular conformation of the azometh-
ineimine's molecule is capable of cyclization upon photoexcitation. At the
same time, this accounts for the comparatively low value of ϕ_{cycl}.

The photoisomerization of diaziridines into azomethineimines proceeds
with a high quantum yield ϕ_{isom} from the fluorescent state [137]. The
fluorescence quenchers proportionately reduce ϕ_{isom} of diaziridines.

The rather low ϕ_f of azomethineimines and diaziridines are a major
drawback from the standpoint of luminescent image recording, despite the
absorption in the visible region of the spectrum and the high ϕ_{isom} in solid
matrices.

COMPOUNDS WITH AN INTRAMOLECULAR HYDROGEN BOND

Many molecules containing acidic and basic groups linked together by a
hydrogen bond in the six-membered chelate ring exhibit intramolecular proton
phototransfer [138-140]. Often the proton phototransfer proceeds adiabatically, leading to the excited state of the photoisomer and its fluorescence.
Thus, upon the excitation of methyl salicylate and other derivatives of

Table 3. Spectral-luminescent and kinetic characteristics of azomethinimines and diaziridines [136,137]

No. of compound	R; Ar	Solvent	λ_{abs}^{max}, nm ($\epsilon \cdot 10^4$)	λ_f^{max}, nm	ϕ_f	τ_f, ns	ϕ_B
52	H; phenyl	toluene	343(4.34);365(4.21)	400	$1.5 \cdot 10^{-4}$	(~ 0.01)	0.11
		dimethylformamide	336(2.96);350(3.08)	400,420	$8 \cdot 10^{-2}$	7.6	0.05
	C_6H_{13}; p-anisyl	toluene	352(4.21);372(4.27)	410	$6 \cdot 10^{-3}$	1.2	0.14
		dimethylformamide	345(4.36);365(4.20)	440	$1.3 \cdot 10^{-2}$	3.2	0.05
	C_6H_{13}; p-bromophenyl	toluene	392(3.46)	483	$1.3 \cdot 10^{-3}$	0.1	0.07
		dimethylformamide	431(3.19);397(2.6)	540	$2.9 \cdot 10^{-2}$	1.6	0.04
	H; p-bromophenyl	toluene	347(3.84);266(3.84)	463	$< 10^{-4}$		0.15
	H; 4-pyrenyl	toluene	438(4.8)	463	$1 \cdot 10^{-3}$	0.2	0.12
		anisole	438(3.75)	459	$1.2 \cdot 10^{-3}$	0.3	0.13
		methyl ethyl ketone	430(4.65)	459	$0.9 \cdot 10^{-3}$	2.3	0.12
		acetonitrile	427(3.5)	457	$0.5 \cdot 10^{-3}$	0.9	0.13
53	H; 4-pyrenyl	toluene	379(0.02–0.07);348(3.5)	385	$3.4 \cdot 10^{-2}$	25	0.65
		anisol	348(3.0)	385	$3 \cdot 10^{-2}$	10.5	0.66
		methyl ethyl ketone	346(2.8)	383	$2.5 \cdot 10^{-2}$	15.7	0.69
		acetonitrile	345(3.7)	385	10^{-3}	11	0.66

salicylic acid, the fluorescence spectrum corresponds to that of the form involving proton transfer, for instance:

54   55

The excitation of the molecule causes a dramatic jump in the acidity of the hydroxyl group ($\Delta pK*$ of about 5-10) and a concurrent increase in the basicity of the carbonyl oxygen atom ($\Delta pK*$ of about 10), which is what induces the transfer of a proton from one oxygen atom to another. One consequence of this is the appearance of very large Stokes' shifts (of up to 10,000 cm^{-1}) characteristic of many salicylic acid derivatives and other analogous chelate compounds with an intramolecular hydrogen bond (hydroxy-azomethines, hydroxyphenylbenzothiazoles, hydroxyaldehyde azines, and others) [141-155]. The phototransfer of a proton occurs very fast ($k > 10^{11}$ s^{-1}), which accounts for the fact that only the emission of the proton transfer form* is observed and a four-level scheme [156] has to be used to describe the fluorescence of such compounds, where one pair of levels corresponds to the initial form and the other to the proton transfer form.

The electronic structure of the latter is described in different ways. Originally, a quinoid structure was proposed [33]. However, it was later noted, quite rightly, that the spectral-luminescent properties of this form contradicted the quinoid structure, for which the lower excited state should have belonged to the n,π*-type, and so a benzoid bipolar structure with a π,π*-type lower excited state was substantiated instead [157].

After the loss of the electronic excitation, the proton can just as easily revert to the original position via the hydrogen bond. That is why the lifetime of such photoisomers is very short (on the order of 10^{-9}-10^{-7} s). To fix the photoisomer, it is necessary that isomerization, breaking the hydrogen bond, take place along with proton phototransfer. Thus, the photochromism of salicylidenanilines observed in frozen solutions and in the crystalline state is attributable to the isomerization relative to the C=N bond [155]. Photoisomers with a lifetime of 10^{-7}-10^{-1} s [158] were detected in the azines of hydroxyaldehydes in fluid solutions.

The rate of the reverse thermal isomerization of the azomethines of 7-hydroxyquinolin-8-aldehyde 60, 61 (Table 4) is low due to the formation of a new intramolecular hydrogen bond with a quinoline nitrogen atom [159], for example:

60   65

Only stable photoisomers could be used for image recording, although the quantum yields of the fluorescence and the photoisomerization of many chelate compounds with an intramolecular hydrogen bond are sufficiently high.

---

*If the system contains molecules with a broken intramolecular hydrogen bond (e.g., in alcoholic media), emission bands of the starting form are also sometimes observed.

Table 4. Spectral-luminescent and kinetic characteristics of compounds with an intramolecular hydrogen bond in different solvents

Compound No.	Name	λ_{abs}^{max}, nm (ϵ^{max})	λ_f^{max}, nm	ϕ_f	τ_f, ns
54	Methyl salicylate [146,148]	305,308(4260) 238(9600)	355,443,460, ethanol		9, ethanol butyronitrile (77°K) 9, heptane (193°K)
56	Phenyl salicylate [146]	305, ethanol	355,450,480, ethanol		
57	Salicylidenaniline [148,150,159]	339(12800), ethanol 340(11200), isooctane	515 (ether-pentane-alcohol) 517 (trimethylpentane)		
58	Salicylaldazine* [153,158]	360, heptane 353, ethanol 362(20000), toluene	535, toluene	$2 \cdot 10^{-3}$, toluene	0.5, heptane
59	o-Hydroxynaphthaldazine [158]	404, ethanol 411, heptane	516, ethanol,heptane		1.0, ethanol
60	8-(Phenyliminomethylene)-7-hydroxyquinoline** [159]			0.06, heptane	0.3, heptane
61	8-(p-Anisyliminomethylene)-7-hydroxyquinoline³* [159]			0.07, heptane	0.3, heptane
62	2-Hydroxyphenylbenzothiazole [154,160]	349,338(16500), trimethylhexane	455–588 (trimethylhexane), 516, solid sample	0.31, solid sample	<10, solid sample
63	2-Deuterohydroxyphenyl-benzothiazole [154,160]		510, solid sample	0.21, solid sample	
64	2-Hydroxy-5-methylphenyl-benzothiazole [154,160]		530, solid sample	0.57, solid sample	

* λ_{abs}^{max} 470 nm [154].   ** ϕ_B 0.7 [155].   ³* ϕ_B 0.4 in heptane, ϕ_B 0.2 in dibromoethane [155].

CONCLUSION

The luminescent properties of the various types of photochromic compounds described above show that substances are now available whose parameters are adequate for the production of highly sensitive, luminescent image recording systems (e.g., dimerizing polyacenes, diarylethylenes, arylmethane derivatives, and so on). But all of these substances require excitation in the ultraviolet region of the spectrum. However, the development of luminescent image recording systems sensitive to the visible and the near-infrared regions call for synthesizing and investigating the luminescent properties of a wider range of compounds. In this respect, thioindigoid dyes, with their good photochromic and luminescent characteristics, are very promising if only dyes with a sufficiently low rate of thermal Z/E-isomerization ($k \sim 10^{-10}$ s^{-1}) could be synthesized. A more careful study of the luminescent properties of the numerous spiropyrans already available could also help find those which meet the requirements for luminescent materials. There is no doubt that in the near future luminescent image recording systems will find practical applications once special equipment for image observation and reading is developed.

REFERENCES

1. E. A. Bukatin, Zh. Nauchn. Prikl. Fotogr. Kinematogr., 10(3):219-220 (1965).
2. A. Zweig, SPSE, III Symposium on Unconventional Photographic Systems, Washington (1979), p. 79.
3. A. Zweig, IUPAC Symposium, Baden-Baden, 1972, Pure Appl. Chem., 33(2-3):389-410 (1973).
4. V. B. Nazarov and M. V. Alfimov, in: Advances in Scientific Photography, Nauka, Moscow (1980), Vol. 20, pp. 13-35.
5. V. B. Nazarov, O. P. Pilyugina, and M. V. Alfimov, Zh. Nauchn. Prikl. Fotogr. Kinematogr., 26(3):205-207 (1981).
6. Patent 3869363, USA (1973).
7. Patent 2446700, FRG (1973).
8. Patent 3892642, USA (1971).
9. G. A. Delzenne, in: Advances in Photochemistry, Wiley, New York (1979), Vol. 11, pp. 1-103.
10. P. A. Kondratenko, M. V. Kurik, and G. A. Sandul, in: Methods of Information Recording on Silverless Carriers, Issue 5, pp. 81-94, Kiev (1974).
11. M. V. Kurik and G. A. Sandul, in: Advances in Scientific Photography, Nauka, Moscow (1978), Vol. 19, pp. 262-269.
12. S. McGlynn et al., Molecular Spectroscopy of the Triplet State, Prentice Hall (1969).
13. R. N. Nurmukhametov, Absorption and Luminescence of Aromatic Compounds, Khimiya, Moscow (1971).
14. J. Bowen and D. W. Tanner, Trans. Faraday Soc., 51:475-481 (1955).
15. A. S. Cherkasov and T. M. Vember, Opt. Spektrosk., 4(2):203-210 (1958).
16. A. S. Cherkasov and T. M. Vember, Opt. Spektrosk., 6(4):503-511 (1959).
17. A. N. Terenin, The Photonics of Dye Molecules, Nauka, Leningrad (1967).
18. A. S. Cherkasov, in: Molecular Photonics, Nauka, Leningrad (1970), pp. 244-264.
19. A. Schoenberg, Preparative Organic Photochemistry, Springer-Verlag (1968).
20. E. A. Chandross, J. Chem. Phys., 43(11):4175-4176 (1965).
21. E. A. Chandross, J. Ferguson, and E. G. McRae, J. Chem. Phys., 45(10):3546-3553 (1966).
22. K. S. Wei and R. Livingston, Photochem. Photobiol., 6(3):229-232 (1967).

23. W. J. Tomlinson, E. A. Chandross, R. L. Fork, et al., Appl. Opt., **11** (3):533-548 (1972).
24. J. Ferguson and A. W.-H. Mau, Mol. Phys., **27**(2):377-387 (1974).
25. J. O. Williams, D. Donati, and J. M. Thomas, J. Chem. Soc. Faraday Trans. II, **9**:1169-1177 (1977).
26. W. J. Tomlinson, Appl. Opt., **11**(4):823-831 (1972).
27. W. J. Tomlinson, Appl. Opt., **13**(10):2456-2467 (1975).
28. J. Bendig and D. Kreysig, Z. Phys. Chem., Leipzig, **259**(3):551-556 (1978).
29. J. Bendig, B. Geppert, and D. Kreysig, J. Prakt. Chem., **320**(5):739-748 (1978).
30. C. A. Coulson, L. E. Orgel, W. Taylor, and J. Weiss, J. Chem. Soc., 2961-2962 (1955).
31. N. P. Shimanskaya, A. P. Kilimov, A. P. Grekov, et al., Opt. Spektrosk., **7**(3):239-370 (1959).
32. V. A. Barachevskii, G. I. Lashkov, and V. A. Tsekhomskii, Photochromism and Its Applications, Khimiya, Moscow (1977).
33. V. L. Broude, A. V. Prikhot'ko, and É. I. Rashba, Usp. Fiz. Nauk, **67**(1):99-117 (1959).
34. R. Schoental and E. J. V. Scott, J. Chem. Soc., 1683-1696 (1949).
35. S. Malkin and E. Fischer, J. Phys. Chem., **66**(12):2482-2486 (1962); **68**(5):1153-1163 (1964).
36. R. H. Dyck and D. S. McClure, J. Chem. Phys., **36**(9):2326-2345 (1962).
37. J. Saltiel, J. D'Agostino, E. D. Megarity, et al., Org. Photochem., **3**(3):1-113 (1973).
38. Ch. Goedicke, et al., Z. Phys. Chem., N.F., **101**:181-196 (1976).
39. L. D. Weis, T. R. Evans, and P. A. Leermakers, J. Am. Chem. Soc., **90**(22):6109-6118 (1968).
40. Y. B. Sheck, N. P. Kovalenko, and M. V. Alfimov, J. Luminescence, **15**(2):157-168 (1977).
41. V. F. Razumov, V. A. Alfimov, G. A. Shevchenko, and N. P. Kovalenko, Dokl. Akad. Nauk SSSR, **238**(4):885-888 (1978).
42. M. V. Alfimov and V. F. Razumov, Dokl. Akad. Nauk SSSR, **241**(3):599-601 (1978).
43. M. V. Alfimov and V. F. Razumov, Mol. Cryst. Liq. Cryst., **49**(3):95-97 (1978).
44. L. Al'der, M. V. Koz'menko, N. A. Sadovskii, et al., Opt. Spektrosk., **46**(1):76-84 (1979).
45. M. N. Pisanias and D. Shulte-Frohlinde, Ber. Bunsenges., **79**(8):662-667 (1975).
46. D. J. S. Birch and J. B. Birks, Chem. Phys. Lett., **38**(3):432-436 (1976).
47. J. Saltiel, J. T. D'Agostino, O. L. Chapman, and R. D. Lura, J. Am. Chem. Soc., **93**(11):2804-2805 (1971).
48. J. Saltiel, J. Am. Chem. Soc., **89**(4):1036-1037 (1967).
49. J. Saltiel, O. C. Zafiriou, E. D. Megarity, and A. A. Lamola, J. Am. Chem. Soc., **90**(17):4759-4760 (1968).
50. J. Saltiel, J. Am. Chem. Soc., **90**(23):6394-6400 (1968).
51. J. Saltiel and E. D. Megarity, J. Am. Chem. Soc., **91**(5):1265-1267 (1969).
52. G. S. Hammond, J. Saltiel, A. A. Lamola, et al., J. Am. Chem. Soc., **86**(16):3197-3217 (1964).
53. J. Saltiel and G. Megarity, Mol. Photochem., **5**(1):227-229 (1973).
54. M. Sumitani, N. Nakashima, K. Yoshihara, and S. Nagakura, Chem. Phys. Lett., **51**(1):183-185 (1977).
55. J. L. Charlton and J. Saltiel, J. Phys. Chem., **81**(20):1940-1944 (1977).
56. L. A. Brey, G. B. Schuster, and H. G. Drickamer, J. Am. Chem. Soc., **101**(1):129-134 (1979).
57. F. Heisel, J. A. Miehe, and B. Sipp, Chem. Phys. Lett., **61**(1):115-118 (1979).
58. A. J. Merer and R. S. Mulliken, Chem. Rev., **69**(5):639-656 (1969).

59. F. Momicchoioly, M. C. Bruni, and J. Baralchi, J. Chem. Soc., Faraday Trans. II, 70(7):1325-1333 (1974).

60. G. Orlandi and W. Siebrand, Chem. Phys. Lett., 30(3):352-354 (1975).

61. P. Tavan and K. Schulten, Chem. Phys. Lett., 56(2):200-204 (1978).

62. N. P. Kovalenko, M. V. Alfimov, A. A. Abunadirov, and Yu. V. Shekk, Izv. Akad. Nauk SSSR, Ser. Khim., No. 6, 1247-1251 (1979).

63. E. Lippert, Angew. Chem., 73(21):695-706 (1961).

64. D. Schulte-Frohlinde, H. Blüme, and H. Gusten, J. Phys. Chem., 66(12): 2486-2492 (1962).

65. T. Kobayaschi, E. O. Degenkolb, and P. M. Rentzepis, J. Appl. Phys., 50(5):3118-3121 (1979).

66. D. Schulte-Frohlinde and H. Görner, Pure Appl. Chem., 51:279-297 (1979).

67. B. M. Krasovitskii, L. A. Kutulya, L. Sh. Afanasiadi, et al., Zh. Prikl. Spektrosk., 29(2):272-277 (1978).

68. J. Saltiel, Al. Marinari, D. W.-L. Chang, et al., J. Am. Chem. Soc., 101(11):2982-2996 (1979).

69. S. Sharafy and K. A. Muszkat, J. Am. Chem. Soc., 93(17):4119-4125 (1971).

70. J. Saltiel and J. T. D'Agostino, J. Am. Chem. Soc., 94(18):6445-6456 (1972).

71. L. A. Brey, G. B. Schuster, and H. G. Drickamer, J. Am. Chem. Soc., 101(1):129-134 (1979).

72. R. N. Nurmukhametov, D. N. Shigorin, and N. S. Dokunikhin, Zh. Fiz. Khim., 34(9):2055-2059 (1960).

73. N. P. Kovalenko, Yu. V. Shekk, L. Ya. Malkes, and M. V. Alfimov, Izv. Akad. Nauk SSSR, Ser. Khim., No. 2, 298-306 (1975).

74. G. Hauke and R. Paetzold, Photophysikalische Chemie indigoider Farbstoffe. Nova Acta Leopoldina, Suppl. 11, Halle (1978), p. 123.

75. D. A. Rogers, J. D. Margerum, and G. M. Wyman, J. Am. Chem. Soc., 79(10):2464-2468 (1957).

76. G. M. Wyman and B. H. Zarnegar, J. Phys. Chem., 77(6):831-837 (1973).

77. G. Hauke, M. Erler, R. Paetzold, et al., Vestn. Mosk. Gos. Univ., Khim., No. 2, 190-194 (1976).

78. J. Blanc and D. L. Ross, J. Phys. Chem., 72(8):2817-2824 (1968).

79. E. A. Gastilovich, K. V. Tskhai, and D. N. Shigorin, Opt. Spektrosk., 41(4):566-572 (1976).

80. K. V. Tskhai, E. A. Gastilovich, and D. N. Shigorin, Zh. Fiz. Khim., 50(10):2694-2696 (1976).

81. A. D. Kirsch and G. M. Wyman, J. Phys. Chem., 79(5):543-544 (1975).

82. C. R. Cuiliano, L. D. Hess, and J. D. Margerum, J. Am. Chem. Soc., 90(3):587-594 (1968).

83. G. V. Gobov, Izv. Akad. Nauk SSSR, Ser. Fiz., 27(1):11-14 (1963).

84. A. N. Sevchenko, K. N. Solov'ev, S. F. Shkirman, and M. V. Sarzhevskaya, Dokl. Akad. Nauk SSSR, 153(6):1391-1394 (1963).

85. N. N. Malykhina and M. T. Shpak, Opt. Spektrosk., 14(6):829-831 (1963).

86. V. A. Levshin and Kh. I. Mamedov, Opt. Spektrosk., 12(5):593-598 (1962).

87. K. A. Muszkat, D. Gegiou, and E. Fischer, J. Am. Chem. Soc., 89(18): 4814-4816 (1967).

88. E. Haas, G. Fischer, and E. Fischer, J. Photochem., 9(2-3):277-279 (1978).

89. B. J. Greene, R. M. Hochstrasser, and R. B. Weisman, Chem. Phys. Lett., 62(3):427-430 (1979).

90. R. N. Nurmukhametov, D. N. Shigorin, and N. S. Dokunikhin, Izv. Akad. Nauk SSSR, Ser. Fiz., 24(6):728-729 (1960).

91. D. N. Shigorin, R. N. Nurmukhametov, and Yu. I. Kozlov, Opt. Spektrosk., 12(5):659-661 (1962).

92. R. N. Nurmukhametov, D. N. Shigorin, and Yu. I. Kozlov, Izv. Akad. Nauk SSSR, Ser. Fiz., 27(5):686-689 (1963).

93. R. N. MacNair, Photochem. Photobiol., 6(11):779-797 (1967).

94. G. H. Brown, ed., Photochromism, Wiley, New York (1971).
95. T. M. Grigor'eva and M. G. Kuz'min, in: Advances in Scientific Photography, Nauka, Moscow (1978), Vol. 19, pp. 177-193.
96. T. M. Grigor'eva, V. L. Ivanov, and M. G. Kuz'min, Khim. Vys. Energ., 13(3):245-249 (1979).
97. T. M. Grigor'eva, V. L. Ivanov, and M. G. Kuz'min, Khim. Vys. Energ., 13(4):325-330 (1979).
98. V. B. Ivanov, Author's Abstract of Candidate's Dissertation, Moscow State University, Moscow (1972).
99. W. R. Longworth and C. P. Mason, J. Chem. Soc., A(9):1164-1167 (1966).
100. Yu. Yu. Kulis and M. G. Kuz'min, Dokl. Akad. Nauk SSSR, 192(5):1079-1082 (1970).
101. L. Harris and J. Kaminsky, J. Am. Chem. Soc., 57(7):1154-1159 (1935).
102. J. Calvert and H. J. L. Rechen, J. Am. Chem. Soc., 74(8):2101-2103 (1952).
103. G. J. Fischer, J. C. Le Blans, and H. E. Johns, Photochem. Photobiol., 6(10):757-767 (1967).
104. T. M. Grigor'eva, V. L. Ivanov, and M. G. Kuz'min, Zh. Org. Khim., 17(2):439-444 (1981).
105. T. M. Grigor'eva, Author's Abstract of Candidate's Dissertation, Moscow State University, Moscow (1979).
106. T. M. Grigor'eva, V. L. Ivanov, and M. G. Kuz'min, Zh. Org. Khim., 17(1):162-167 (1981).
107. U. K. A. Klein and F. W. Hafner, Chem. Phys. Lett., 43(1):141-145 (1976).
108. T. M. Grigor'eva, V. L. Ivanov, N. Nizamov, and M. G. Kuz'min, Dokl. Akad. Nauk SSSR, 232(5):1108-1111 (1977).
109. T. M. Grigor'eva, V. L. Ivanov, and M. G. Kuz'min, Dokl. Akad. Nauk SSSR, 238(3):603-606 (1978).
110. J. Ferguson and A. W. H. Mau, Chem. Phys. Lett., 17(4):543-546 (1972).
111. M. L. Hers, J. Am. Chem. Soc., 97(23):6777-6785 (1975).
112. A. H. Sporer, Trans. Faraday Soc., 57(6):983-991 (1961).
113. A. Siemiarczuk, Z. R. Grabowski, A. Krówczyn'ski, et al., Chem. Phys. Lett., 51(2):315-320 (1977).
114. D. J. Cowley, P. J. Healy, and A. H. Peoples, J. Photochem., 9(2-3): 240-242 (1978).
115. E. Kirkor-Kaminska, K. Rotkiewics, and A. Grabowska, Chem. Phys. Lett., 58(3):379-384 (1978).
116. V. B. Ivanov, V. L. Ivanov, E. M. Kaplan, and M. G. Kuz'min, Teor. Eksp. Khim., 9(2):250-254 (1973).
117. J. Szychlinski, Rocz. Chem., 41(12):2123-2134 (1967).
118. N. Tyer and K. Becker, J. Am. Chem. Soc., 92(5):1289-1294; 1295-1302 (1970).
119. G. I. Lashkov, M. V. Sevast'yanova, A. V. Shablya, and T. A. Shakhverdov, in: Molecular Photonics, Nauka, Leningrad (1970), p. 299.
120. G. I. Lashkov, Author's Abstract of Candidate's Dissertation, Lensovet Institute of Technology, Leningrad (1971).
121. R. Becker and J. Roy, J. Phys. Chem., 69(4):1435-1436 (1965).
122. C. Balny, P. Douzon, T. Bercovici, and E. Fischer, Mol. Photochem., 1(2):225-233 (1969).
123. I. L. Belaits and T. D. Platonova, Opt. Spektrosk., 35(2):218-223 (1973).
124. G. I. Lashkov and A. V. Shablya, Izv. Akad. Nauk SSSR, Ser. Fiz., 32(9):1569-1574 (1968).
125. C. Balny, M. Mossé, C. Audic, and A. Hinnen, J. Chim. Phys., Phys.-Chim. Biol., 68(7-8):1078-1083 (1971).
126. P. Appriou, J. Brelivet, C. Trebaul, and R. Guglielmett, J. Photochem., 6(1):47-54 (1976-1977).
127. G. I. Lashkov and A. V. Shablya, Opt. Spektrosk., 19(5):821-824 (1965).
128. V. Hirshberg and E. Fischer, J. Chem. Soc., No. 1, 297-303 (1954).
129. V. Hirshberg and E. Fischer, J. Chem. Soc., No. 9, 3129-3137 (1954).

130. A. V. Shablya, K. B. Demidov, and Yu. N. Polyakov, Opt. Spektrosk., **20**(4):738-740 (1966).
131. J. Arnaud, J. Chim. Phys., **66**(1):159-160 (1969).
132. R. S. Becker, E. Dolan, and D. E. Balke, J. Chem. Phys., **50**(1):239-245 (1969).
133. M. Schulz and G. West, J. Prakt. Chem., **312**(1):161-164 (1970).
134. M. Schulz and G. West, J. Prakt. Chem., **315**(4):711-716 (1973).
135. M. Hippius, Dissertation, Hümboldt-Universität, Berlin (1978).
136. P. Rot, G. Tomashevskii, M. V. Koz'menko, et al., Khim. Vys. Energ., **13**(1):66-69 (1979).
137. A. Klimakova, M. V. Koz'menko, G. Tomashevskii, et al., Khim. Vys. Energ., **14**(2):149-155 (1980).
138. W. Klöpffer, in: Advances in Photochemistry, Wiley, New York (1977), Vol. 10, pp. 311-358.
139. A. Weller, in: Progress in Reaction Kinetics, Pergamon Press, London (1961), Vol. 1, p. 187.
140. Yu. I. Martynov, A. B. Demyashkevich, B. M. Uzhinov, and M. G. Kuz'min, Usp. Khim., **46**(1):3-31 (1977).
141. J. D. Margerum and L. J. Miller, in: Photochromism, ed. by G. H. Brown, Wiley, New York (1971), Chap. 6, p. 557.
142. M. B. Stryukov, A. E. Lyubarskaya, and M. I. Knyazhanskii, Zh. Prikl. Spektrosk., **27**(6):1055-1060 (1977).
143. M. I. Knyazhanskii, O. A. Osipov, A. E. Lyubarskaya, et al., Izv. Sev. Kavk. Nauchn. Tsentra Vyssh. Shkola, Ser. Estestv. Nauk, No. 2, 20-24 (1973).
144. M. I. Knyazhanskii, P. V. Gilyanovskii, and O. A. Osipov, Khim. Geterotsikl. Soedin., No. 11, 1455-1474 (1977).
145. E. Hadjuodis, Mol. Cryst. Liq. Cryst., **13**(3):233-241 (1971).
146. I. Yu. Martynov, B. M. Uzhinov, and M. G. Kuz'min, Khim. Vys. Energ., **11**(6):443-447 (1977).
147. D. S. Lo, Appl. Opt., **13**(4):861-865 (1974).
148. J. D. Margerum and J. A. Sousa, Appl. Spectr., **19**(3):91-97 (1965).
149. E. Fischer, Fortschr. Chem. Forsch., **7**:605-641 (1967).
150. W. Richey and R. Becker, J. Chem. Phys., **49**(5):2092-2101 (1968).
151. D. L. Williams and A. J. Heller, J. Phys. Chem., **74**(26):4473-4480 (1970).
152. M. Ottolenghi and D. S. McClure, J. Chem. Phys., **46**(12):4613-4620 (1967).
153. A. Weller and H. Wolf, Lieb. Ann., **657**:44 (1962).
154. M. D. Cohen and S. J. Flavian, J. Chem. Soc., B(4):317-321 (1967).
155. A. P. Simonova, R. N. Nurmukhametov, and A. L. Prokhoda, Dokl. Akad. Nauk SSSR, **230**(4):900-903 (1976).
156. B. S. Neporent, Izv. Akad. Nauk SSSR, Ser. Fiz., **37**(2):236-247 (1973).
157. R. N. Nurmukhametov, O. I. Betin, and D. N. Shigorin, Dokl. Akad. Nauk SSSR, **230**(1):146-149 (1976).
158. D. Fassler, V. L. Ivanov, and M. G. Kuz'min, Khim. Vys. Energ., **10**(2):187-189 (1976).
159. K. Franke, V. Rikhter, and G. Tomashevskii, Khim. Vys. Energ., **13**(5):442-447 (1979).
160. M. D. Cohen and S. Flavian, J. Chem. Soc., B(4):321-328 (1967).
161. B. S. Neporent, Izv. Akad. Nauk SSSR, Ser. Fiz., **37**(2):236 (1973).
162. M. G. Kuz'min, in: Problems of Biophotochemistry, Nauka, Moscow (1973), pp. 79-91.
163. W. Klöpffer, Adv. Photochem., **10**:311-358 (1977).

Aci-nitro acid, 185, 198

Actinometry, 46, 135

Acyl migrations, 218
  O-acylated isoamides, 219
  2-(N-acyl-N-arylaminomethylene)-
    3(2H)benzo[b]furan(thiophen)ones,
    219
  2-(acyloxybenzal)anilines, 220
  1-(acyloxymethylene)-
    2-indanones, 219
  photomechanism, 223

Ana-quinones, 212

Anils, 28

Anthracene dimerization
  Huckel method, 22
  luminescence, 246
  Longuet-Higgins treatment, 22
  solar energy conversion, 239
  Woodward-Hoffmann, 5

Aryl diazo compounds, 136
  cyanides, 137
  sulfides, 150
  sulfonates, 152
  triazenes, 143

Azamerocyanines, 29, 179
  kinetics of cyclizations, 196
  quantum yields of formation, 201

Azines, 26

Azo compounds, 112
  actinometry, 135
  aliphatic azo compounds, 126
  azobenzene $\Pi-\Pi^*$, $n-\Pi^*$ transitions,
    113
  azoheterocycles, 123
  inversion mechanism, 106, 134
  liquid crystals, 136
  luminescence, 123
  photocyclization, 133
  photoinitiators, 135
  polymer bound, 135
  polymer solutions, 136

Azo compounds, cont'd
  quantum yields, cis-trans
    isomerizations, 116
  solar energy conversion, 135, 235
  substituent effects, 124, 131

Bianthrones, 23, 36

Chromenes, 19

Chromone photodimerization, 6

Copper salt sensitization, 237

Diazo compounds, 152

Dihydronaphthalenes, 19, 21

Dihydroquinolines, 19, 21

2,4-Dinitrophenylacetic acid, 184

Dissociation-recombination
    mechanism, 107

2-DNBP, acronym for 2-(2,4-dinitro-
    benzyl)pyridine, 178
  charge transfer complex with·
    poly(N-vinylcarbazole), 210
  E.P.R., 186
  Huckel method, 192
  photodegradation, 207
  polymeric films, 209
  PPP analysis, 192

4-DNBP, 187

Electrocyclic reactions, 5

3-Ethylrhodanine derivatives, 193

Fluorescence, 246
  O-acetylsalicylalanines, 223
  aryloxyquinones, 212
  thioindigoids, 222

Fulgides, 20

General acid catalysis, 198

Heavy atom effects, 159

Heterolytic cleavage, 2

Hydrazides, 156

Hydrogen bonding, 157

Hydrogen transfer tautomerism, 37, 177

Indigoids, 24, 25

Inversion mechanism, 33, 106, 136

Liquid crystals, 136

Luminescence, 246
 azo derivatives, 123
 azomethines, 158
 merocyanines, 256
 spiropyrans, 256

Merocyanine electronic structure, 10
 luminescence, 256
 PPP analysis, 17

Nitrones, 236

Norbornadiene, 236

Optical data storage, 46, 146, 152, 217, 259

Oxime ethers, 161

PEPIPS method (Russian acronym), 47

Peri-aryloxy-p-quinones, 210

Photocatalytic isomerization, 111, 134, 148

Photochromism, definition, 1

Photoconductivity, indigo/zinc oxide films, 98
 poly(vinylcarbazole)/2-DNBP, 209

Photocyclization of azobenzenes, 133

Photodegradation
 aryldiazocyanides, 151
 aryldiazosulfates, 151
 aryldiazosulfides, 150
 aryloxy-p-quinones, 212, 217
 azo-alkanes, 127
 azobenzenes, 133
 azooxybenzenes, 134
 oximes, 160
 thioindigoids, 47
 triazenes, 150

Photodissociation, 253

Photoisomerizations, 30, 105

Photo oxides, 248

pH dependence of photochromes, 143, 194

Picosecond laser photolysis, 93, 98

Polar rotational mechanism, 107

Polymer-bound photochromics
 azobenzenes, 135

Polymer-bound photochromics, cont'd
 dinitrobenzyl derivatives, 210, 179
 thioindigoids, 98

Polymer solutions of photochromics
 aryloxy-p-quinones, 217
 azobenzenes, 136
 2-DNPB, 209
 thioindigoids, 97

PPP method
 anils, 28
 chromenes, 19, 20
 description, 7
 fulgides, 20
 merocyanines, 17
 spiropyrans, 9, 12, 34
 thioindigoids, 15, 47

Potential energy surfaces, 29

Prismones, 236

Q-factor, 232

Quadricyclene, 236

Rhodamine B, 253

Rotational mechanism, 33, 106

Salicylidene anils, 27

Selenochromenes, 19

Selenoindigo, 92

Semicarbazides, 163

Sensitization (triplet), 108, 110
 diazo compounds, 128
 spiropyrans, 11
 triazenes, 147

Solar energy conversion, 229
 O-acetylsilicylanines, 224
 advantages over conventional
   solar energy conversions, 223
 anthracenes, 239
 azobenzenes, 135, 235
 indigoids, 46, 234
 nitrones, 236
 prismanes, 236
 Q-factor, 232
 quadricyclene, 236

Solvatochromism
 azomerocyanines, 184
 3-ethylrhodanines, 193
 triazenes, 143

Solvent effects, 38
 activation energies, 149
 fluorescence, 213
 photochromic imines, 154
 photoreduction suppression, 149
 oxime stereoisomer stabilities, 160

Spiropyrans, CNDO method, 12
  color/structure, 17
  energy transfer, 10
  luminescence, 256
  PPP method, 9
  thermal bleaching, 13
  triplet sensitization, 11

Stilbene electrocylic ring closure, 5
  adiabatic Z–E conversion, 108

Tautomerism, 2

Thermal E/Z isomerizations, 160

Thioindigoids, 45
  cis-trans isomerizations, 58
  color contrast, 49, 72
  fluorescence, 222
  luminescence, 252
  PEPIPS diagram, 48
  PPP method, 25, 47
  quantum efficiencies in
    photoreactions, 80
  substituent effects, 50

Viscosity effects, 97